AF615291

MACHINING HARD MATERIALS

Dr. Roy Williams
Editor

Published by

Society of Manufacturing Engineers
One SME Drive
P.O. Box 930
Dearborn, Michigan 48128

MACHINING HARD MATERIALS

First Edition

First Printing

Library of Congress Catalog Card Number 82-50538

International Standard Book Number: 0-87263-083-8

Manufactured in the United States of America

SME wishes to express its acknowledgement and appreciation to the following publications for supplying the various articles within the contents of this book.

Industrial Diamond Review
Charters, Sunninghill, Ascot
Berkshire, SL5 9 PX
England

Manufacturing Engineering
Society of Manufacturing Engineers
One SME Drive
P.O. Box 930
Dearborn, Michigan 48128

Philips Technical Review
N.V. Philips Gloeilampenfabrieken
Eindhoven,
The Netherlands

Production Engineering
Penton/IPC Publication
Penton Plaza
Cleveland, Ohio 44114

Stanki I Instrument
PERA
Melton Mowbray
Leicestershire LE13 OPB
England

The Carbide And Tool Journal
The Society of Carbide and Tool Engineers
P.O. Box 437
Bridgeville, Pennsylvania 15017

Tooling & Production
Huebner Publications Inc.
5821 Harper Road
Solon, Ohio 44139

Vestnik Mashinostroeniya
PERA
Melton Mowbray
Leicestershire LE13 OPB
England

Grateful acknowledgement is also expressed to:

General Electric
P.O. Box 568
Worthington, Ohio 43083

Industrial Diamond Association of America Inc.
1215 Lady Street
P.O. Box 11187
Columbia, South Carolina 29211

Machinability Data Center
3980 Rosslyn Drive
Cincinnati, Ohio 45209

Cover illustration courtesy of the Specialty Materials Department, General Electric

PREFACE

The manufacture of parts from hard brittle materials originates in the history of man. Crude chipping and rubbing methods of material removal have given way to ultrasonic machining, laser beam, diamond grinding, cubic boron, and diamond compact machining, to name a few.

Historically, the abrasive machining processes have been used to machine hard materials. Aluminum oxide, silicon carbide, and diamonds have been the primary abrasives used in material removal. Recent developments in traditional cutting tools have significantly improved the role of the turning and milling processes in machining hard materials. Traditionally one selects a material harder than the product for the cutting tool. Development of oxides, compact diamonds, and cubic boron have provided extremely hard materials for cutting tools. The development of non-traditional machining methods have contributed significantly to machining of hard materials. Methods such as EDM, ECM, LBM, and EBM are unaffected by material hardness.

The future of machining hard materials looks bright because more new materials and methods have been developed in the past 25 years than any other time. During the same period we have seen a tremendous growth of hard materials applied in product design. This realization should encourage increased emphasis on research and development of new materials and methods. In addition, renewed emphasis should be placed on developing new applications for existing methods and materials.

The papers in this book are representative of several hundred papers which were reviewed. The material is divided into three chapters: (1) Nontraditional Machining, (2) Cutting and (3) Abrasive Machining. The papers were selected on the basis of application to industrial problems and application to a variety of work materials.

An attempt was made to include as many different tool materials, work materials, and processes as possible. Where several papers were available on the same topic, the paper with the most data and the broadest application was selected. There are many excellent papers covering theoretical aspects of machining hard materials which were not included.

Chapter One of this volume discusses nontraditional machining of hard materials. The chapters include discussions of Electrical Discharge Machining, Photochemical Machining, Laser Beam Machining and Ultrasonic Machining. Material removal rates and tolerances are explored. The surface integrity of nontraditional material removal processes is discussed.

Chapter Two is related to material removal using conventional machine tools. Recent developments in cutting tool materials have enabled lathes and milling machines to be utilized in machining hard materials. This development can dramatically improve the cubic inch per minute removal rate, thus reducing the machining cost. Tool materials of polycrystalline diamonds and cubic boron have been responsible for a revolution in machining hardened steels and jet aircraft superalloys. Several papers have been included with machining data. Several other papers have been included which cover machining ceramics and glass.

Chapter Three is related to material removal using the more traditional method of abrasive machining. These papers cover metallic and non-metallic materials. Principle in the metallics are carbide and tool steel; in the non-metallics aluminum oxide (ceramic) is most prominent. Traditionally, diamond wheels with natural industrial diamonds have been the primary abrasive used for grinding hard materials. Recent advances in synthetic diamonds and cubic boron abrasives have been responsible for significant improvement in grinding hard brittle materials.

Additional papers have been included on surface integrity and economics of grinding. Surface integrity is very important because property requirements extend to the surface of the material and grinding damage and can significantly change the material properties close to the surface. Economics is always a significant factor in choosing any process.

I wish to thank the authors of each of these articles for their contributions. (Company affiliations are those that the author held when they wrote the journal article or paper.) I also wish to express my gratitude to the publications who supplied the material in this volume. They include: *Industrial Diamond Review, Manufacturing Engineering, Philips Technical Review, Production Engineering, Stanki I Instrument, The Carbide and Tool Journal, Tooling & Production* and *Vestnik Mashinostroeniya.* In addition, my thanks is also extended to General Electric, Industrial Diamond Association of America Inc. and the Machinability Data Center, Finally, my thanks to the staff of the SME Marketing Services Department for their help in assembling this volume.

Roy Williams
University of Houston
Editor

ABOUT THE EDITOR

Roy L. Williams is Professor and Chairman of the Mechanical Technology Department at the University of Houston. He holds a Bachelor of Science degree in Industrial Engineering and received an M.S. in Industrial Engineering from the University of Houston.

Roy L. Williams

Mr. Williams has held the positions of Project Engineer with Metcut Research Association in Cincinnati and was a Development Specialist at the Union Carbide Nuclear Division in Oak Ridge, Tennessee.

A noted lecturer and author, Mr. Williams is a member of several professional organizations and is a Registered Professional Engineer in Texas. He was presented with the SME Gold Medal in 1973. He has been an active member of the Society of Manufacturing Engineers authoring several technical papers and serving as co-editor of the Machining Fundamentals, programmed learning course. He is Vice Chairman of the SME Professional Registration Committee.

SME

The informative volumes of the Manufacturing Update Series are part of the Society of Manufacturing Engineers' many-faceted effort to provide its members with the latest information and developments in engineering.

Technology is constantly evolving. To be successful, today's engineers must keep pace with the torrent of information that appears each day. To meet this need, SME provides, in addition to the Manufacturing Update Series, many opportunities in continuing education for its members.

These opportunities include:

- Monthly meetings through three associations and their more than 230 chapters which provide a forum for member participation and involvement.
- Educational programs including seminars, clinics, programmed learning courses, as well as videotapes and films.
- Conferences and expositions which enable engineers and managers to examine the latest manufacturing concepts and technology.
- Publications including the periodicals *Manufacturing Engineering, Robotics Today* and *CAD/CAM Technology*, the *SME Newsletter*, the *Technical Digest*, and a wide variety of text and reference books covering everything from the basics to manufacturing trends.
- The SME Manufacturing Engineering Certification Institute formally recognizes manufacturing engineers and technologists for their technical expertise and knowledge acquired through experience and education.
- The Manufacturing Engineering Education Foundation, was created by SME to improve productivity through education. The foundation provides financial support for equipment development, laboratory instruction, fellowships, library expansion, and research.

SME is an international organization with more than 60,000 members in 60 countries worldwide. The Society is a forum for engineers and managers to share ideas, information, and accomplishments.

The Society works continuously with organizations such as the American National Standards Institute, the International Organization for Standardization, and others, to establish and maintain the highest professional standards.

As a leader among professional societies, SME assesses industry trends, then interprets and disseminates the information. SME members have discovered that their membership broadens their knowledge and experience throughout their careers. The Society is truly industry's partner in productivity.

MANUFACTURING UPDATE SERIES

Published by the Society of Manufacturing Engineers, the Manufacturing Update Series provides significant, up-to-date information on a variety of topics relating to the manufacturing industry. This series is intended for engineers working in the field, technical and research libraries, and also as reference material for educational institutions.

The information contained in this volume doesn't stop at merely providing the basic data to solve practical shop problems. It also can provide the fundamental concepts for engineers who are reviewing a subject for the first time to discover the state-of-the-art before undertaking new research or application. Each volume of this series is a gathering of journal articles, technical papers and reports that have been reprinted with expressed permission from the various authors, publishers or companies identified within the book.

SME technical committees, which are made up of educators, engineers and managers working within industry, are responsible for the selection of material in this series.

We sincerely hope that the information collected in this publication will be of value to you and your company. If you feel there is a shortage of technical information on a specific manufacturing area, please let us know. Send your thoughts to the Manager of Educational Resources, Marketing Services Department at SME. Your request will be considered for possible publication by SME—the leader in disseminating and publishing technical information for the engineer.

MANUFACTURING UPDATE SERIES

Published by the Society of Manufacturing Engineers, the Manufacturing Update Series provides significant, up-to-date information on a variety of topics relating to the manufacturing industry. This Series is intended for engineers working in the field, technical and research libraries, and also as reference material for educational institutions.

The information contained in this volume doesn't stop at merely providing the basic data to solve a [illegible] problem. [illegible] fundamental concepts [illegible] of better [illegible] new research [illegible] experts [illegible] with [illegible] permission [illegible] identified within the book.

SME [illegible] who [illegible] engineers and managers [illegible]

[illegible] values [illegible] technical [illegible] Manufacturing [illegible] publication by SME [illegible] dissemination and [illegible]

TABLE OF CONTENTS

CHAPTERS

1 NONTRADITIONAL MACHINING

Nontraditional Machining Guide
By *Guy Bellows*
Reprinted from Nontraditional Machining Guide: 26 Newcomers for Production, MDC 76-101 by Guy Bellows, by permission of the Machinability Data Center, Metcut Research Associates Inc. Copyright 1976 by Metcut Research Associates Inc. 3

Surface Integrity Of Nontraditional Material Removal Processes
By *Guy Bellows and John B. Kohls*
Presented at SME's Nontrad '76 Conference, October 1976 13

Applications For Industrial Lasercutter Systems
By *Willard E. Hanson*
Presented at SME's Lasers in Manufacturing Conference, December 1976 53

NC/EDM—"Cutting" Costs With NC And A Wire Electrode
By *C. Michael Mercincavage*
Presented at SME's NC Machining Conference, April 1975 64

Electronic Bandsaw Sparks A Revolution
By *Donald E. Hegland*
Reprinted from Production Engineering, March 1977. Copyright Penton/IPC 1977 79

Rotary Ultrasonic Machining
By *William R. Tyrrell*
Presented at SME's Nontraditional Machining Seminar, January 1970. Updated for *Machining Hard Materials* 83

2 CUTTING

The Application Of Polycrystalline Tooling
By *S. K. Bhattacharyya and D. Aspinwall*
Presented at the Industrial Diamond Association of Japan's 30th Anniversary Meeting and Seminar, May 1978. Copyright General Electric Company 95

Machining Sintered Carbides With Polycrystalline Diamond Tools
By *G. F. Wilson, R. Alworth and S. Ramalingam*
Presented at the International Conference on Manufacturing Engineering August 1980. Copyright General Electric Company 104

Machining Of Hard Ceramics With Sintered Diamond Tools
By *Hidehiko Takeyama*
Presented at the Industrial Diamond Association of Japan's 30th Annual Meeting and Seminar, May 1978. Copyright General Electric Company 110

Now: Turn Hardened Steels And Tough Superalloys As Easily As Mild Steels
By *Daniel E. Herzog*
Reprinted from Manufacturing Engineering, October 1975 114

BZN Case Histories And Application Briefs
Reprinted courtesy of the General Electric Company 117

Diamonds In Glass Industry
By *William M. DeAngelis*
Presented at the Industrial Diamond Association of America's
Diamond—Partners in Productivity Conference, 1974 125

3 ABRASIVE MACHINING

Grinding Brittle Materials
By *A. Broese van Groenou and J. D. B. Veldkamp*
Reprinted from the Philips Technical Review, Volume 38, 1979 131

Grinding Damage In Ceramics
By *B. G. Koepke and R. J. Stokes*
Presented at SME's International Tool And Manufacturing
Engineering Conference, May 1977 145

Cutting With Wires And Polishing With Diamonds
By *Howard B. McLaughlin*
Presented at SME's Westec '79, March 1979 158

CDA—The New High Performance Diamond Abrasive For Grinding Carbide
By *A. T. Notter and M. O. Nicolls*
Reprinted from Industrial Diamond Review, December 1976 168

Dry Grinding Of Carbides
By *Roger J. Loecy*
Reprinted from The Carbide And Tool Journal, March-April 1979 174

High-Performance Carbide Grinding
By *Laszlo M. Zsolnay*
Reprinted from The Carbide Journal, September-October 1976 179

Cost-Effective Surface Grinding
By *Paul Grieb*
Reprinted from Tooling & Production, June 1977 182

High Precision Grinding With Superabrasives
By *Hank Deutschendorf*
Presented at SME and GE's Abrasives Workshop, March 1975 187

Grinding High Speed Tool Steels
By *R. A. Moir and W. N. Ault*
Presented at SME's South Atlantic Conference, February 1977 198

Application Of Superhard Materials In Thread Grinding
By *A. D. Pokhorovskii and A. M. Belikov*
Reprinted from Stanki I Instrument, Volume 48, Issue 2, 1977 212

Machining Turbine Parts—Problem Solving With Diamond And CBN Tooling
By *H. Hujer*
Reprinted from Industrial Diamond Review, December 1978 217

Grinding Hardenable Steel Bonded Carbide Alloys With Diamond Wheels
By *F. Frehn and A. T. Notter*
Reprinted from Industrial Diamond Review, June 1979 219

Diamond Tools As A Means Of Rationalising Production In The Optics Industry
By *E. Loh*
Reprinted from Industrial Diamond Review, February 1977 227

Diamond Honing Of Hard Non-metallic Materials
By *V. R. Kangun and R. Z. Tsypkin*
Reprinted from Vestnik Mashinostroeniya, Volume 55, Issue 6, 1975 231

INDEX 241

CHAPTER 1

NONTRADITIONAL MACHINING

Nontraditional Machining Guide

By Guy Bellows
Machinability Data Center

ACTIVE NONTRADITIONAL MATERIAL REMOVAL PROCESSES

Since the early 1940's, many new machining processes have emerged which are generally nonmechanical, layless and involve new energy modes. The 26 nontraditional machining (NTM) processes listed below are those that are in regular productive use and are commercially available — circa 1976.*

MECHANICAL

AFM - Abrasive Flow Machining
AJM - Abrasive Jet Machining
HDM - Hydrodynamic Machining
LSG - Low Stress Grinding
USM - Ultrasonic Machining

ELECTRICAL

ECD - Electrochemical Deburring
ECDG - Electrochemical Discharge Grinding
ECG - Electrochemical Grinding
ECH - Electrochemical Honing
ECM - Electrochemical Machining
ECP - Electrochemical Polishing
ECT - Electrochemical Turning
ES - Electro-stream
STEM - Shaped Tube Electrolytic Machining

THERMAL

EBM - Electron Beam Machining
EDG - Electrical Discharge Grinding
EDM - Electrical Discharge Machining
EDS - Electrical Discharge Sawing
EDWC - Electrical Discharge Wire Cutting
LBM - Laser Beam Machining
LBT - Laser Beam Torch
PBM - Plasma Beam Machining

CHEMICAL

CHM - Chemical Machining
ELP - Electropolish
PCM - Photochemical Machining
TCM - Thermochemical Machining

*See Machinability Data Center publication no. MDC 76-100, *MACHINING, A Process Checklist*[3] for a complete listing of material removal processes.

TYPICAL SURFACE FINISH RANGES

Some material removal processes are more adapted to rough initial cuts whereas others are more suited for use as fine finishing cuts. Typical ranges of attainable surface finishes for some material removal processes are shown in Table I. Additional comparative information is contained in the *Machining Data Handbook*[4] and in American National Standards Institute (ANSI) Standard B46.1-1962, *Surface Texture*.[5] Note that the following chart is an updated version of the chart found in the Second Edition of the *Machining Data Handbook*.

MATERIAL REMOVAL RATES

Material removal rates provide a general guideline only as an initial step in selection among competing processes. Table II compares the material removal rates of some NTM processes with both conventional (mechanical) turning and abrasive grinding. For those NTM processes which simultaneously cut all exposed surfaces, the penetration rate is a more significant value. Part of this data was extracted from *Electrochemical Machining* by DeBarr and Oliver.[6]

Table I

TYPICAL SURFACE FINISHES FROM NONTRADITIONAL MATERIAL REMOVAL PROCESSES

Average Application ■

Less Frequent Application ░

Surface Finish, arithmetic average in microinches

Process		500–250	250–125	125–63	63–32	32–16	16–8	8–4	4–2	
AFM					░	■	■	░		
LSG				░	■	■	■	░		
USM				░	■	■	░			
CHM/PCM		░	■	■	░	░	░			
ELP					░	■	■	■	░	░
EBM/LBM			■	■	■	░				
EDG				░	■	■	░			
EDM roughing	░	■	■	░	░					
EDM finishing			░	■	■	░	░	░	░	
PBM	░	■	░							
ECD/ECP					░	■	■	░		
ECG				░	░	■	■	░		
ECH					░	■	■	░	░	
ECM frontal cut; ECT				░	■	■	░	░	░	
ECM side cut		░	■	■	░	░	░			
ES				░	■	■	░	░		
STEM			░	■	■	░	░			

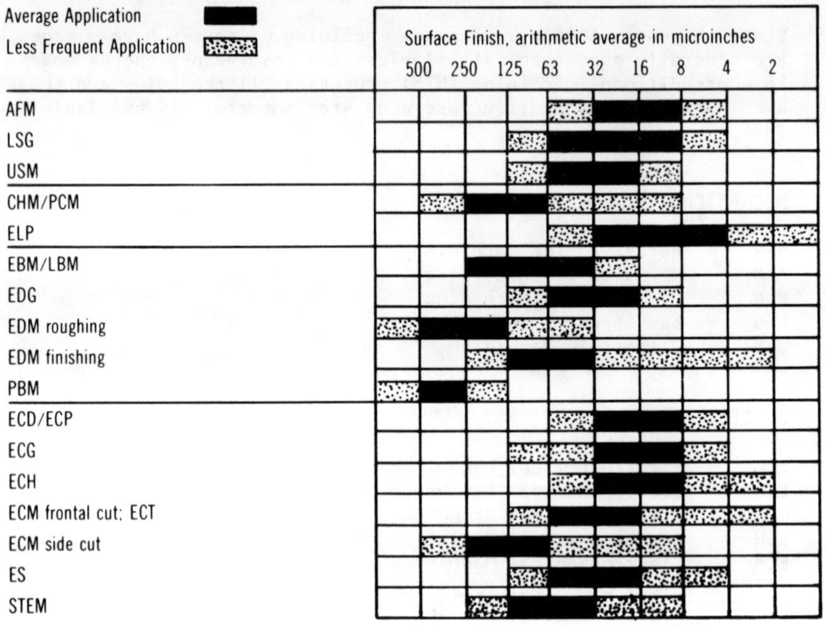

Table II

METAL REMOVAL RATES

Process	Maximum Rate of Metal Removal, inch³/minute	Power Consumption, hp/inch³/minute	Cutting Speed, feet/minute	Penetration Rate, inches/minute	Accuracy, inch: Attainable	Accuracy, inch: At Maximum Metal Removal Rate	Typical Machine Input, horsepower
Conventional Turning	200	1	250	---	0.0001	0.005	30
Conventional Grinding	50	10	10	---	0.0001	0.003	25
PBM	10	20	50	10	0.01	0.1	200
EDM	0.3	40	---	0.5	0.0005	0.005	15
ECM	1	160	---	0.5	0.0005	0.005	200
USM	0.05	200	---	0.02	0.0002	0.001	15
EBM	0.0005	10000	200	6	0.0002	0.001	10
LBM	0.0003	60000	---	4	0.0005	0.005	20
CHM	30	---	---	0.001	0.0005	0.002	---

ULTRASONIC MACHINING

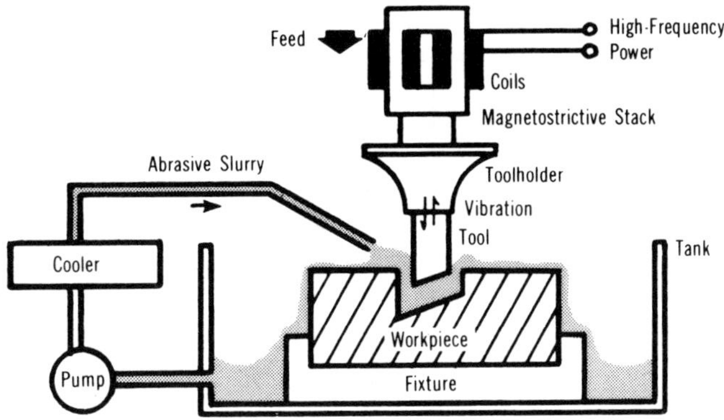

Ultrasonic machining (USM) is the removal of material by the abrading action of a grit-loaded slurry circulating between the workpiece and a tool vibrating at a frequency above the audible range. A high-frequency power source activates the stack of magnetostrictive material, which produces the vibrating motion in the toolholder. The tool forms a reverse image in the workpiece as the grit-loaded slurry abrades the material. Frequencies of 20 to 40 kilohertz with amplitudes of 0.0005 to 0.004 inch are most commonly used. Boron nitride, aluminum oxide and silicon carbide are the most frequently used grits with grit sizes ranging from No. 100 to No. 800. It is important to maintain a full flow of slurry (usually cooled) in the 0.001- to 0.005-inch cutting gap. Overcut is approximately equal to twice the size of the abrasive grit. Surface roughness increases with the size of the grit. The tools — made from brass, tungsten carbide, mild steel or tool steel — will wear from the action of the grit with a ratio that ranges from 1:1 to 200:1 (workpiece to tool wear), depending on the materials involved. The tool must be designed to resonate at the desired frequency for best results and must be strong enough to resist fatigue failure.

PRACTICAL APPLICATIONS

While USM can cut any material, conductive or nonconductive, metallic, ceramic or composite, it is most effective on materials harder than 40 R_c. Holes, slots and irregular shapes can be produced in delicate ceramics. Over 2,000 holes, 0.031-inch square, were "drilled" simultaneously in 0.040-inch-thick carbon in less than 10 minutes. A 1/4-inch-diameter hole can be drilled 5 inches deep in glass in 130 seconds with a rotating ultrasonic tool. To achieve resonance the tool size is limited to about 1-1/2 inches diameter currently. Tool wear and taper in the cut can be limiting with practical depth-to-width ratio being 2.5:1. Threading of ceramics can be accomplished with a rotating tool and workpiece. Coining, lapping, broaching and deburring are also done with the ultrasonic process. Ultrasonic assist is sometimes helpful as added energy to conventional drilling and drawing. Ultrasonic welding has been applied to plastic assemblies.

MATERIAL REMOVAL RATES AND TOLERANCES

The removal rate is slow and depends on the ability to circulate the slurry; the rate is inversely proportional to area of cut and proportional to grit size and the square of the amplitude of vibration. For glass, a penetration rate of 0.150 inch per minute with a 1/2-inch-diameter tool proceeds with a 100:1 wear ratio when using a low carbon steel tool with a 35 percent slurry of 200-grit boron carbide vibrating at 25 kHz and 0.0015 inch amplitude. Finish improves with smaller grit size and can be in the 10- to 40-microinch AA range. Accuracy is typically ± 0.001 inch and can achieve ± 0.0005 inch. Break-out and chipping of exits can be a problem. Thin parts are often cemented to a sacrifice plate. Surface integrity is good and the compressive layers can enhance fatigue strength.

AVAILABILITY
Equipment for cavity and hole sinking is available as are rotary heads with axial ultrasonic vibration. A portable ultrasonic drill is produced and used on aircraft assembly lines. Currently, a 3-1/2-inch-diameter tool is the largest in use.

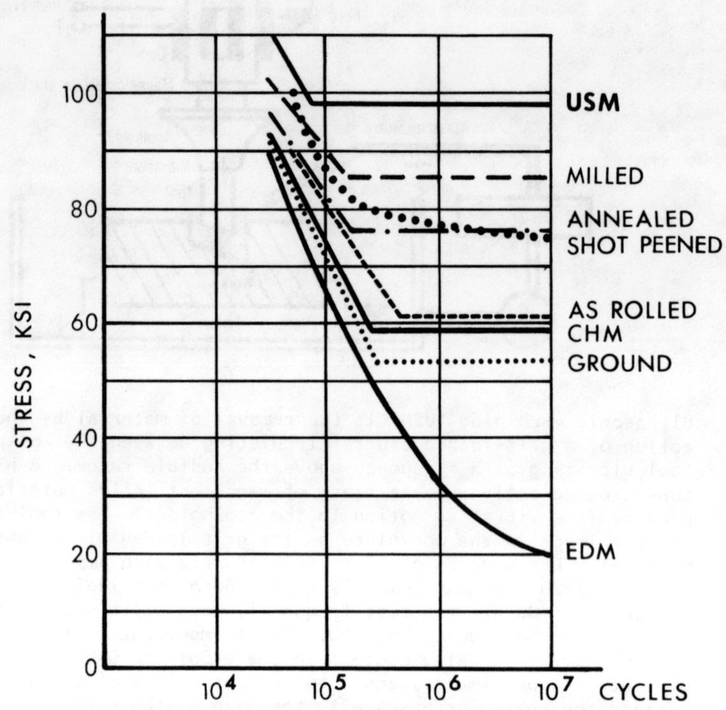

Fatigue Endurance Strength of 6Al-2½V Titanium Alloy. USM compared to other processes. (Source: AFML report WADC-TR-57-310)

ELECTROCHEMICAL MACHINING

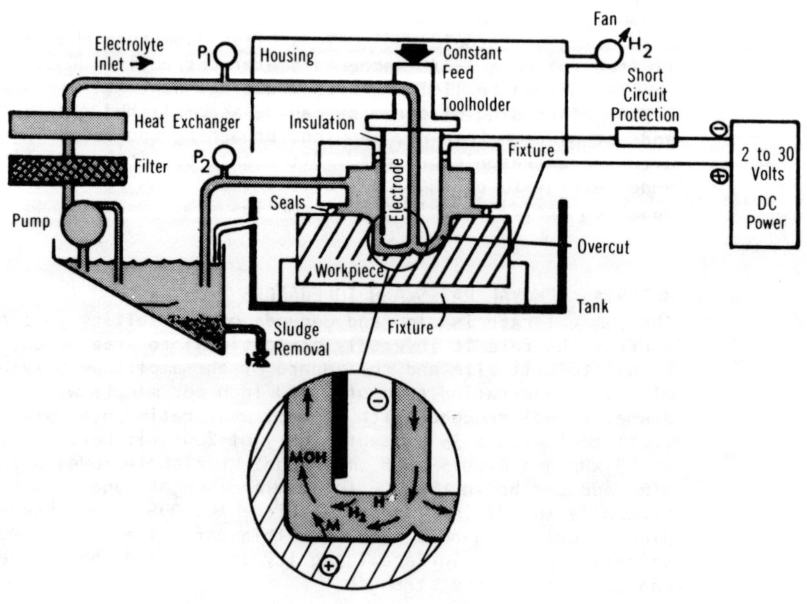

Electrochemical machining (ECM) is the removal of electrically conductive material by anodic dissolution in a rapidly flowing (50 to 200 fps) electrolyte, which separates the workpiece from a shaped electrode. The electrolyte is pumped under pressure to bring a freshly filtered solution with controlled conductivity and controlled temperature to the cutting area. The shape of the workpiece is nearly a mirror image of the shape of the electrode, which is advanced into the workpiece at a constant feed that exactly matches the rate of dissolution of the material. The working gap ranges from 0.003 to 0.030 inch with 0.010 inch being typical. The current density (50 to 1000 amperes per square inch) is the chief factor in setting feed rates and attaining smoothness; higher feed rates create better finishes. Hydraulic pressures and tool/workpiece separating forces increase with smaller gaps and affect workpiece tolerances. Tools are insulated with epoxies and other plastics. They must be securely attached to the tool walls to withstand the temperature, hydraulic pressure and electrolyte flow. Copper and brass electrodes are common and must be designed to carry the high currents. The highly conductive electrolytes are aqueous solutions of inorganic salts such as NaCl, KCl, $NaNO_3$ or mixtures with proprietary additives operating at 100 to 120°F. Acids are used in some applications. A sludge of considerable volume is generated with salt electrolytes and must be removed by filtration, settling, or centrifuge. The electrolyte must be cooled to control conductivity. The inlet pressure (to 300 psi) and outlet pressure (to 30 psi) are controlled to assure uniform, generally turbulent, flow of about 0.25 gpm per 100 amperes.

To obtain tight tolerances, tool design must compensate for the variable current density that occurs with shape and electrolyte variations. Exact control of all critical parameters is needed for the best results. In "cutting", the metal ions are removed from the workpiece surface; hydrogen ions from the electrode. These combine quickly in the electrolyte to form metal hydroxides and hydrogen. The hydrogen gas which is released at the electrode must be adequately vented.

PRACTICAL APPLICATIONS

ECM is best suited for mass production of complex shapes in difficult-to-machine materials. Small, odd-shaped and deep holes down to 1/8-inch diameter can be "drilled" individually or in multiple. The stress-free material removal eliminates distortion from machining (but not necessarily from prior stress-inducing operations). The ability to cut on the entire surface simultaneously aids productivity. Tool design and development, except for the most simple shapes, is time consuming and may require several "cut and try" cycles. Process control must be exact. Sludge or effluent disposal should be environmentally planned. Concentration of current density at edges of the workpiece provides automatic rounding and absence of burrs. The workpiece must be thoroughly cleaned after electrochemical machining to prevent corrosion.

MATERIAL REMOVAL RATES AND TOLERANCES

Material removal rate is independent of material hardness and is approximately 0.1 cubic inch minute per 1000 amperes. Accuracy to ±0.002 inch is usual for cavities and to ±0.0005 inch for frontal cuts or with highly refined tools. Internal radii of 0.007 inch and external radii of 0.002 inch are attainable. Deep cuts will have taper; 0.001 inch per inch is common with a 0.005-inch overcut gap. Tolerance capabilities are dependent on part geometry, tool design and particular shop practices and need careful checking with experienced people for particular applications. Surface finishes of 16 to 63 microinches AA are normal and improve with higher cutting rates. Mirror finishes in frontal cuts of nickel alloys are easily obtained. The side-gap areas are generally much rougher because of the lower current densities in these areas. Incorrect electrolytes, different workpiece material heat treatment, or low current densities can produce selective etching pits or intergranular attack at the grain boundaries. The absence of residual stress (as contrasted to the compressive stresses from many mechanical processes) produces a reduced high cycle fatigue strength (by comparison) but these values more nearly represent the pure material values. Any physical blemish in the tool will be reproduced on the workpiece and poor flow conditions can produce striations in the surface. Hydrogen embrittlement is not a problem because the hydrogen is liberated at the electrode.

AVAILABILITY

The available equipment ranges from small, 50-ampere bench models to 40,000-ampere models with a five-foot-cube working space. Power sources from 4 to 30V dc are available with special, fast acting, short circuit protective devices. Ancillary equipment such as pumps, tanks, plastic piping, mixing vats and sludge-disposal devices must be engineered as a complete system. Equipment and tools should be rigid enough to withstand the forces from the high hydraulic pressures.

ELECTRICAL DISCHARGE MACHINING

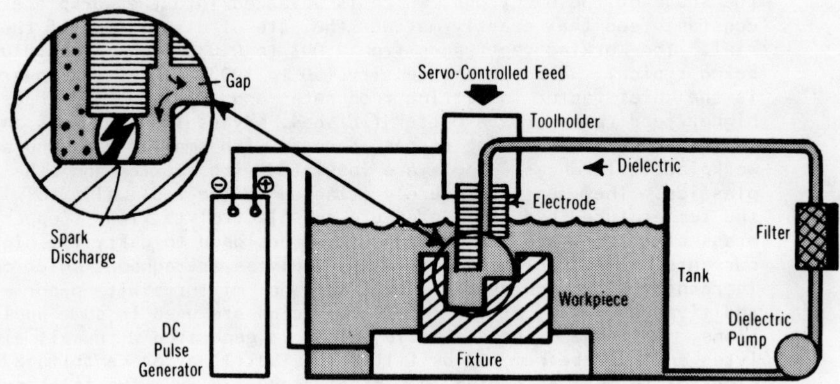

Electrical discharge machining (EDM) removes electrically conductive material with rapid repetitive spark discharges from a pulsating d-c power supply with a dielectric flowing between the workpiece and the tool. The shaped tool (electrode) is fed into the workpiece under servo control until a spark discharge breaks down the dielectric fluid. The frequency (200 to 500,000 sparks per second) and energy per spark are set and controlled with a d-c power source. The servo control maintains a constant gap between the tool and the workpiece while advancing the electrode. The dielectric oil flushes out the vaporized and condensed material while reestablishing insulation in the gap. Surface finish improves with increased frequency and reduced current. Material removal rate, surface roughness and overcut all increase with a current increase or with a frequency decrease (or longer "on" time cycles). Electrode materials frequently used are brass, copper, copper tungsten, tungsten wire and graphite. Spark gaps range from 0.0005 to 0.020 inch with closer tolerance control and slower cutting rates associated with the smaller gaps. Erosion occurs on the tool as well as the workpiece with wear ratios ranging from 0.5:1 to 100:1, depending on spark wave-shape from the power source, electrode material and workpiece material. A nearly "no wear" combination of operating parameters can be found for electrical discharge machining steel when using reverse polarity, as opposed to "standard" polarity (positive on the workpiece).

PRACTICAL APPLICATIONS

EDM cuts any electrically conductive material regardless of its hardness and is particularly adapted for machining small irregular slots or cavities. Because of the absence of physical contact, delicate structures can be cut successfully. Cutting is three-dimensional as the shaped electrode is fed into the workpiece. Because the sparks focus first on peaks and corners, burr-free cutting occurs. Multiple electrode, automatic dressing, automatic positioners and NC motion control all contribute to electrical discharge machining's versatility. Tool and die work is frequent but mass production and even transfer line applications exist. Small and/or shaped holes at shallow angles to the workpiece surface are commonplace. A recast and heat-affected layer occurs on all materials cut with EDM and needs to be removed or modified on critical or fatigue-sensitive surfaces.

MATERIAL REMOVAL RATES AND TOLERANCES

Feed rates and material removal rates range from 0.06×10^{-4} to 6.6×10^{-4} cubic inches per minute per ampere. Corner radii to 1/64 inch are common. Production tolerances to ± 0.001 inch are normal; tolerances of ± 0.0002 inch are repeatable with careful selection of cutting conditions. Finish levels are typically in the 63- to 125-microinch AA range, but deluxe equipment can attain 2 to 4 microinches AA. The recast layer can be controlled and is repeatable to a few ten thousandths of an inch. Material removal rates, surface roughness, recast layer and heat-affected zones all increase as spark intensity increases.

AVAILABILITY

EDM equipment is available in a wide range of sizes from a bench model with a few amperes capacity to 4 ft. x 6 ft. die sinkers with 3,000 amperes capacity. Automatic or NC controls are common. Automatic feed, interchangeable electrode holders and rotary turret electrode holders are available to aid electrode changing and automation of the EDM process. Multiple electrodes and multi-lead power supplies enhance the productivity of many current equipment types. Integrated systems are the usual order; thus, EDM machines can be placed almost anywhere in the normal shop. Fume vents are needed and tooling should provide for venting the gases liberated. The pulse power supply usually contains full control of "on and off" times for each discharge as well as discharge energy. Good safety practice makes it desirable to operate with the spark fully submerged in the dielectric.

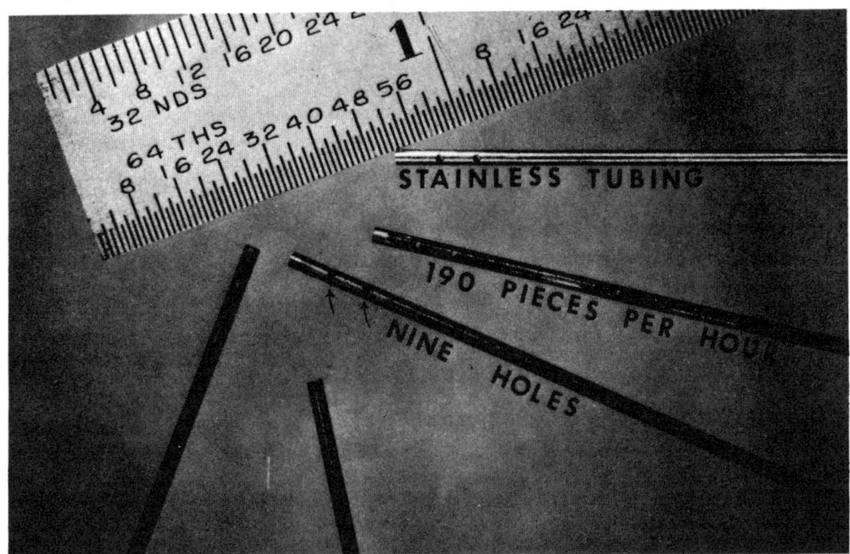

EDM of stainless steel tubing. (Courtesy of Raycon Corp.)

LASER BEAM MACHINING

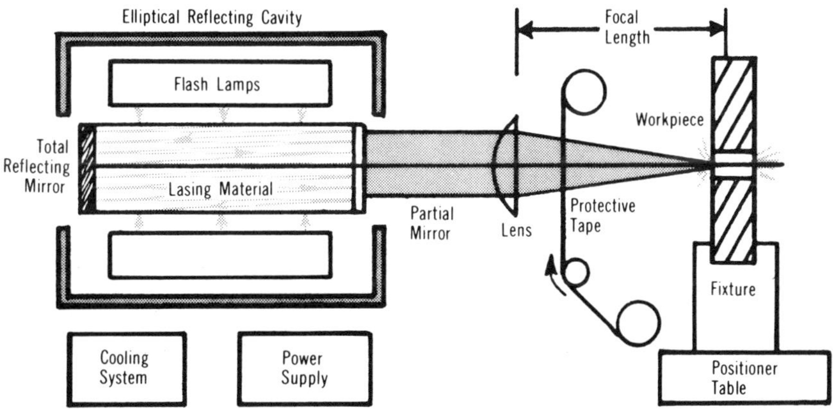

Laser beam machining (LBM) removes material by melting and vaporizing the workpiece at the point of impingement of a highly focused beam of coherent monochromatic light. Laser is an acronym for "light amplification by stimulated emission of radiation." The electromagnetic radiation operates at wavelengths from 0.3 to 300 microns. The most common wavelengths for material removal using solid state lasers are 0.69 micron for ruby, 1.06 for neodynium glass (Nd:Glass), 1.06 for neodymium yttrium aluminum garnet (Nd:YAG), and 10.6 for carbon dioxide gas laser. Of these four, only the 0.69 micron radiation of the ruby is in the visible light range. The Nd:Glass and Nd:YAG wavelengths are in the near infrared and the CO_2 wavelength is in the far infrared range. The optical characteristics of the workpiece determine which wavelength should be used.

Most of the solid state lasers operate only in a pulsed mode with greater than 10 pulses per second repetition rate common with adequate cooling systems. An exception to this is the Nd:YAG laser, which like the CO_2 laser, can be operated either pulsed or in a continuous wave to produce a continuous power output.

For pulsed operation of solid state lasers, the power supply produces a short intense pulse into the flash lamps which concentrate their light flux on the lasing material. The resulting energy from the excited atom is released at a constant frequency. The monochromatic light is amplified by successive reflections from the mirrors. The thoroughly collimated light exits through the partially reflecting mirror to the lens, which focuses it on or just below the surface of the workpiece. The small beam divergence, high peak power and single frequency provide excellent small-diameter spots of light with up to 3×10^{10} watts per square inch of power that can sublime almost any material.

During laser beam machining, the expulsed material solidifies to a dust; therefore, a cleaning system is needed in addition to protection for the lens from the molten particles. Adequate eye protection is needed for both direct and reflected laser light.

PRACTICAL APPLICATIONS

Small precision cuts or holes in thin materials can be produced by LBM. Scribing of ceramics can be done since there is no massive heat shock or mechanical contact or large forces between the tool and workpiece. LBM is not a mass material removal process; however, its operation in air at rapid repetitive rates and its ease of electrical control commend it for mass micromachining production. Multiple pulses permit hole drilling up to 50:1 depth-to-diameter ratios on 0.005-inch-diameter holes while 0.050-inch-diameter holes can be drilled through 0.100-inch-thick material. Shallow angles (15 degrees) to the surface can be drilled. Other applications include engraving, resistor trimming, sheet metal trimming and blanking. The same equipment can be used to weld, surface heat treat or machine, which makes the laser a "universal" machine tool.

MATERIAL REMOVAL RATES AND TOLERANCES

Removal rate is slow, 4×10^{-4} cubic inch per minute; however, 0.020-inch-diameter holes can be drilled in milliseconds with accuracies to ±0.001 inch in thin stock of tungsten, brass or ceramics. The hole walls will be irregular and tapered as well as having a recast structure from the heat-affected surface. Repeatability is good and yields can be improved over conventional micromachining processes. The heat-affected zones with a hardened recast layer are typical of laser machined surfaces. This recast layer can be detrimental to material properties and should be removed or modified on applications where high stresses or fatigue life is a concern.

AVAILABILITY

Several sources of laser components or systems exist. The principal equipment concern is workpiece positioning and control. Integration of an NC table with focus, beam intensity or standoff distance is common. Safety interlocked enclosures are commonly used. Bench-top equipment with a few watts capacity to computer controlled systems of several kilowatts capacity are commercially available.

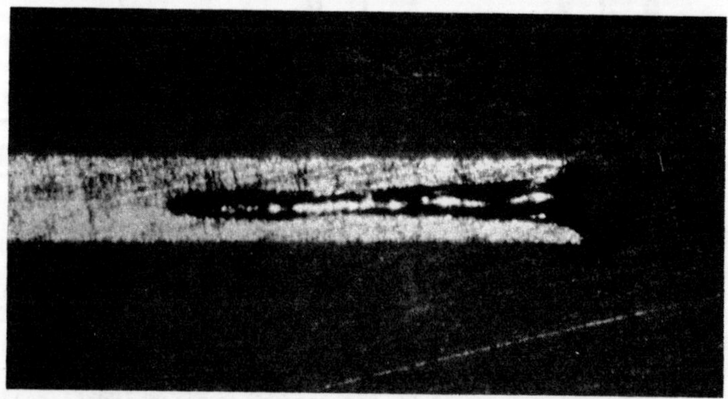

Surgical needle (0.013 inch diameter) with laser drilled 0.006-inch-diameter hole, 0.060 inch deep. (Courtesy of Holobeam Laser Inc.)

PHOTOCHEMICAL MACHINING

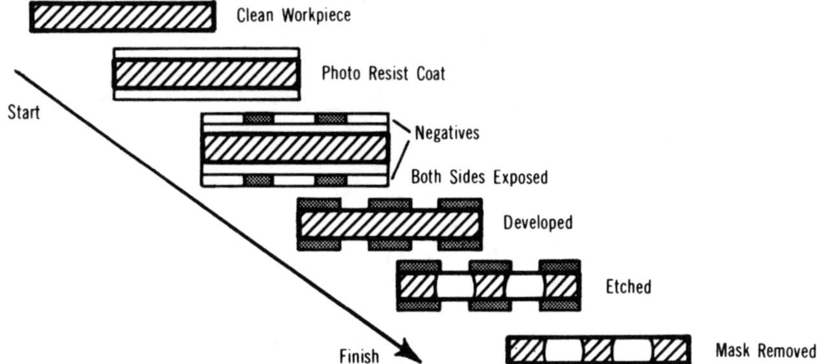

Photochemical machining (PCM) is a variation of CHM where the chemically resistant mask is applied to the workpiece by a photographic technique. A photographic negative, often a reduced image of an oversize master print, is applied to the workpiece and developed. Precise registry of duplicate negatives on each side of the sheet is essential for accurately blanked parts. Immersion or spray etching is used to remove the exposed material. The chemicals used must be active on the workpiece but inactive against the photoresist mask. There will be some undercutting behind the mask, however, which limits PCM to thin materials (3/32 inch).

PRACTICAL APPLICATIONS
Photochemical blanking is burr free and is capable of intricate designs. A wide range of materials can be etched; brittle materials can be "blanked" since there is no mechanical strain. Small lot sizes can be produced at lower cost than with conventional mechanical presses and there is a short cycle time from design to finished parts. Cost of tooling is low and design changes can be quickly effected. Care in handling the corrosive chemicals is needed.

MATERIAL REMOVAL RATES AND TOLERANCES
Cutting rates range from 0.0004 to 0.0020 inch per minute, depending on the material and its metallurgical state. Thin materials can be held to a tolerance of 10 percent of sheet thickness with rigid control, and 0.001- to 0.002-inch tolerance for routine production. Shallow pockets can be held to the same tolerance.

AVAILABILITY
Batch and automatic equipment is available for both dip and spray etching strips up to two feet wide. Some installations include an electrical assist to the light, low-concentration chemicals in order to obtain a more straight cut. DC power, with low current density, is required.

Typical parts blanked with PCM. (Courtesy of Chemcut Corp.)

Presented at SME's Nontrad '76, October 1976.

Surface Integrity Of Nontraditional Material Removal Processes

By Guy Bellows
And John B. Kohls
Metcut Research Associates Inc.

Nontraditional material removal processes now total 26 with several in each principal energy mode. These processes have emerged in the last 35 years and are now in regular production use. In the same time period the principles of surface integrity have also evolved with their contribution to greater product reliability. While surface integrity principles affect conventional machining to an equal extent, this paper presents the current status of surface integrity of the principal nontraditional machining processes. Much new data has been generated since the last comprehensive summary, nearly 5 years ago. An updated selection of the most pertinent data, typical illustrations of the various surface effects and guidelines for processing are presented herein. An awareness of the impact of processing on material properties is essential for all design, manufacturing and quality control engineers.

INTRODUCTION

Almost four years has elapsed since the International Conference on Surface Technology[1,2] was held at Carnegie-Mellon University and eight years since the first summary of surface integrity effects on nontraditional machining (NTM).[3] In 1968, there were 12 commercially available NTM processes. The 18 processes of 1973 have now grown to 26 (Table 1).

Surface integrity has changed from a concept in the mid 60's to today's disciplined approach to enhance quality assurance. It is backed by collatable comparative data generated in several government programs[4,5,6,7] and in industrial laboratories.[8,9] From an early heavy emphasis on the effects from NTM it is now well established that all material removal processes have some impact on material properties - the conventional mechanical processes can have effects as severe as those seen on the NTM processes.[10,11] Only the application and use of the workpiece can determine whether the surface integrity effects are beneficial or detrimental.

This paper is an update of the previous summaries[2,3] to include the latest available data in a consolidated manner. The depths of the altered material zones has been reviewed to include information from Metcut Research Associates surface integrity encyclopedia. Typical surface section photomicrographs have been selected from among the several thousands reviewed. While many of the words remain unchanged, the former guidelines have been reviewed and amended to reflect our experiences to date.

TABLE 1

COMMERCIALLY AVAILABLE NONTRADITIONAL MACHINING PROCESSES (circa June 1976)[19]

MECHANICAL	-	AFM	-	Abrasive Flow Machining
	-	AJM	-	Abrasive Jet Machining
	-	HDM	-	Hydrodynamic Machining
	-	LSG	-	Low Stress Grinding
	-	USM	-	Ultrasonic Machining
CHEMICAL	-	CHM	-	Chemical Machining
	-	ELP	-	Electropolish
	-	PCM	-	Photochemical Machining
	-	TCM	-	Thermochemical Machining
THERMAL	-	EBM	-	Electron Beam Machining
	-	EDG	-	Electrical Discharge Grinding
	-	EDM	-	Electrical Discharge Machining
	-	EDS	-	Electrical Discharge Sawing
	-	EDWC	-	Electrical Discharge Wire Cutting
	-	LBM	-	Laser Beam Machining
	-	LBT	-	Laser Beam Torch
	-	PBM	-	Plasma Beam Machining
ELECTRICAL	-	ECD	-	Electrochemical Deburring
	-	ECDG	-	Electrochemical Discharge Grinding
	-	ECG	-	Electrochemical Grinding (sometimes ELG)
	-	ECH	-	Electrochemical Honing
	-	ECM	-	Electrochemical Machining (Milling)
	-	ECP	-	Electrochemical Polishing
	-	ECT	-	Electrochemical Turning
	-	ES	-	Electro-stream
	-	STEM	-	Shaped Tube Electrolytic Machining

SURFACE INTEGRITY

Surface integrity is a subject covering the *description* and *control* of the many possible alterations produced in a surface layer during manufacturing including their *effects* on the material properties or on the performance of hardware. The objective of surface integrity is the development of unimpaired or enhanced surface conditions in hardware by controlled manufacturing processes to the extent needed in specific applications. The principal causes of alterations from material removal operations are:

1. high temperatures or high temperature gradients

2. plastic deformation

3. chemical reactions on the nascent machined surface

4. excessive heating from electrical conductance.

The distortion, inaccuracies or change in material properties due to fabrication processes are of concern to the designer *and* the manufacturer as a reduction of quality or a source of loss and cost. Some of the effects can also contribute to the initiation of early component failures. Table 2 is a tabulation of the surface integrity effects that have been observed. These effects on material properties can in turn influence the mechanical properties of which fatigue life and stress corrosion resistance are of the most concern.

A well disciplined sequence and method for evaluating surface integrity exists today[4,12] and data is slowly accumulating upon which to draw guidelines and uncover trends or patterns. These guidelines,[2] however, must be considered only general or starting recommendations. Each material-process combination is unique and can have variable effects depending on the metallurgical state of the material and the energy intensity level used during processing. The designer should assess the critical areas on the workpiece and apply surface integrity specifications to these spots only - or the costs of the component can be excessive. The manufacturing engineer also should realize that maintaining the proper sequence of operations is as important to surface integrity as the selection and precise maintenance of the process operating conditions.

Processes are usually operated over a range of intensities. The roughing or finishing modes are reflections of these differing intensities - or energy densities. Sometimes they are labeled abusive or gentle or similar words. From a surface integrity standpoint it is necessary to consider the change in effects on the workpiece material over the full range of energy levels expected to be used. Actual shop practice, however, results in a planned set of operating conditions or process parameters. These typical, usual or conventional values are the "standard" values or "standard" operating intensity for a given process on a specific item of hardware. If these are "standard" conditions, then there are "off-standard" conditions that represent deviations from standard. A convenient "rule of thumb" claims that a deviation of more than 10 percent from a planned operating parameter is sufficient to be labeled "off-standard." A guideline for quality assurance

TABLE 2

SURFACE INTEGRITY ALTERED MATERIAL ZONES
(by principal energy mode)

Mode	Altered material zones
Mechanical:	Plastic deformations Tears, laps and protuberances Hardness alterations Cracks (macroscopic & microscopic) Residual stress Inbedded processing materials
Metallurgical:	Transformation of phases Grain size and distribution Precipitate size, shape and distribution Foreign inclusions in material Recrystallization
Chemical:	Intergranular attack (IGA) Intergranular corrosion (IGC) Intergranular oxidation (IGO) Contamination Embrittlement Pits or selective etch Corrosion Stress corrosion
Thermal:	Heat-affected zone (HAZ) Recast or redeposited material Resolidified material
Electrical:	Conductivity change Magnetic change

states that the necessity to operate with two or more parameters off-standard is sufficient cause to shut down a process until corrective action is taken to bring the process back to "standard" operating conditions. The surface integrity of components produced with off-standard operating conditions is frequently less than that produced under standard conditions. While NTM processes are becoming better understood, it is essential that they (as well as the conventional processes) be operated under relentless insistence upon maintaining the standard, preplanned operating conditions. This paper will use the terms "typical process parameters" and "off-standard conditions" to delineate the surface integrity effects. Table 3 lists some of the descriptors of this terminology which as yet, unfortunately, means different things to different people.

TABLE 3
ROUGHLY EQUIVALENT TERMINOLOGY

Typical Process Parameters	Off-Standard Conditions
standard	abusive
optimum	roughing
gentle	atypical
finishing	high stress
conventional	accidental
average	extra-ordinary
controlled	irregular
preplanned	
low stress	
ordinary	

Perhaps, with sufficient data, surface integrity effects could ultimately be related to process energy levels for particular materials. This could bring precision and eliminate the need for the collective general terms of Table 3 by creating an energy level that extends from gentle to finishing to normal to roughing to abusive processing.

With a consistent material metallurgical state, carefully selected process operating parameters and relentless quality monitoring, the surface integrity of workpieces can be assured at the level needed in today's reliability conscious market. For further details the reference list contains sources for very specific data.[4,5,6,7,10,11,15,16]

The family of chemical nontraditional material removal processes is characterized by an absence of stress introduced into the workpiece as a result of the processing. The gentle chemical action of removing material by dissolution molecule by molecule is sometimes aided by the addition of small amounts of electricity. The principal varieties of nontraditional chemical machining today are:[13]

CHM	Chemical Milling (or Machining)
CHP	Chemical Polishing
EGM	Electro Gel Machining
ELP	Electropolish (electrical assist)
PCM	Photochemical Machining
TCM	Thermal Chemical Machining (thermal assist)

The CHM surface roughness values are influenced by the roughness present in the initial workpiece surface and the degree of adherence to the preplanned conditions. Typically very smooth surfaces (32 microinches AA or less) are slightly roughened when chemically machined while rough surfaces (125 microinches AA or more) can be slightly smoothed. On a microscale the uniformity of the workpiece material and the fineness of the grain structure are other principal influences on surface roughness. Selective etching or intergranular attack (IGA) are the most worrisome of subsurface effects. A slight softening of the surface (Rebinder Effect)[14] is frequently (but not always) found. Surface discontinuties frequently result in different surface texture or cutting rates with welded joints a particular difficulty because of its different metallurgical state. There is no lay pattern to CHM surface - the surface texture is mottled on a microscale with each grain chemically etched to its individual level.

Good surface integrity from chemical material removal begins with a careful matching of the reagents to the metallurgical state and composition of the workpiece material. Proper process controls of reagent composition, temperature and adequate stirring to assure a constant stream of fresh chemicals impinging on the surfaces to be cut is needed to attain good processing. Off-standard conditions include deviations from the above and reagents "too old" or too loaded with metallic ions. The typical or off-standard conditions can produce some of the effects shown in Table C1. The altered material zones (AMZ) will not necessarily be present or be present to the depths displayed in Table C1. These are the maximum values that have been observed in reviewing data on many cases of CHM on many different materials. The values represent the maximum extent that has been observed to date of particular effects at the two levels of processing.

Figures C1 and C2 are 1000X photomicrographs of typical and off-standard CHM surfaces. Figure C1 is CHM of Rene' 41 of hardness 45 R_c in the fully solution treated condition. Its original 8 to 12 microinch AA finish became 50 to 65 microinch AA finish and shows an unleveling of the surface in a random pattern with no pronounced selective etch. One insignificant spot of IGA, .0003 in. deep, was found on this specimen. Figure C2 is off-standard CHM of fully solution treated and aged Rene' 41 with a 100 to 130 microinches AA finish and filled with selective etch and IGA. The selective etch proceeded down the twinning planes and the IGA had a maximum crack .0025 inch deep in this section.

Figure C3 shows a typical "Rebinder Effect" softening in a solution treated and aged specimen of Rene' 41 that had been chemically machined. The depth of hardness alteration was .00075 inch. This is the same specimen as Figure C2 only chemically machined to a greater depth. Its finish was 190 to 290 microinches AA.

Static material properties (tensile strength, yield) have not been found to change in specimens prepared by CHM. Dynamic properties (fatigue and corrosion) have been found to have some diminution as compared to conventionally prepared specimens. There is a strong indication that the principal mechanism for this difference is the absence of residual stress in chemically machined surfaces as compared to the fatigue enhancing compressive residual stress usually found in conventionally prepared surfaces. The chart in Table C2 illustrated some of the data currently available on high cycle fatigue (HCF) endurance strength. The range shown covers experience with both typical process parameters and off-standard conditions. Table C1 relates HCF to conventionally prepared specimens. These data are for room temperature, HCF, in full reverse cantilever bending and are the values at 10^7 cycles. Obviously, these reductions in fatigue strength are of concern to the designer and post machining processing is essential if CHM is applied to fatigue critical or stress sensitive surfaces - but not of concern if applied to ordinary nonfatigue critical surfaces.

Guidelines for CHM can only be a starting point for surface integrity practices at this time. They represent a concensus of current data and experience reduced to the series of statements in Table C3.

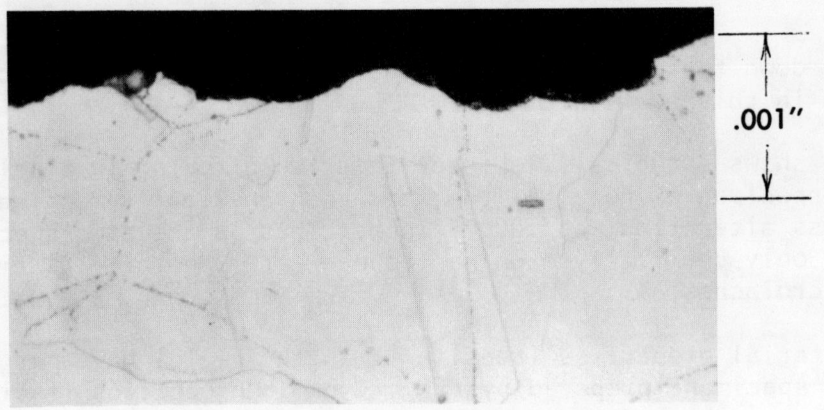

FIG. C1. TYPICAL CHM SURFACE RENE' 41st, 45 R_C
50-65 Microinches AA

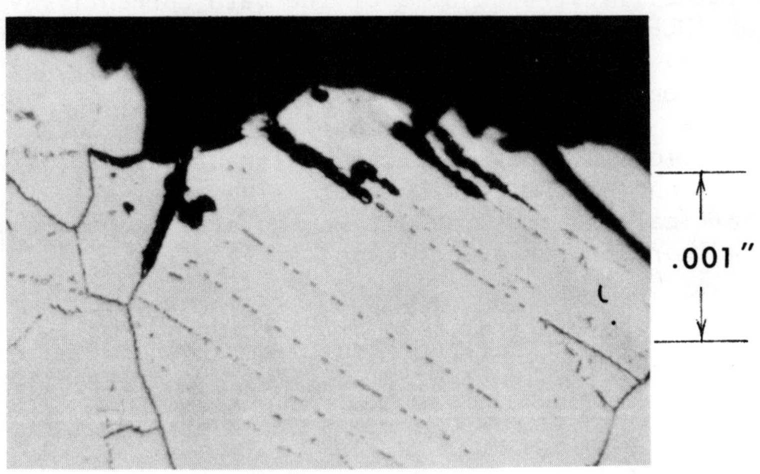

FIG. C2. OFF-STANDARD CHM SURFACE RENE' 41 STA, 45 R_C
100-130 Microinches AA
Note: both IGA and selective etch

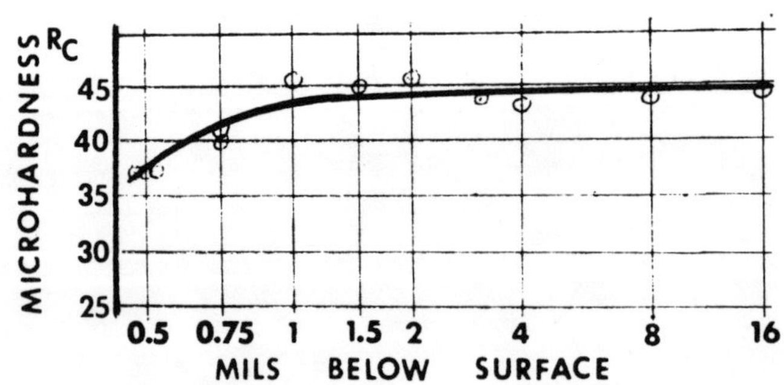

FIG. C3. REBINDER EFFECT ON CHM
OF RENE' 41, STA, 190-290 MICROINCHES
AA FINISH

TABLE C1

SURFACE INTEGRITY EFFECTS OBSERVED IN CHEMICAL NONTRADITIONAL MATERIAL REMOVAL PROCESSES

Type of Effect		Typical Process Parameters	Off-Standard Conditions
Surface Roughness:	Average Range - Microinches AA	63 - 250	125 - 500
	Less Frequent	8 - 500	63 - 500
Mechanical* Altered Material Zones:	Plastic Deformation (PD)	---	---
	Plastically Deformed Debris (PD^2)	---	---
	Hardness Alteration †	1.0	3.1
	Microcracks or Macrocracks		
	Residual Stress §	1.0	1.0
Metallurgical* Altered Material Zones:	Recrystallization	---	---
	Intergranular Attack (IGA)	0.3	6.0
	Selective etch, pits, protuberances	0.6	1.5
	Metallurgical Transformations		
	Heat-Affected Zone (HAZ) or Recast Layer	---	---
High Cycle Fatigue (HCF):	% change from "handbook" values at room temperature #	+18 to -39	-22 to -37

Source: Guy Bellows, *NONTRADITIONAL MACHINING GUIDE, 26 Newcomers for Production*, MDC 76-101, Cincinnati, OH: Machinability Data Center, 1976.

Note: A blank in the table indicates no or insufficient data.
A --- in the table indicates no occurrences or not expected.

*Maximum observed depths in thousandths of an inch, normal to the surface.

†Depth to point where hardness becomes less than ±2 points R_c (or equivalent) of bulk material hardness (hardness converted from Knoop microhardness measurements).

§Depth to point where residual stress becomes and remains less than 20 ksi or 10% of tensile strength, whichever is greater.

#"Handbook" values from HCF testing are frequently generated from low stress ground specimens, hand or gentle machine polishing or occasionally electropolishing. These values are based upon LSG as 100% (with its minor amount of retained but enhancing compressive residual stress).

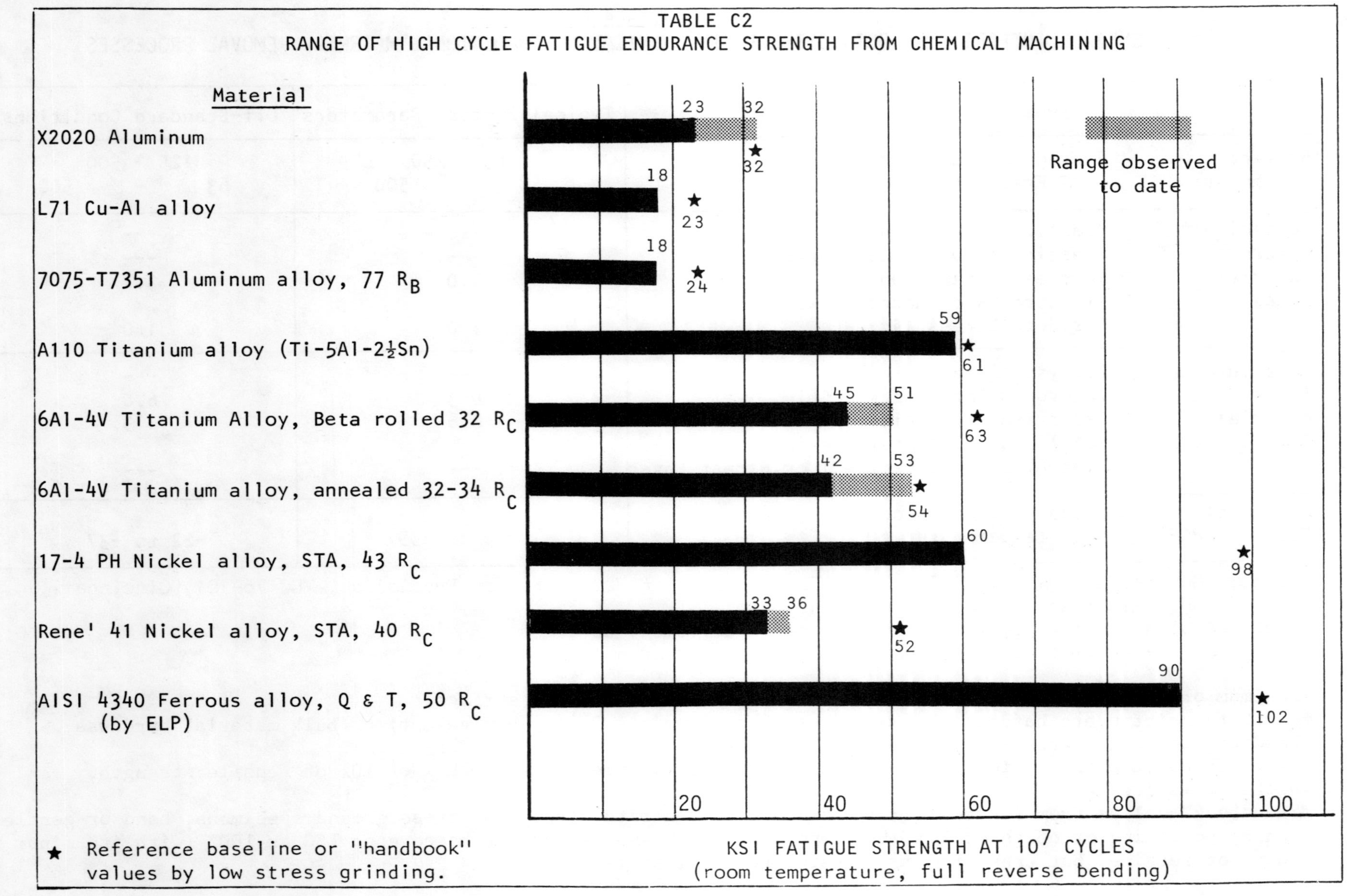
TABLE C2
RANGE OF HIGH CYCLE FATIGUE ENDURANCE STRENGTH FROM CHEMICAL MACHINING
Material
X2020 Aluminum
23
32
32
L71 Cu-Al alloy
18
23
7075-T7351 Aluminum alloy, 77 R B
18
24
A110 Titanium alloy (Ti-5Al-2½Sn)
59
61
6Al-4V Titanium Alloy, Beta rolled 32 R C
45
51
63
6Al-4V Titanium alloy, annealed 32-34 R C
42
53
54
17-4 PH Nickel alloy, STA, 43 R C
60
98
Rene' 41 Nickel alloy, STA, 40 R C
33 36
52
AISI 4340 Ferrous alloy, Q & T, 50 R C (by ELP)
90
102
Range observed to date
20
40
60
80
100
★ Reference baseline or "handbook" values by low stress grinding.
KSI FATIGUE STRENGTH AT 10 7 CYCLES
(room temperature, full reverse bending)

TABLE C3
CHEMICAL PROCESSES GUIDELINES

1. Chemical machining does not induce any significant stress in the machined surfaces.

2. Surface roughness standards should be reassessed when applying CHM due to the different texture and absence of lay pattern.

3. The chemical reagents must be matched to the expected metallurgical state of the workpiece material.

4. Surface roughness variations during processing can be a good indicator of changing processing conditions.

5. Careful rinsing of the solutions from the finished workpiece is essential.

6. The metallurgical and heat treat state of the workpiece is essential.

7. Selective etching, intergranular attack and pitting can result from off-standard conditions such as high temperature in the solutions, incomplete stirring, depleted or unbalanced solutions or contaminated solutions, and variations in the metallurgical state of the workpiece material.

8. Weld areas usually show a different rate of cutting with an increase in surface roughness.

9. Steel and nickel base alloys, susceptible to hydrogen embrittlement, should have a post heat treatment of a few hours at a low temperature (375-400° F). It should be applied immediately after chemical processing.

10. Room temperature fatigue endurance strength generally is lower when compared to conventionally prepared low stress ground specimens.

11. The use of post treatments to add a compressively stressed surface layer may be desirable to enhance component fatigue strength.

12. A test coupon for metallurgical evaluation should be made at least at 90 to 180 day intervals, or whenever the solutions are changed.

THERMAL MATERIAL REMOVAL

The thermal nontraditional material removal processes utilize a variety of heat sources to melt, vaporize or sublime the workpiece surface: electrical sparks, plasma, electrons or light.[13] While details of the surface effects differ, the surface and the subsurface heat-affected zones are similar. The principal difference is the energy density level. The speed of the thermal wave differs and the presense of other energy modes can be significant in the surface integrity effects. The temptation to extend the more abundant EDM data to the other processes should be resisted.[2] The principal varieties of nontraditional thermal machining today are:[13]

EBM - Electron Beam Machining
EDG - Electrical Discharge Grinding
EDM - Electrical Discharge Machining
EDS - Electrical Discharge Sawing
EDWC - Electrical Discharge Wire Cutting
LBM - Laser Beam Machining
LBT - Laser Beam Torch
PBM - Plasma Beam Machining

The surface texture reflects the impingement of the heat source and the molten state that occurred. Figure T1 shows a scanning electron microscope (SEM) view of a typical EDM surface taken at 1250X magnification and 45 degrees. A regular surface profile trace, Figure T2, can be made of such a surface, however, for reasonably accurate, repeatable AA readings an average of five readings is recommended with each reading in a different direction so as to accommodate the absence of a lay pattern. Special gages have been generated for EDM to better illustrate the texture as shown on the example in Figure T3. The texture roughens and exhibits more splatter and globules as the operating energy level increases for all of the thermal processes. The particular surface finish values vary with material and the grain structure. Figure T4 shows the range of values to be expected from EDM, using material removal rate (MRR) in cubic inches per minute as a measure of process intensity.

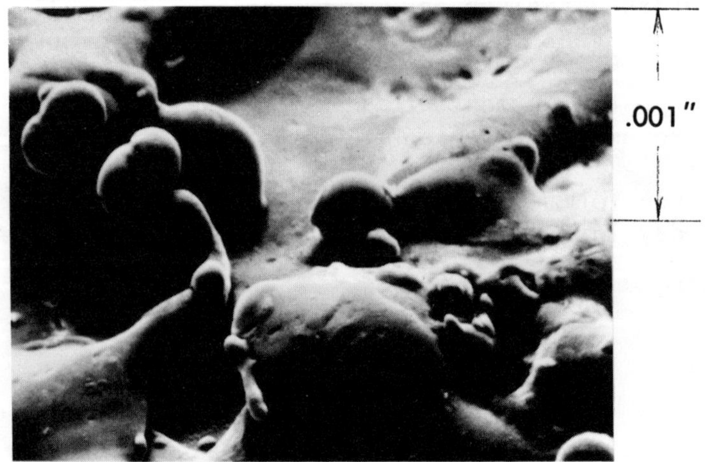

FIG. T1. TYPICAL EDM SURFACE AT 1250X BY SEM AT 45° (Hastelloy X)

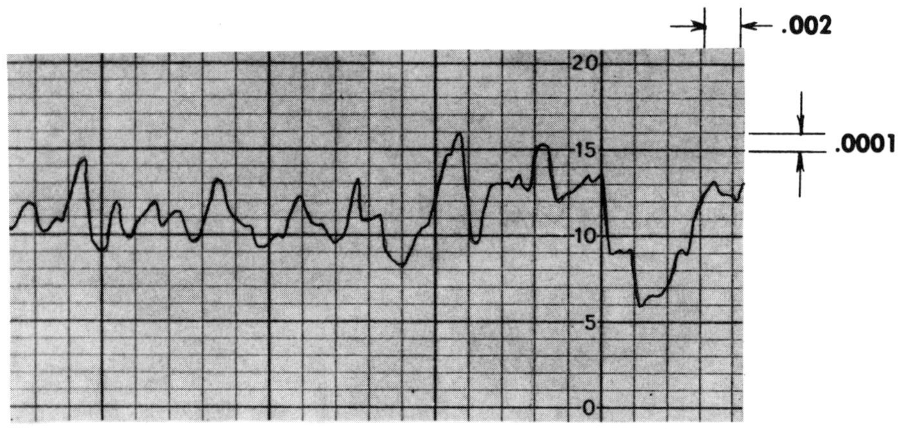

FIG. T2. TYPICAL SURFACE PROFILE TRACE OF EDM SURFACE
(162 Microinches AA of Hastelloy X)

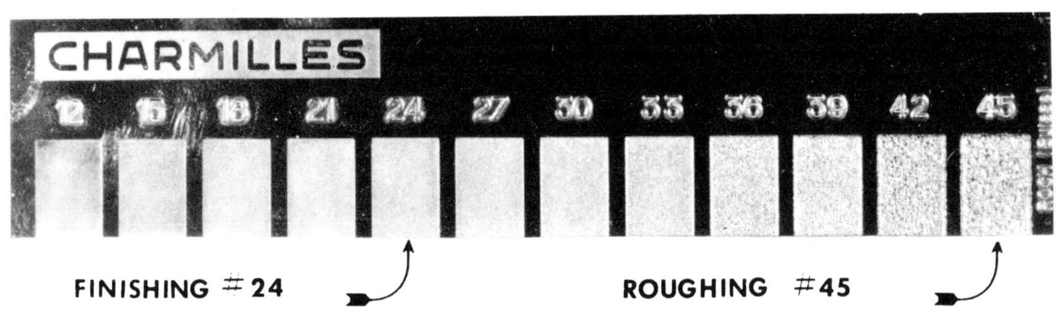

FIG. T3. SURFACE TEXTURE GAGE FOR EDM
(Courtesy of Charmilles Co.)

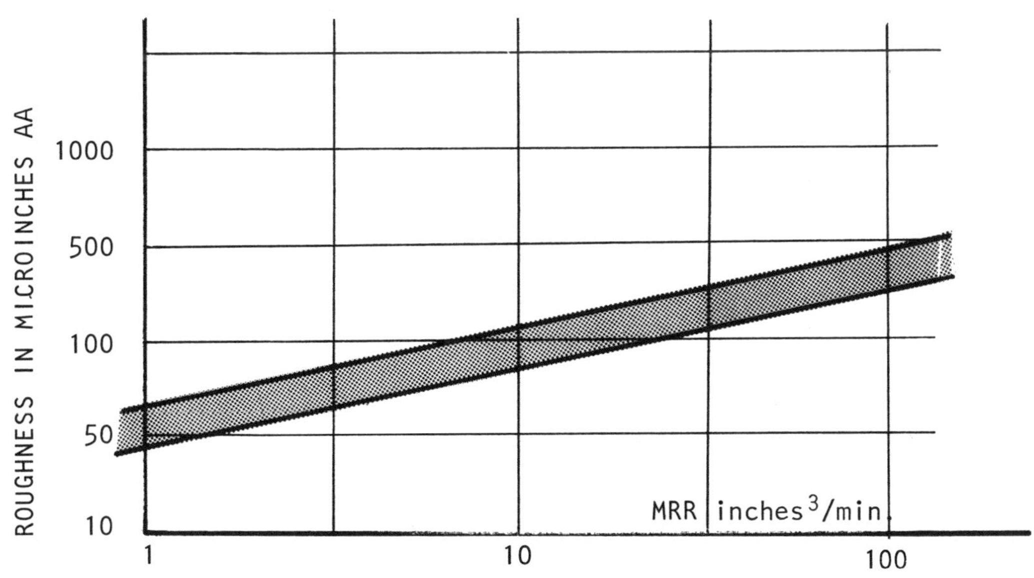

FIG. T4. APPROXIMATE RELATION OF SURFACE ROUGHNESS
TO MATERIAL REMOVAL RATE FOR EDM

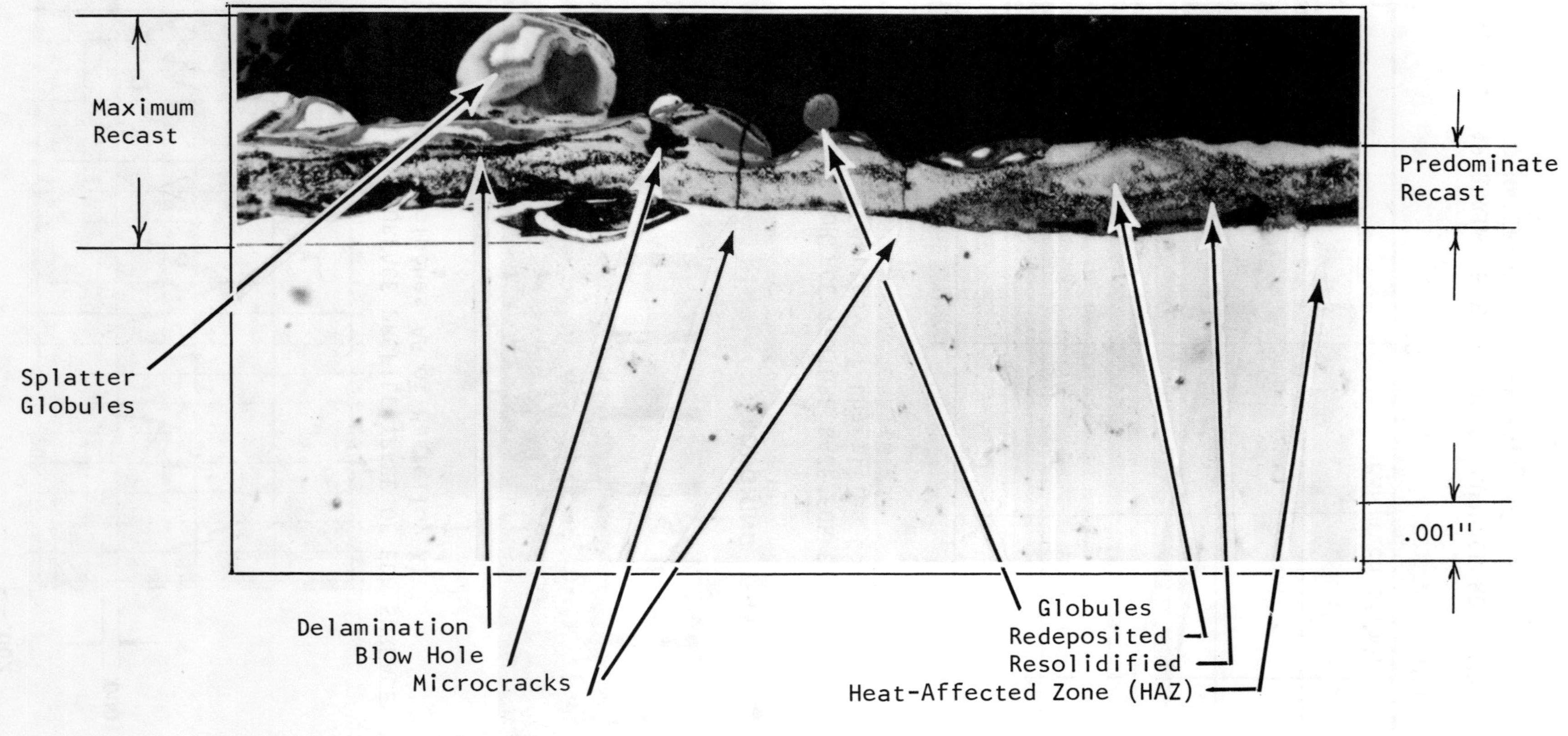

FIG. T5. RECAST SURFACE WITH TYPES OF ALTERATIONS (HASTELLOY X AT 300X)

The shallow thermal impact of these processes on the workpiece is accompanied by the very rapid quench rate from the heat sink of the bulk of the material. These transient thermal waves produce a recast and/or resolidified layer on the surface with a heat-affected zone (HAZ) below the recast. With higher energy levels, splatter and globules appear - usually lightly attached to the surface. Thermally sensitive materials can have microcracks in the recast layers that sometimes penetrate the base metal. Often the microcracks will stop at the depth of the recast or the depth of the HAZ but it is *not* established that these transitions in the metallurgical transformations act as a crack propagation barrier. Figure T5 illustrates many of these effects in a 300X photomicrographic section of EDM of Hastelloy X. Two terminologies have come into general use in assessing the recast layer: predominate recast thickness and maximum recast. The latter is measured over the globules and the former by multiple measurements over a significant length of surface (or by "eyeballing" a typical spot).

Figure T6 illustrates the significantly different recast appearance frequently found in laser generated surfaces.[16] The much greater rapidity of the laser thermal wave can generate epitaxial growth in the deposited layers with nucleation occurring from the base metal grain structure. Columnar structures have been seen even in successive layers generated by successive laser pulses. Microcracks also are frequent as seen on this 1000X view of Inconel 718 STA. This was the worst crack seen on the specimen. Multiple layers will occasionally show delamination type cracks in which case spalling might be expected.

Precise preplanning and relentless insistence upon operation at the planned level will repeatedly produce surfaces with well controlled recast and HAZ. Most modern equipment can produce surfaces with a minimum of altered material zones (AMZ). Table T1 lists the values for typical process parameters and for off-standard conditions. Figure T7 illustrates the surface of well controlled EDM. It is a 1000X view of Udimet 700 with a predominate recast

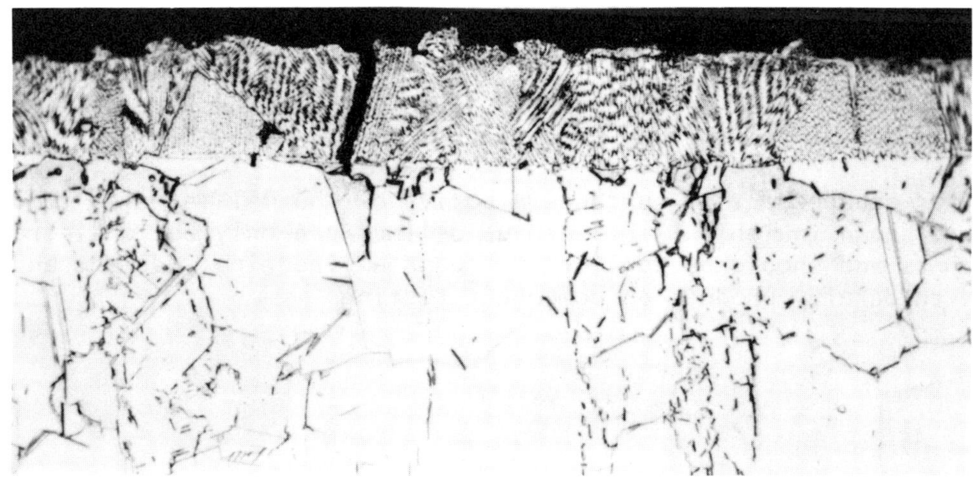

FIG. T6. SECTION OF LASER DRILLED INCONEL 718 STA

of 0.0002 inch. There is a HAZ about another 0.0002 inch below the recast. One rule of thumb for EDM is that the depth of the HAZ below the recast is equal to the recast thickness. This figure has been used for a number of years as the reference appearance for well controlled EDM processing. The extent of the AMZ depends upon the material and its metallurgical state as well as the particular process intensity.

Careful studies of the recast and microcracking from EDM has shown that the HAZ is always present but decreases as the processing intensity decreases. Figure T8 shows the shape of these curves with the processing intensity represented by the material removal rate (MRR). By the nature of these processes, there is no removal rate or process intensity so low as to have no HAZ. These data are typical for superalloys.

The HAZ below the recast is usually detected by a microhardness traverse on carefully mounted and prepared metallographic section with adequate edge retention and back up. Knoop indentation readings can repeatedly be made to within 0.0005 inch of the surface. Figure T9 is a typical example with the harder recast layer above the softened deeper HAZ. (The Knoop readings have been converted to equivalent Rockwell readings.)

Residual stresses in the EDM surface of a workpiece are shallow but can be of high tensile values. Bending fatigue endurance strengths from EDM are predictably low thereby. This is a consistent pattern over a large number and variety of material as shown on Table T2. The comparisons are made to low stress ground conventional specimens or modified conventional ground specimens from a variety of sources.[6,11,15] It should be noted that the reduction in room temperature HCF at 10^7 cycles is in many cases as severe for gentle, low intensity EDM as it is for high intensity EDM. Even the presence of a thin, almost invisible recast layer is just as detrimental as a thick recast layer. Where the EDM surface is to be used in a highly stressed application, adequate steps of post processing must be used to alleviate the reduction in fatigue strength.

Data on fatigue strengths from other thermal processes is still lacking. Caution should be exercised in using the thermally stressed surfaces without adequate checking. Direct application of EDM results to other processes would also be unwise.

Only in rare special cases have the bulk static material properties been altered by the use of the thermal processes.

Guidelines for enhancement of surface integrity of thermal material removal processes are shown in Table T3. These guidelines can only be considered starting points and should be applied with good engineering judgment and backed up with careful product on component testing.

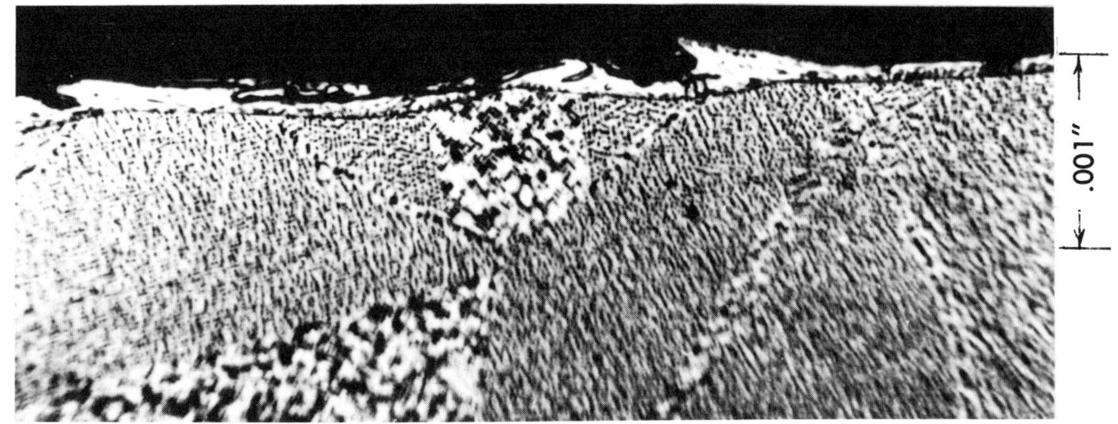

FIG. T7. TYPICAL WELL CONTROLLED EDM HEAT-AFFECTED ZONE
(0.0002 in. predominate recast on EDM of Udimet 700)

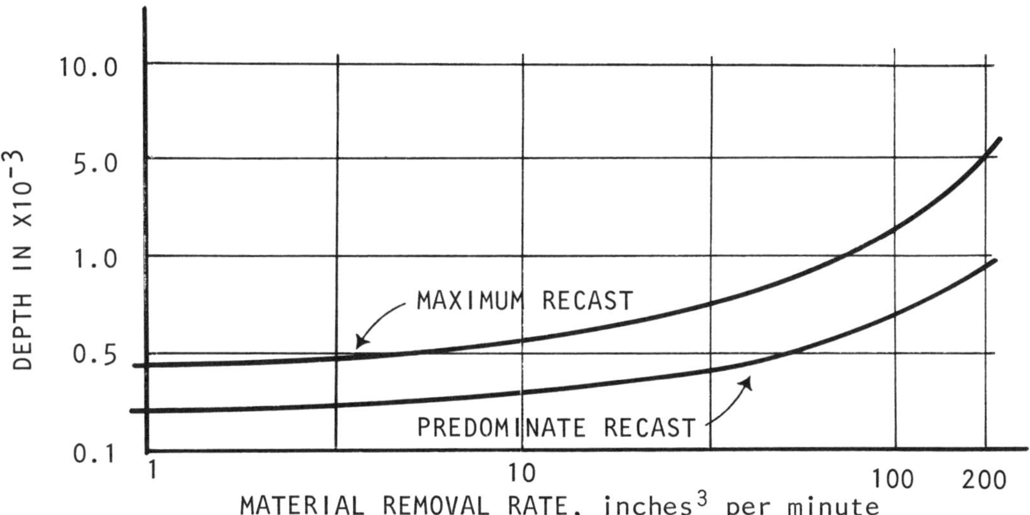

FIG. T8. EDM RECAST VS. PROCESS INTENSITY

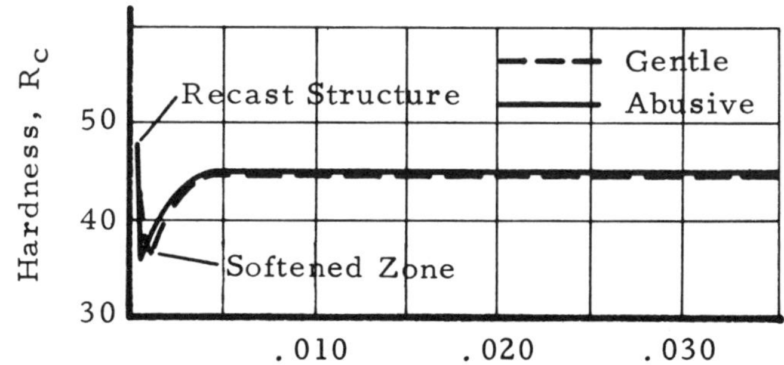

FIG. T9. MICROHARDNESS TRAVERSE ON EDM SURFACE
(Inconel 718 STA)

TABLE T1
SURFACE INTEGRITY EFFECTS OBSERVED IN THERMAL NONTRADITIONAL MATERIAL REMOVAL PROCESSES

Type of Effect		Typical Process Parameters	Off-Standard Conditions
Surface Roughness:	Average Range - Microinches AA	32 - 125	125 - 500
	Less Frequent	2 - 250	32 - 1000
Mechanical* Altered Material Zones:	Plastic Deformation (PD)	---	---
	Plastically Deformed Debris (PD^2)	---	---
	Hardness Alteration †	1.1	8.0
	Microcracks or Macrocracks	0.5	7.0
	Residual Stress §	2.0	3.0
Metallurgical* Altered Material Zones:	Recrystallization	---	---
	Intergranular Attack (IGA)	---	---
	Selective etch, pits, protuberances		0 - 1.6
	Metallurgical Transformations		
	Heat-Affected Zone (HAZ) or Recast Layer	0.6	10.0
High Cycle Fatigue (HCF):	% change from "handbook" values at room temperature #	-17 to -96	-48 to -64

Source: Guy Bellows, *NONTRADITIONAL MACHINING GUIDE, 26 Newcomers for Production*, MDC 76-101, Cincinnati, OH: Machinability Data Center, 1976.

Note: A blank in the table indicates no or insufficient data.
A --- in the table indicates no occurrences or not expected.

*Maximum observed depths in thousandths of an inch, normal to the surface.

†Depth to point where hardness becomes less than ±2 points R_c (or equivalent) of bulk material hardness (hardness converted from Knoop microhardness measurements).

§Depth to point where residual stress becomes and remains less than 20 ksi or 10% of tensile strength, whichever is greater.

#"Handbook" values from HCF testing are frequently generated from low stress ground specimens, hand or gentle machine polishing or occasionally electropolishing. These values are based upon LSG as 100% (with its minor amount of retained but enhancing compressive residual stress).

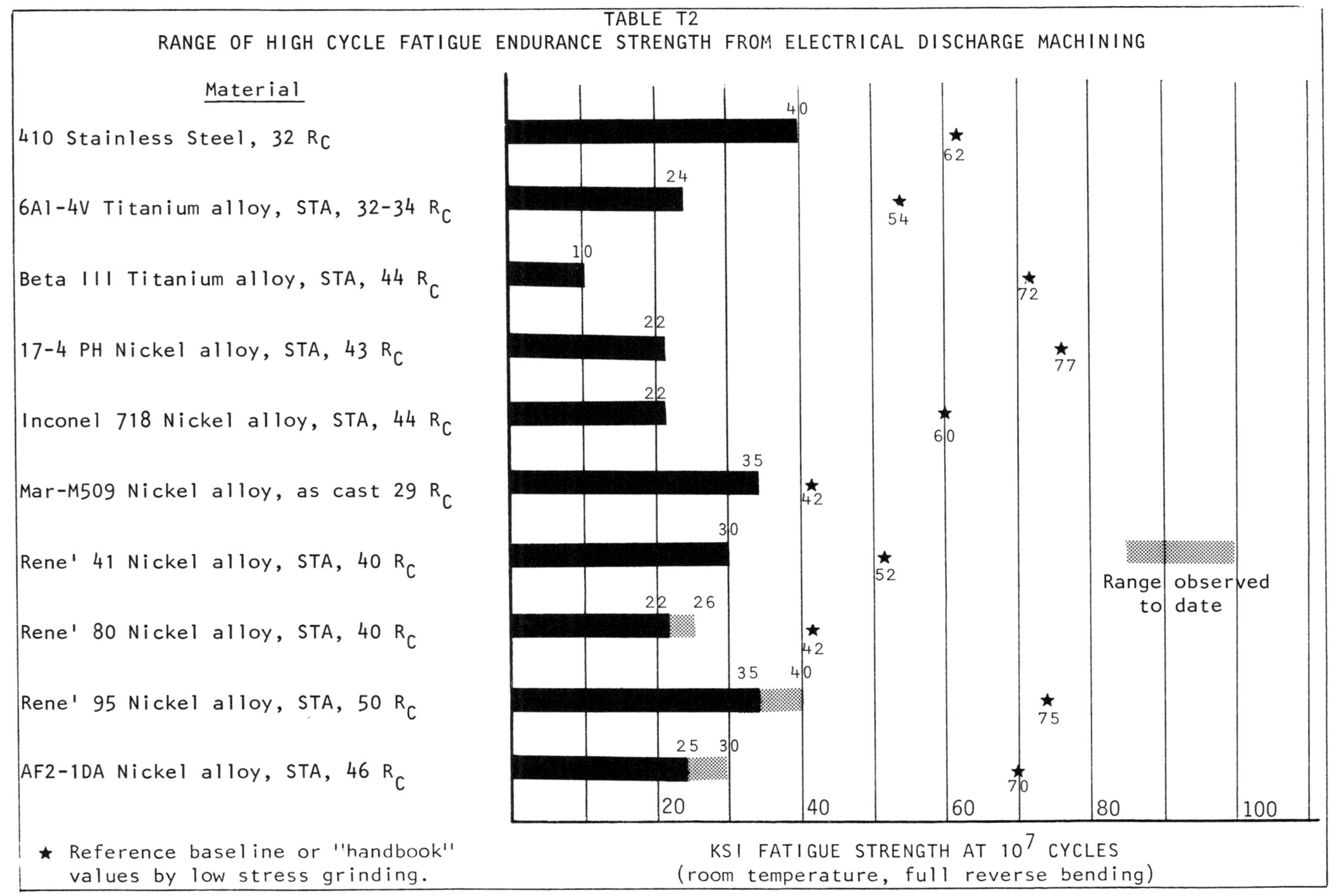
TABLE T2
RANGE OF HIGH CYCLE FATIGUE ENDURANCE STRENGTH FROM ELECTRICAL DISCHARGE MACHINING
Material
410 Stainless Steel, 32 R_C
6Al-4V Titanium alloy, STA, 32-34 R_C
Beta III Titanium alloy, STA, 44 R_C
17-4 PH Nickel alloy, STA, 43 R_C
Inconel 718 Nickel alloy, STA, 44 R_C
Mar-M509 Nickel alloy, as cast 29 R_C
Rene' 41 Nickel alloy, STA, 40 R_C
Rene' 80 Nickel alloy, STA, 40 R_C
Rene' 95 Nickel alloy, STA, 50 R_C
AF2-1DA Nickel alloy, STA, 46 R_C
40
62
24
54
10
72
22
77
22
60
35
42
30
52
22
26
42
35
40
75
25
30
70
Range observed to date
20
40
60
80
100
★ Reference baseline or "handbook" values by low stress grinding.
KSI FATIGUE STRENGTH AT 10^7 CYCLES
(room temperature, full reverse bending)

TABLE T3

SURFACE INTEGRITY GUIDELINES FOR THERMAL MATERIAL REMOVAL PROCESSES

1. The surface texture is composed of a random array of craters or cusps, sometimes with macrocracks at the roughing level of processing.

2. Surface roughness standards should be reassessed when using these processes due to the different texture and absence of lay patterns, as well as different methods of checking.

3. The prior condition of the workpiece is as important to generating surface quality as is the selection of operating parameters.

4. Careful monitoring of the preselected operating parameters is necessary. Some older types of equipment do not maintain their initial settings throughout a long day or run.

5. The depth of the heat-affected zone (HAZ) or recast structure on the surface is approximately proportional to the magnitude of the energy impinging on that surface. It is always present to some degree.

6. The HAZ can induce a substantial tensile residual stress on the surface along with hardness variations.

7. The depth of the HAZ below the recast layers is usually equal to the depth of the recast.

8. Fatigue strength is frequently severely reduced by the HAZ.

9. Highly stressed or critical surfaces should have the heat-affected zones (HAZ) produced by thermal material removal processes removed (or given a post-processing treatment).

10. Removal of the altered material zones (AMZ) may not be necessary if component or laboratory tests result in meeting design requirements.

11. Microholes (those less than one millimeter in diameter) produced by thermal processes may not be detrimental to fatigue endurance strength. A check by fracture mechanics of the critical hole size for a specific material is desirable.

12. On thin components, high current densities may overheat the workpiece.

13. Thorough cleaning to remove dielectric fluids, beads and vapor residue is desirable.

14. The microcracks per inch of cross section can be a valuable clue to the relative thermal sensitivity of materials.

ELECTRICAL MATERIAL REMOVAL

All the commercially available electrical nontraditional material removal processes are based upon the principles of electrolysis set forth by Faraday. The molecule by modecule dissolution of the electrically conductive workpiece is a forceless material removal that introduces no residual stresses into the workpiece. Removal occurs without concern for material hardness and at rates that are substantially uniform for all materials - and ECM can "cut" almost any conductive material. The principal varieties of nontraditional electrical machining today are:[13]

- ECD - Electrochemical Deburring
- ECDG - Electrochemical Discharge Grinding
- ECG - Electrochemical Grinding
- ECH - Electrochemical Honing
- ECM - Electrochemical Machining (Milling)
- ECP - Electrochemical Polishing
- ECT - Electrochemical Turning
- ES - Electro-stream
- STEM - Shaped Tube Electrolytic Machining

The surface texture of these process reflects the molecular dissolution and sometimes even displays individual grain patterns. The finishes typically are in the 8 to 63 microinch AA range with the smoother finishes associated with the higher current densities and the fastest cutting rates -- quite a contrast to mechanical cutting tools. Figure E1 shows a typical surface taken by SEM at 45° and Figure E2 is the conventional surface profile trace of this surface and Figure E3 is the 1000X metallographic cross section of the same surface of 17-4 PH. The difference in texture and the absence of a regular lay pattern are apparent. Figure E4 shows one comparator gage that was made to aid in the application of ECM.[4] The roughness ranges from 600 microinches AA on the left spot (F) to 8 microinches AA on the right (A). The roughest areas on ECM are frequently the side walls of cavities produced from electrodes with insulated sides.

The absence of residual stress from the ECM process is illustrated in Figure E5 taken from a deflection etching residual stress specimen of 17-4 PH material. This residual stress profile was prepared by electrochemically machining one side of a low stress ground blank then measuring the deflection as successively deeper layers were removed.[4] The conversion to residual stress shows practically zero stress at all depths and is almost below the practical limits of measurement. The absence of residual stress provides distortionless machining that is useful for thin workpieces.

Figure E6 illustrates another common phenomenon in ECM - the slight softening of the surface. This "Rebinder Effect" occurs on most ECM surfaces, but significantly, not all materials exhibit it nor do all metallurgical states of some materials or all process intensity levels produce it. Individual application checks should be made if surface hardness is of concern.

Surface integrity altered material zones (AMZ) for ECM are characterized by shallow metallurgical type alterations as shown on Table E1. The key to good EDM surfaces is a careful match of the electrolyte composition with the

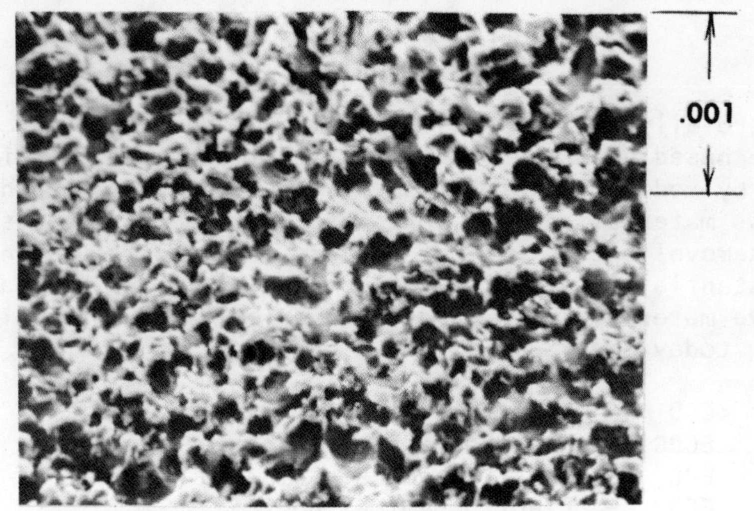

FIG. E1. TYPICAL ECM SURFACE AT 1000X BY SEM AT 45° ANGLE (17-4 PH)

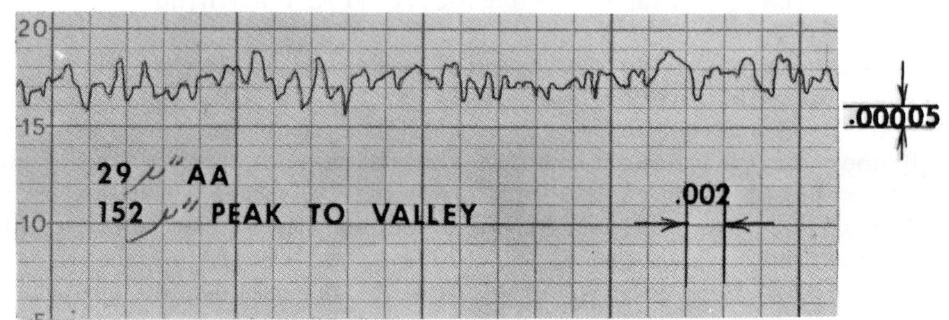

FIG. E2. TYPICAL SURFACE PROFILE TRACE OF ECM SURFACE (29 Microinches AA of 17-4 PH)

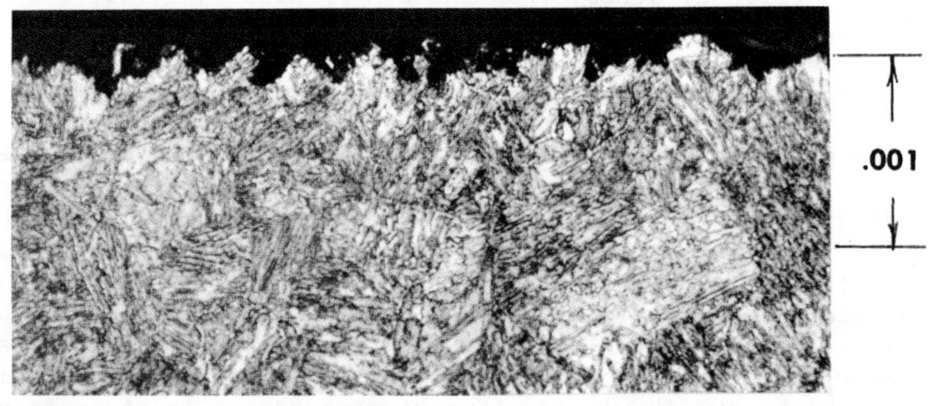

FIG. E3. TYPICAL CROSS SECTION OF ECM SURFACE (17-4 PH)

FIG. E4. MASTER IMPRESSIONS FOR ECM SURFACE FINISH COMPARATOR GAGE (1/4 x 1/2 pockets in 6Al-4V Titanium)

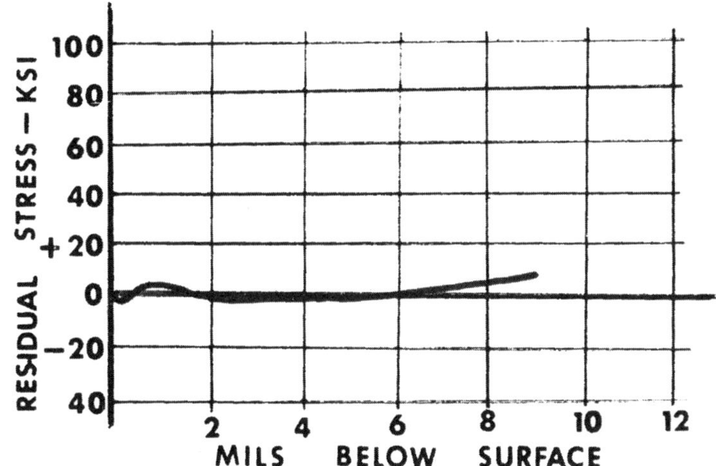

FIG. E5. TYPICAL RESIDUAL STRESS PROFILE FROM ECM (17-4 PH)

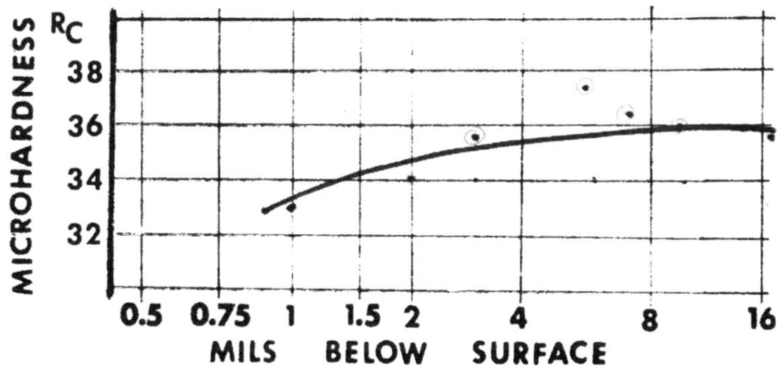

FIG. E6. TYPICAL MICROHARDNESS TRAVERSE FROM ECM SHOWING SURFACE SOFTENING OR REBINDER EFFECT (17-4 PH)

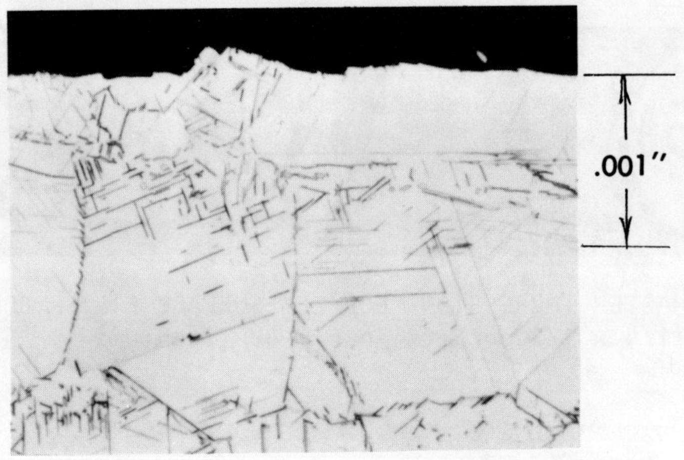

FIG. E7. TYPICAL ECM SURFACE WITH GOOD PROCESS PARAMETERS (43 Microinches AA roughness Inconel 178 STA)

FIG. E8. TYPICAL OFF-STANDARD CONDITIONS DURING ECM (74 Microinches AA roughness Inconel 718 STA)

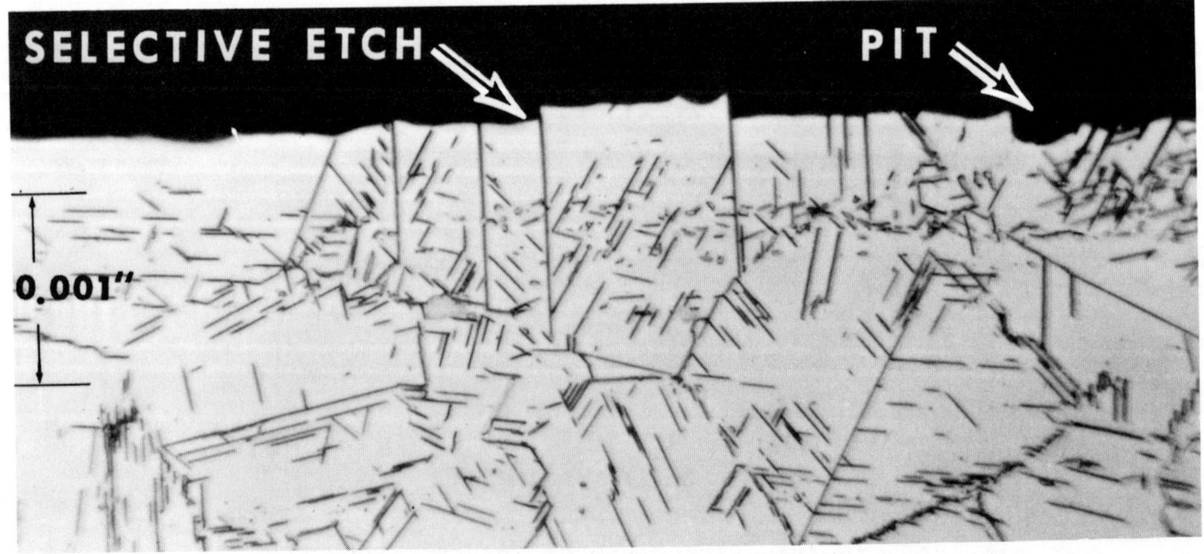

FIG. E9. SELECTIVE ETCHING AND PITS ON OFF-STANDARD ECM OF INCONEL 718 STA

metallurgical state of the workpiece material plus careful control of the operating parameters,[7,14] particularly the current density. Figure E7 shows a typical surface section at 1000X of Inconel 718 (solution treated and aged to 44 R_c). This surface measured 43 microinches AA in roughness value. Departure from preplanned conditions will produce a rougher surface with irregularities as shown on Fig. E8, which was 74 microinches AA in roughness value. This beginning of selective etch can become pits as shown on Figure E9 if conditions are too far off-standard.[19] Infrequently, with poorly matched operating conditions, a minor amount of IGA (up to 0.002 inch deep) will appear. However, frequently IGA will come or go depending on the heat treat state of the workpiece.[3] Figure E10 illustrates a typical and minor amount of IGA on cast 17-4 PH when operated under off-standard ECM conditons. This intergranular attack is a preferential etching of one phase of this alloy. It is, therefore, most important to control the material metallurgical state as carefully as the process parameters are controlled. With modern equipment short circuits during ECM are rare and of minute duration. However, each such case should be carefully checked for severity and the damaged areas removed.[19]

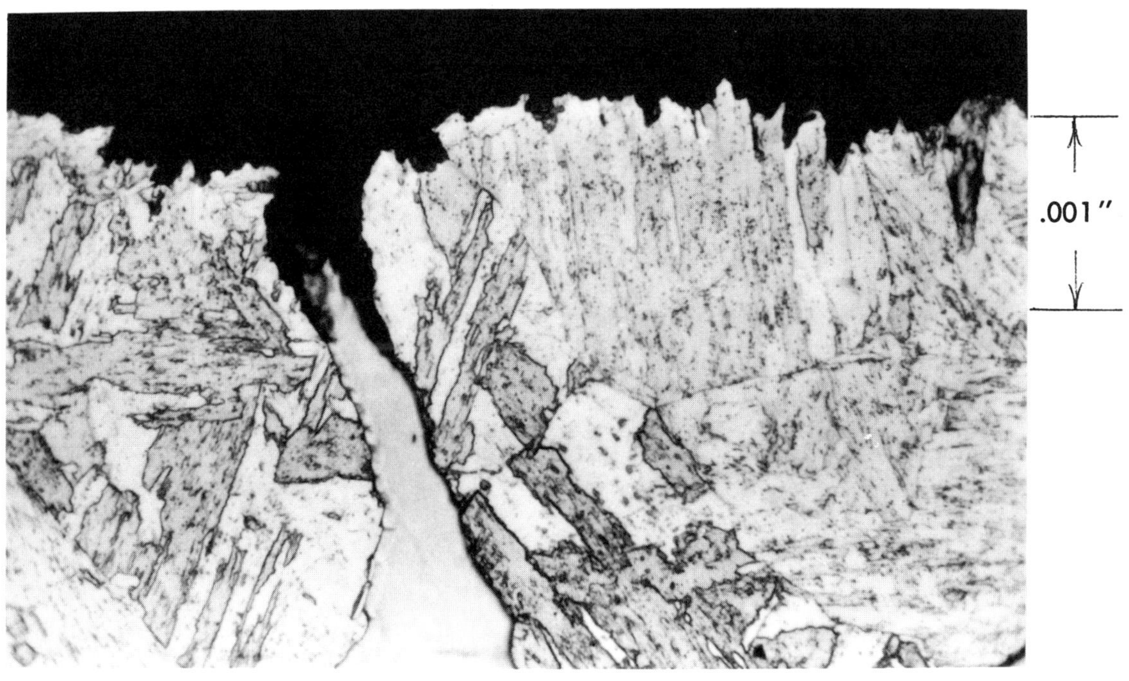

FIG. E10. TYPICAL IGA FROM ECM UNDER OFF-STANDARD CONDITIONS (0.001 in. on cast 17-4 PH of 37 R_c hardness and 85 microinches AA roughness)

A summary of the full reverse bending high cycle fatigue (HCF) endurance strength of surfaces prepared by ECM is shown in Table E2.[6,7,19] These values are the 10^7 cycle endurance at room temperature. Most of the HCF tests indicate a reduction in fatigue strength from "handbook" values secured by use of low stress ground specimens. ECM, however, is of lower residual stress and therefore more nearly represents the "true" or unaffected state of a material. The "reduction" is a change from values generated by engineering techniques of long standing. A shot peening post machining operation will usually more than restore the endurance strength to handbook values.[2] The HCF data collected to date does not permit categorizing a reduction range by alloy groups.[21] For example, the nickel alloys listed in Table E3 range from 8 percent reduction for 17-4 PH to 30 percent reduction for Inconel 718.

The static material properties of materials have not been found to be affected by ECM. Hydrogen embrittlement from ECM has also not been attributed to ECM. The hydrogen ions form and combine at the electrode to hydrogen gas and never reach the workpiece surface.[20]

Guidelines as a starting point for high surface integrity ECM surfaces are given in Table E3. Careful, individuals checks and tests should be made when applying ECM (or any other process) to critically stressed surfaces or components exposed to severe environments.

TABLE E1
SURFACE INTEGRITY EFFECTS OBSERVED IN ELECTRICAL NONTRADITIONAL MATERIAL REMOVAL PROCESSES

Type of Effect		Typical Process Parameters	Off-Standard Conditions
Surface Roughness:	Average Range - Microinches AA	8 - 32	63 - 250
	Less Frequent	2 - 125	8 - 500
Mechanical* Altered Material Zones:	Plastic Deformation (PD)	---	---
	Plastically Deformed Debris (PD^2)	---	---
	Hardness Alteration †	1.4	2.0
	Microcracks or Macrocracks	0.3	1.5
	Residual Stress §	0	0
Metallurgical* Altered Material Zones:	Recrystallization	---	---
	Intergranular Attack (IGA)	0.3	2.0
	Selective etch, pits, protuberances	0.4	2.5
	Metallurgical Transformations	---	---
	Heat-Affected Zone (HAZ) or Recast Layer	---	---
High Cycle Fatigue (HCF):	% change from "handbook" values at room temperature #	+6 to -33	-20 to -45

Source: Guy Bellows, *NONTRADITIONAL MACHINING GUIDE, 26 Newcomers for Production*, MDC 76-101, Cincinnati, OH: Machinability Data Center, 1976.

Note: A blank in the table indicates no or insufficient data.
A --- in the table indicates no occurrences or not expected.

*Maximum observed depths in thousandths of an inch, normal to the surface.
†Depth to point where hardness becomes less than ±2 points R_c (or equivalent) of bulk material hardness (hardness converted from Knoop microhardness measurements).
§Depth to point where residual stress becomes and remains less than 20 ksi or 10% of ultimate strength, whichever is greater.
#"Handbook" values from HCF testing are frequently generated from low stress ground specimens, hand or gentle machine polishing or occasionally electropolishing. These values are based upon LSG as 100% (with its minor amount of retained but enhancing compressive residual stress).

TABLE E2

RANGE OF HIGH CYCLE FATIGUE ENDURANCE STRENGTH FROM ELECTROCHEMICAL MACHINING

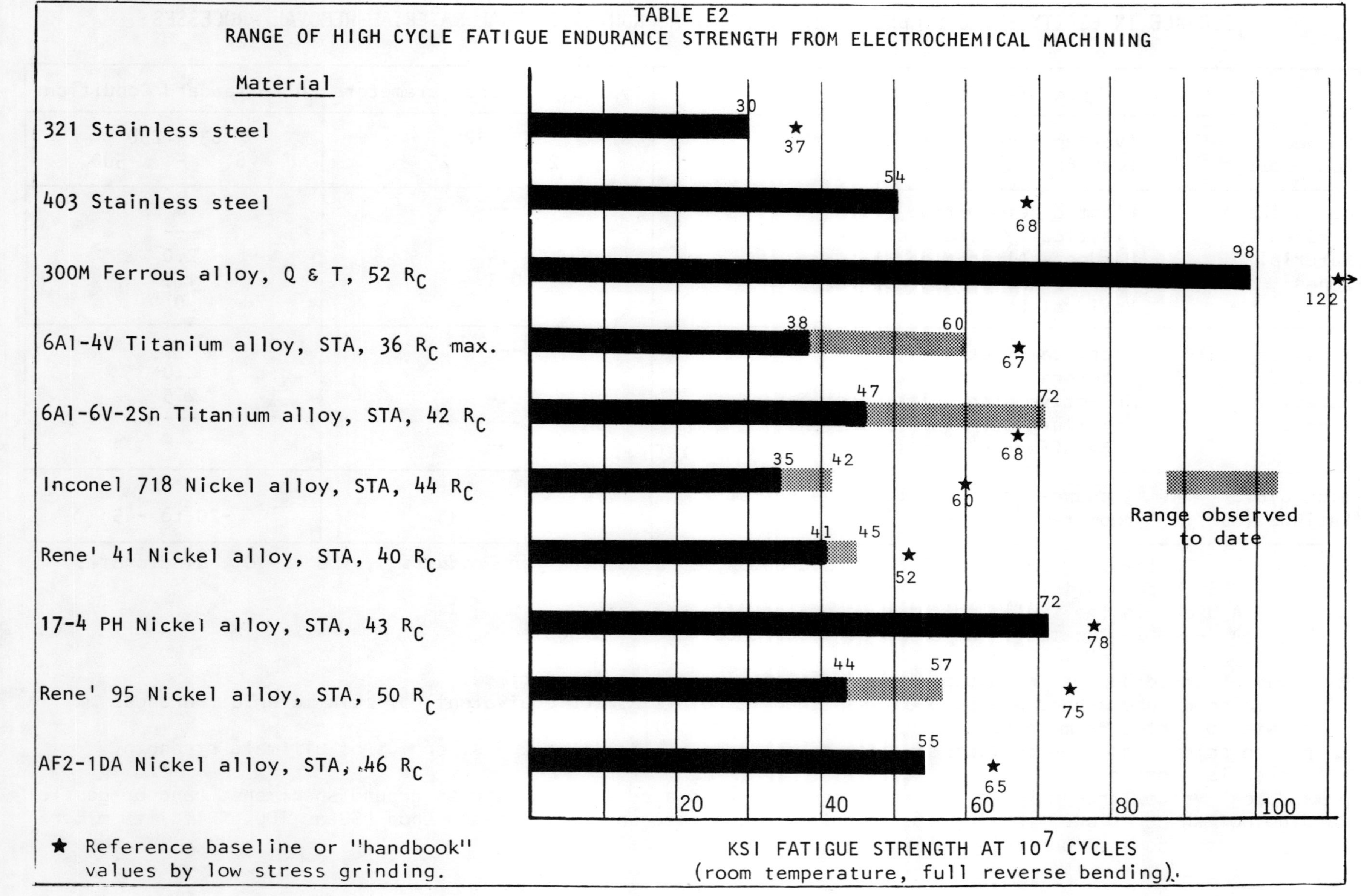

TABLE E3
ELECTROCHEMICAL PROCESSES GUIDELINES

1. Surface roughness standards should be reassessed when applying ECM due to the different textures and absence of lay pattern.

2. Well controlled ECM does not induce any stresses into the surface.

3. High current densities are desirable in the cutting gap for good finishes and rapid metal removal.

4. The current density in the workpiece should be carefully planned to prevent overheating in the material.

5. Relentless process controls should be supplemented with periodic metallographic coupon checks from representative surfaces.

6. The workpiece material heat treat state should be precisely known and controlled to assure the best repeatability of ECM surfaces.

7. Contact between the electrolyte and workpiece without current flow should be minimized.

8. Localized overheating due to poor connections or short circuits should be individually and carefully examined for the extent of any damage. Removal of the surface discoloration on the surface is insufficient.

9. Careful rinsing of the electrolyte from the workpiece is essential.

10. Selective etching, excessive roughness, intergranular attack or pitting are evidences of less than optimum operating conditions or tooling practices. These conditions can occur on surfaces subjected to low current density in the presence of the electrolyte or in areas adjacent to the main electrode cutting face.

11. Room temperature fatigue endurance strength is generally lower when compared to low stress conventionally ground specimens and represent the unblemished or unaffected material properties state.

12. The use of a post-ECMing shot peening treatment may be desirable to enhance component fatigue strength.

13. Anomalies in the workpiece surface or microvariations in material composition can influence the surface quality and may be exposed by ECM.

14. Some surface softening will occur in many, but not all, materials.

15. An increase in roughness of the ECM surface can be a significant alert to departure from optimum or preplanned operating conditions.

16. Hydrogen embrittlement is not known from ECM.

MECHANICAL MATERIAL REMOVAL

The commercially available mechanical nontraditional material removal processes are all abrasive in nature and utilize a multipoint approach for removal. The mechanical forces between the "tool" and the workpiece affects the surface integrity by introducing residual stresses, plastic deformation and sometimes hardness alterations to the surface. The multipoint approach aids in reducing the intensity of these effects as well as aiding productivity. The principal varieties of mechanical nontraditional machining today are:[13]

AFM - Abrasive Flow Machining
AJM - Abrasive Jet Machining
HDM - Hydrodynamic Machining
LSG - Low Stress Grinding
USM - Ultrasonic Machining

Only LSG has a significant amount of surface integrity data available. This data has been generated as LSG is a principal method for manufacturing high cycle fatigue (HCF) test specimens. All of the mechanical NTM processes have some degree of compressive residual stress with the "micro shot peening' effects from USM yielding the most enhancement of HCF endurance strength. Excessive abrading action, or dull grit, can produce excessive heat in the workpiece, hardness alteration or excessive plastic deformation. These alterations to the surface can provide the mechanism for recrystallized layers if a heat treatment follows the material removal.

The advent of LSG in the 1950's provided an additional energy level, or gentleness, to the grinding process. Several studies have shown that the parameters must be carefully controlled in order for LSG to yield the low residual stressed, reduced distortion, freedom from burning and reduction of plastic deformation.[17] The heavier thermal impact from conventional grinding (CGS) or high stress grinding (HSG) can produce surface burns or cracking particularly on the high strength alloys. There is a level of energy density below which cracking is not found and this is desirable to reduce costs by reducing manufacturing losses. The LSG studies have also shown the order of importance of the grinding parameters. In decending order of importance to good surface integrity the LSG parameters are:

- -frequent coarse dressing to maintain sharpness
- -lower wheel speeds (under 3500 sfpm)
- -lower infeed rates (0.0002 to 0.0005 inch per pass)
- -oil-base lubricants with good flow control
- -soft wheels (H, I or J grades)
- -higher table speeds (50 spfm or more)
- -solid fixtures and well-maintained equipment.

These conditions reduce the heat shock to manageable levels. If it is not practical because of machine limitations to control all of these then *one* can be increased if *all* of the remaining parameters are tightened.

The surface textures vary considerably among the mechanical NTM processes.[13] Most are layless with only LSG, as seen in Figure M1, having a uniform lay direction but with variable frequency and amplitude. Surface profile readings are a valid measure of surface roughness for LSG (when taken perpendicular to the lay) as shown on the trace in Figure M2. This trace is of the same specimen as Figure M1. The cross section of this specimen as shown in Figure M3 was taken after aging heat treatment so it shows the minor recrystallized layer of about 0.0004 inch depth that is an artifact from the plastic deformation/stressed layer. This plastic deformation is seen in the "before" view of Figure M4. Microhardness alteration will also frequently (but not always) accompany LSG. Figure M5 illustrates the range sometimes encountered on the same material with different samples and slightly different LSG parameters. While these effects are shallow in depth, they are important for some applications and should be recognized. They can be quite repeatable with good material and process control but if excessive or out of control can be failure initiation sites.[6] Figure M6 illustrates an undesirable surface condition if appearing on a highly stressed component part.

Low stress grinding can be obtained from several degrees of cross feed, wheel widths and dressing as long as the other parameters are maintained. Figure M7 shows the residual stress profiles from surface grinding AISI 4340 (50 R_c) steel with conditions to yield from 8 to 127 microinches AA surface roughness.[5,6] The stresses are very shallow as well as low in magnitude and mostly compressive.

Surface integrity altered material zones (AMZ) for mechanical nontraditional material removal processes are shown in Table M1. Most of this data is based upon grinding as the other processes have much more shallow altered material zones. Table M2 is a summary of the full reverse bending high cycle fatigue (HCF) data. The range shown for grinding includes the reduction in endurance strength usually associated with conventional grinding (CGS) or high stress grinding (HSG) or abusive grinding (ASG). (These latter, more intense grinding values are usually accompanied by high tensile residual stress that extend deeper below the workpiece surface.) The LSG values (★) are the same as those used in Tables C2, T2 and E2.

Guidelines for abrasive material removal processes are given in Table M3. These apply to the conventional as well as the nontraditional material removal processes. For either, they represent good starting points towards achieving good surface integrity but are not a substitute for careful testing or checks on the actual workpiece/component.

FIG. M1. TYPICAL LSG SURFACE AT 1250X BY SEM AT 45° ANGLE (Rene' 80 ST)

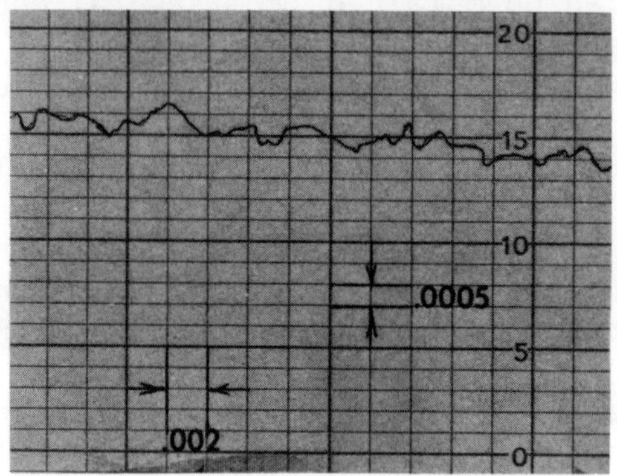

FIG. M2. TYPICAL SURFACE PROFILE TRACE OF LSG SURFACE (29 Microinches AA perpendicular to lay on Rene' 80 ST)

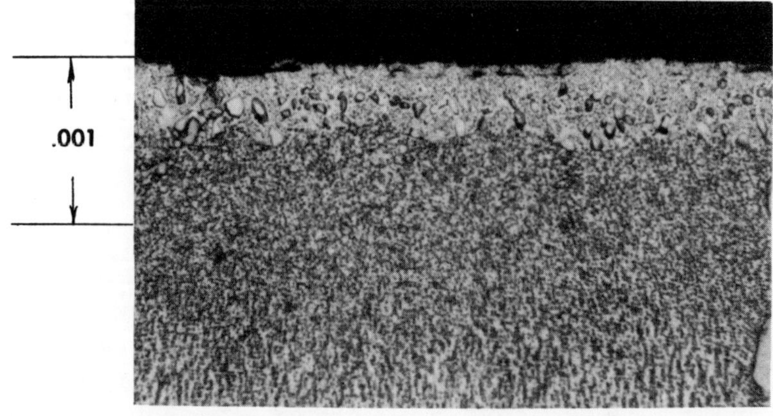

FIG. M3. TYPICAL CROSS SECTION OF LSG SURFACE SHOWING RECRYSTALLIZED LAYER FROM SUBSEQUENT AGE HEAT TREATMENT (Rene' 80 STA)

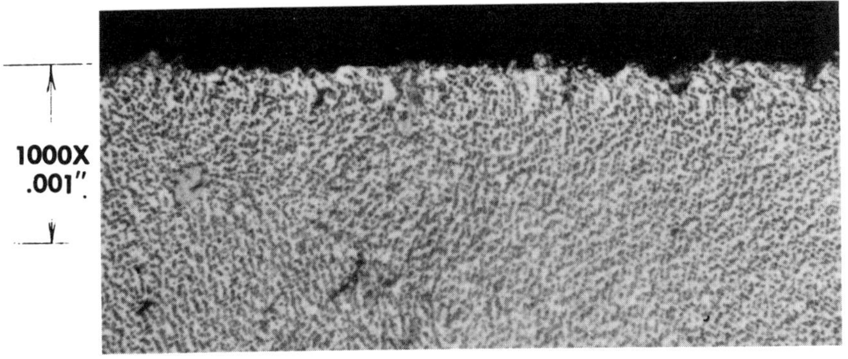

FIG. M4. TYPICAL PLASTIC DEFORMATION FOUND ON LSG GRINDING (Rene' 80 ST)

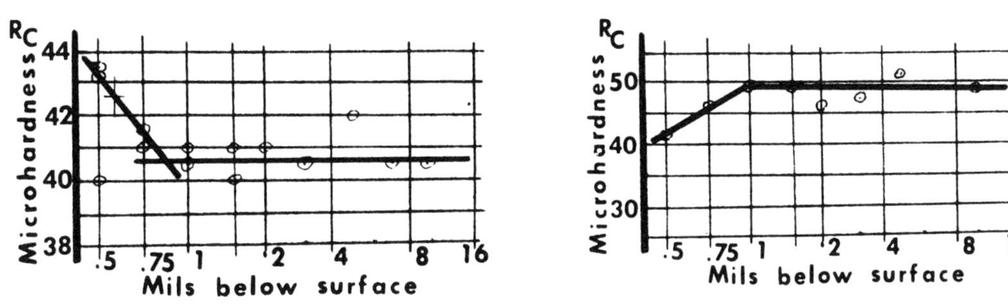

FIG. M5. TYPICAL MICROHARDNESS TRAVERSES FROM LSG OF RENE' 80 STA (R_C values converted from Knoop measurements)

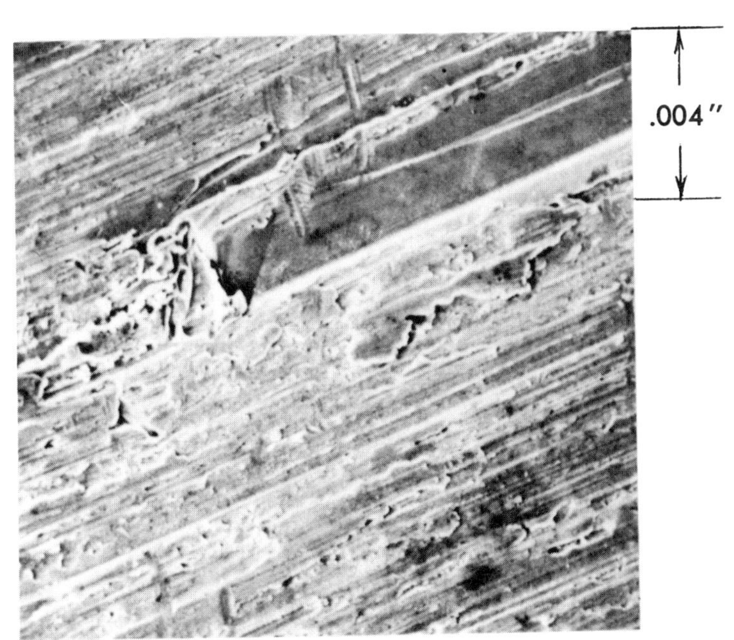

FIG. M6. SEM MICROGRAPH OF "TORN" SURFACE FROM BUILT-UP EDGE, WORN GRIT OR PLASTICALLY DEFORMED DEBRIS ON GROUND SURFACE OF 6Al-4V TITANIUM

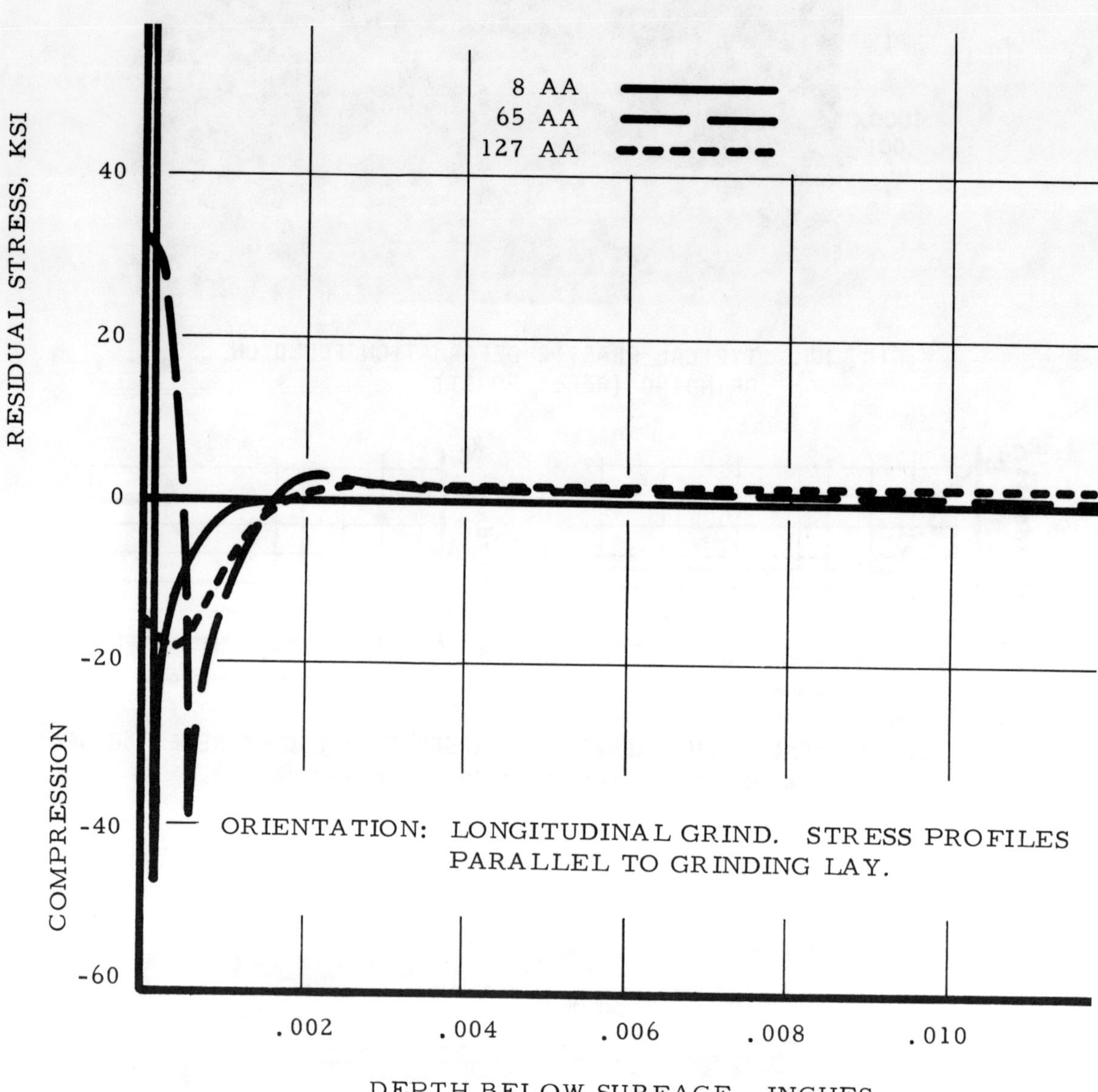

FIG. M7. RESIDUAL STRESS PROFILE FROM LSG OF AISI 4340 STEEL (50 R_c) OVER A RANGE OF SURFACE FINISHES.

TABLE M1

SURFACE INTEGRITY EFFECTS OBSERVED IN MECHANICAL NONTRADITIONAL MATERIAL REMOVAL PROCESSES

	Type of Effect	Typical Process Parameters	Off-Standard Conditions
Surface Roughness:	Average Range - Microinches AA	4 - 63	32 - 150
	Less Frequent	1 - 125	8 - 250
Mechanical* Altered Material Zones:	Plastic Deformation (PD)	0.3	1.7
	Plastically Deformed Debris (PD^2)	0.5	0.3
	Hardness Alteration †	1.5	8.0
	Microcracks or Macrocracks	0.5	9.0
	Residual Stress §	0.5	12.5
Metallurgical* Altered Material Zones:	Recrystallization	0.5	
	Intergranular Attack (IGA)	---	---
	Selective etch, pits, protuberances	0.2	0.4
	Metallurgical Transformations	0.5	6.0
	Heat-Affected Zone (HAZ) or Recast Layer	0.7	12.5
High Cycle Fatigue (HCF):	% change from "handbook" values at room temperature #	0 to 62	-95 to 0

Source: Guy Bellows, *NONTRADITIONAL MACHINING GUIDE, 26 Newcomers for Production*, MDC 76-101, Cincinnati, OH: Machinability Data Center, 1976.

Note: A blank in the table indicates no or insufficient data.
A --- in the table indicates no occurrences or not expected.

*Maximum observed depths in thousandths of an inch, normal to the surface.

†Depth to point where hardness becomes less than ±2 points R_c (or equivalent) of bulk material hardness (hardness converted from Knoop microhardness measurements).

§Depth to point where residual stress becomes and remains less than 20 ksi or 10% of ultimate strength, whichever is greater.

#"Handbook" values from HCF testing are frequently generated from low stress ground specimens, hand or gentle machine polishing or occasionally electropolishing. These values are based upon LSG as 100% (with its minor amount of retained but enhancing compressive residual stress).

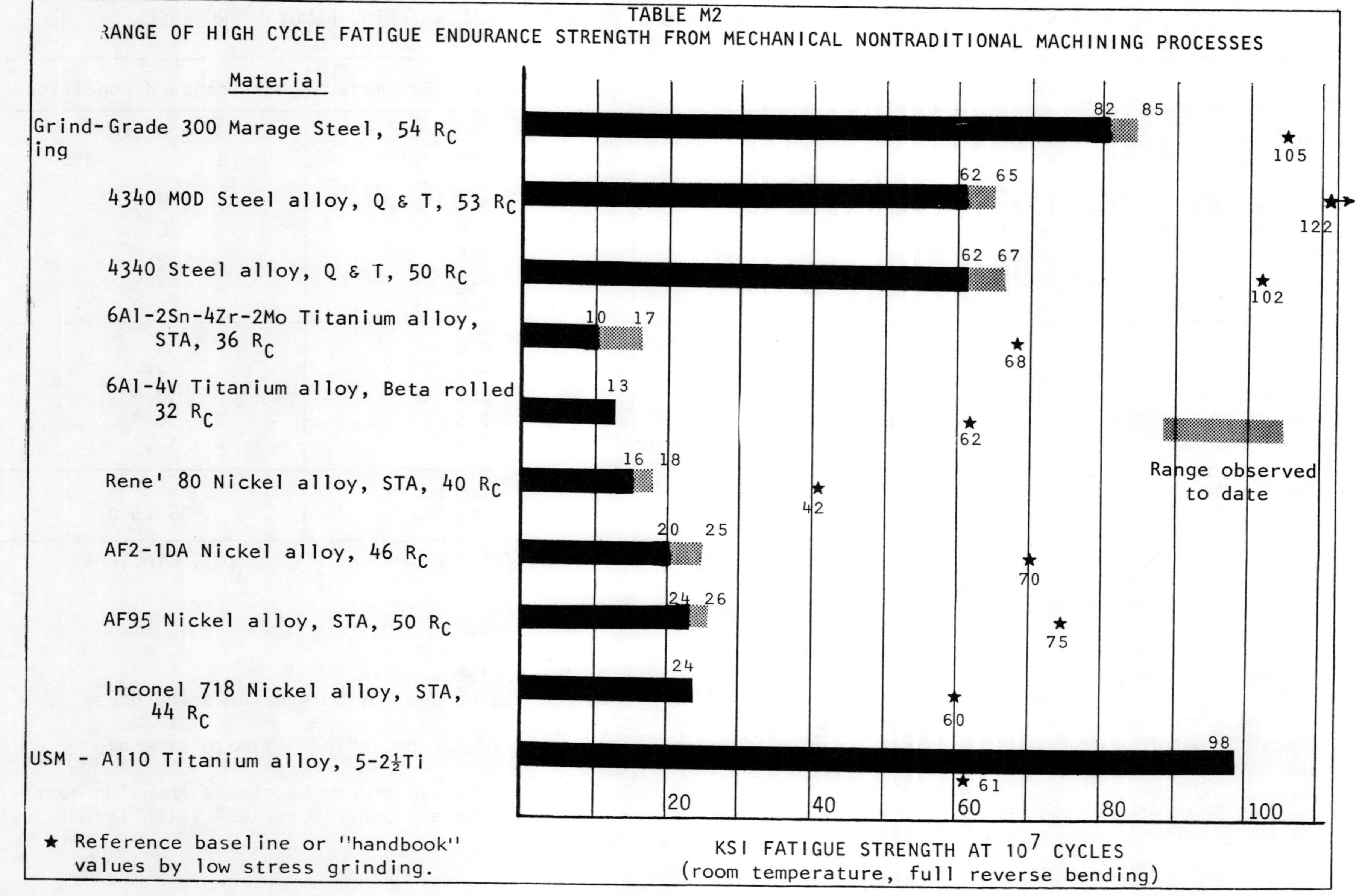
TABLE M2
RANGE OF HIGH CYCLE FATIGUE ENDURANCE STRENGTH FROM MECHANICAL NONTRADITIONAL MACHINING PROCESSES
Material
Grind-ing
Grade 300 Marage Steel, 54 R_C
82 85
105
4340 MOD Steel alloy, Q & T, 53 R_C
62 65
122
4340 Steel alloy, Q & T, 50 R_C
62 67
102
6Al-2Sn-4Zr-2Mo Titanium alloy, STA, 36 R_C
10 17
68
6Al-4V Titanium alloy, Beta rolled 32 R_C
13
62
Range observed to date
Rene' 80 Nickel alloy, STA, 40 R_C
16 18
42
AF2-1DA Nickel alloy, 46 R_C
20 25
70
AF95 Nickel alloy, STA, 50 R_C
24 26
75
Inconel 718 Nickel alloy, STA, 44 R_C
24
60
USM - A110 Titanium alloy, 5-2½Ti
98
61
20
40
60
80
100
★ Reference baseline or "handbook" values by low stress grinding.
KSI FATIGUE STRENGTH AT 10^7 CYCLES
(room temperature, full reverse bending)

TABLE M3
ABRASIVE PROCESSES GUIDELINES

1. Low stress grinding techniques can reduce distortion and surface damage. Fresh, coarse dressing of softer wheel grades, lower infeeds and reduced grinding wheel speeds with lubricating cutting fluids are the principle elements of low stress grinding. Very low tensile and, in some cases, compressive residual stress conditions near the surface are attained.

2. Higher work table speeds can aid both productivity and surface integrity. Lower wheel surface speeds aids surface integrity without sacrifice to productivity.

3. Frequent coarse dressing of the grinding wheels can reduce surface damage by keeping wheels open and sharp, thus helping to reduce temperatures at the wheel-workpiece interface.

4. Modifications to low stress grinding procedures should not be attempted unless testing confirms that compromises can be tolerated.

5. If low stress grinding is specified for the finishing operation, then conventional grinding can be used to within 0.010 inch of the finished size providing materials are not sensitive to cracking.

6. Cutting fluids and coolants should have a good and positively directed flow and be checked for their chemical action on the specific material being ground.

7. Conventional grinding conditions should not be used for grinding highly sensitive alloys such as the high strength steels (4340, D6ac), high temperature nickels (Rene' 41, Inconel 718, L605) cobalt, titanium or molybdenum.

8. The heat-affected zone from rough grinding can be much deeper than the surface discoloration. Microhardness traverses can generally detect the depth of HAZ alterations.

9. Abusive grinding in martensitic steels can create untempered martensite zones and over-tempered zones below them, both of which can limit fatigue strength.

10. Crack detection tests should be made when rough grinding material. Surface cracks can propagate ahead of the finish grinding. LSG should remove more than the AMZ's from any prior conventional grinding.

11. Hand grinding of sensitive alloys should be discouraged, unless under very careful control.

12. Controls for hand power sanders should be maintained.

13. Abrasive cut-off frequently has a harsh and deep surface effect and these altered layers should be removed.

SUMMARY

The fear of the unknown has been reduced for the nontraditional material removal processes as more experience and data has accumulated. The increased realization that the surface effects from conventional machining can be as extensive as those from nontraditional machining has introduced more calm consideration to the application of nontraditional machining. The knowledge that the variabilities in materials has an impact on surface integrity as great as the variabilities in processes has led to better failure analysis. Both materials and processes must be controlled with equal degrees of relentlessness if the best surface integrity is to be maintained.

Good value engineering will insist that surface integrity practices be applied only to those stress or environmentally critical surfaces of a workpiece - not all over. The guidelines provided in this paper are just that - guidelines - and can never substitute for careful analysis and testing of a specific part for its specific application and operating conditions. Consideration of the total sequence of operations is as important as the effects from any one process. It is the "as shipped" state of a surface that must have a level of surface integrity suited to its purpose.

While this paper has grouped the altered material zones under the four headings - chemical, thermal, electrical and mechanical - it is a fact that each process has its unique effects on specific materials. With fine tuning of the testing, it is also possible to distinguish differences in surface integrity effects from different process energy intensities and from different material metallurgical states presented for processing.

The above summary remarks apply equally to conventional or nontraditional processes. As a matter of fact, some of the nontraditional processes are now becoming quite "conventional" in many instances.

REFERENCES

1. *Proceedings of the International Conference on Surface Technology.* Dearborn, MI: Society of Manufacturing Engineers, 1973.

2. Surface Integrity of Nontraditional Machining Processes. G. Bellows, Paper No. MR73-633, *Proceedings of International Conference on Surface Technology*, Dearborn, MI: Society of Manufacturing Engineers, 1973.

3. Impact of Nonconventional Material Removal Processes on the Surface Integrity of Materials. G. Bellows, Paper No. MR68-518, American Society of Tool and Manufacturing Engineers, Dearborn, MI, 1968.

4. Surface Integrity of Machined Structural Components. W. P. Koster, M. Field, L. J. Fritz, et al, AFML-TR-70-11, Metcut Research Associates Inc.

5. Manufacturing Methods for Surface Integrity of Machined Structural Components. W. P. Koster, et al, AFML-TR-71-258, Metcut Research Associates Inc., April 1972.

6. Surface Integrity of Machined Materials. W. P. Koster, et al, AFML-TR-74-60, Metcut Research Associates Inc., April 1974.

7. Electrochemical Machining. J. Cross, et al, General Electric Company, WPAFB report for program MMP 723-9, October 1972.

8. Surface Integrity of Machined Inconel 718 Alloy. G. Bellows and R. M. Niemi, Paper No. IQ71-239, Society of Manufacturing Engineers, Dearborn, MI April 1971.

9. Surface Integrity of Electrical Discharge Machining. G. Bellows, General Electric Company report, R72AEG236, July 1972.

10. *Machining Data Handbook.* 2d. ed., Cincinnati, OH: Machinability Data Center, Metcut Research Associates Inc., 1972.

11. Surface Integrity - Update '72. G. Bellows and W. P. Koster, General Electric Company, TM72-384, July 1972.

12. Introduction to Surface Integrity. G. Bellows and D. N. Tishler, General Electric Company pamphlet, TM70-974, October 1970.

13. *Nontraditional Machining Guide, 26 Newcomers for Production.*
G. Bellows, MDC 76-101, Cincinnati, OH: Machinability Data Center, Metcut Research Associates Inc., 1976.

14. Surface Effects in Metal Deformation.
W. A. Glaeser, Paper No. MF69-101, Society of Manufacturing Engineers, Dearborn, MI, 1969.

15. Electrical Discharge Machining (EDM).
D. J. Moracz, TRW Company, AFML-TR-73-66, May 1973.

16. Laser Drilling.
L. M. Heglin, General Electric Company, AFML-TR-75-44, April 1975.

17. Machining of New Materials.
N. Zlatin, et al, Metcut Research Associates Inc., AFML-TR-73-165, July 1973.

18. *Machining, A Process Checklist.*
G. Bellows, MDC 76-100, Cincinnati, OH: Machinability Data Center, Metcut Research Associates Inc., 1976.

19. Surface Integrity of ECM.
G. Bellows, General Electric Company report, R69AEG172, July 1969.

20. *Principles of ECM.*
J. A. McGeough, New York, NY: Wiley, 1974.

21. *Practice and Theory of ECM.*
J. F. Wilson, New York, NY: Wiley, 1971.

Presented at SME's Lasers in Manufacturing Conference, December 1976

Applications For Industrial Lasercutter Systems

By Willard E. Hanson
Hughes Aircraft Company

The application of lasers to industrial processes has spanned the gamut from the mundane to the esoteric. One application that has received fairly widespread notoriety has been that of using a laser beam to cut fabrics in the manufacturing of apparel. As a result of this application, the cutting of many other types of materials has been and is currently being investigated.

This paper will address current applications as well as the status of scheduled and projected applications of lasercutting for industrial purposes.

APPAREL

Hughes' initial efforts in designing and building lasercutting systems was in the area of cutting the individual pieces, both shell and lining for men's suits. For a variety of reasons (fabric utilization, manufacturing flexibility, inventory control, response to fashion change, etc.), it is reasonable to want to cut suit material "one-high", or one layer at a time. A schematic of the system required to accomplish this process is illustrated in Figure 1. The action begins with the designer creating styles for the new season. Geometry of the base size patterns plus growth instructions are entered into the system. As orders are received from the field sales force, a lasercutting plan is concieved which takes into account delivery schedules, piece goods availability, shop loading, and other factors. The lasercutting plan is entered via magnetic tape or punched cards. For example, this information may consist of the following:

MODEL NUMBER:	1278
SIZE:	44 long
PIECE GOODS WIDTH:	59 inches
TYPE OF MATERIAL	Texturized polyester

Please refer to Figure 2.

Since it is the industry standard that patterns are based on size 40 regular, grading of the patterns to the desired size of 44 long is the first step. Now, the patterns, approximately 40 in number for a suit, are efficiently arranged by a nesting program which takes into consideration piece goods width, grain and pattern rotation constraints, and plaid and stripe matching requirements. The result of this activity is a cutting tape which then is read into the system's controller in a format which controls conveyor advance, laser switching and positioning of the optics.

Figure 3 shows a pictorial representation of a typical lasercutting system. Major elements include:

- ... Laser source
- ... Cutting areas
- ... Spreader and Conveyor
- ... Control unit

A suit contains nearly 2000 inches of peripheral cutting. A standard lasercutting system will produce more than 40 suits an hour which is more than 1/2 million inches of peripheral cutting per shift.

Lasercutting in the apparel industry is no longer in the prototype and trial stages. To date, sixteen systems are in daily use both in the United States and abroad.

ADVANCED COMPOSITE MATERIALS

As a result of the exposure gained with the apparel industry, it was brought to the attention of Hughes that cutting problems existed within the general area of advanced composite materials — boron epoxy, graphite epoxy, Kevlar, etc. These cutting problems were much the same as those which existed in apparel, to wit, making fast clean and accurate intricate cuts on a variety of shaped parts in an efficient, automatic manner. It has been realized that many of the production problems associated with processing composites are similar to that of apparel especially if the advanced composite material is in broadgoods form, resembling a bolt of cloth. The ability to nest parts efficiently while maintaining grainline (fiber) constraints is a must. Automatic dispensing of the goods to the cutting area followed by efficient cutting and movement to the pick-up area is also desirable.

Hughes is currently in the process of delivering a system to McDonnell Douglas, St. Louis, Missouri. This system will be mainly used to profile single layers of boron epoxy broadgoods. The individual layers will then be collated, cured and used as integral skin surface members on the empennage of the F-15.

Figure 4 is a pictorial representation of the advanced composites lasercutter. Major elements include:

... Laser source

... Boron broadgoods despooler

... Boron epoxy spreader and conveyor

... Cutting area

... Control unit

It is anticipated that systems of this general configuration will be used in larger numbers as applications of advanced composites become widespread. One of the current drawbacks in using these materials has been that the manufacturing processes associated with advanced composites have been highly labor intensive. For example, boron epoxy is currently provided in 3-inch strips and is layed up on a mylar master as depicted in Figure 5. A "pizza" knife device is used to fracture the boron epoxy along the pattern outline. This process is expensive in both labor and material. It is anticipated that extensive savings will be realized by effecting an efficient pattern nest and by eliminating the manual cutting operation.

METAL CUTTING

The use of a laser beam to cut metals has been a popular topic of discussion in metalworking circles for a number of years. Of special interest during our supersonic transport era was that of cutting titanium since machining by other means was difficult. Titanium, possessing low conductivity and reflectivity, can be easily and quickly cut by relatively low-powered lasers. On the other end of the spectrum are the easily machined materials such as aluminum which, due to high reflectivity and conductivity, are more difficult to cut, but can be cut at slower speeds. Cutting speeds and quality are, of course, a function of laser power, mode, and thickness of the material being cut. Figure 6 tabularizes some recent cutting tests performed using a United Technologies high-power laser.

Of some concern in the application of lasercutting to metals has been the extent and nature of heat affected zone, commonly referred to as the H.A.Z. This concern is intensified if the cutting process is relatively slow, e.g. in aluminum. The Air Force Materials Laboratory will soon be funding an intensive investigation to determine the laser and process parameters which minimize this effect. Preliminary findings have been made that acceptable cuts in aluminum can be made that will meet or exceed the standards currently used in the aircraft industry. Figure 7 illustrates that there is almost no H.A.Z. when using the cutting parameters listed thereon.

One application currently being examined in depth by a number of companies is that of profiling sheet aluminum using such a laser. Figure 8 illustrates a possible manner in which such a system could be implemented. Master pattern shapes (tens of thousands, if desired) are entered into the computer system data bank through a digitizing process which transforms geometric shapes into an information format understood by the computer. As manufacturing requirements are generated, pattern identifiers are entered into the system along with the size of the sheet aluminum blank from which the pattern shapes are to be cut. Special software programs are then used to nest the patterns in an efficient manner insuring that proper grain orientation, if appropriate, is maintained. This requirement is analogous to that of the apparel industry. After the pattern layout has been accomplished within the computer, this information can be transferred to lasercutter controller either directly or through the use of magnetic tape. When the appropriate aluminum blank has been loaded, cutting will commence and will continue until all parts have been cut. Cut parts will then be conveyed from the cutting zone to an area for pickup and distribution.

Many benefits accrue to the manufacturer using this concept. These include:

... Improved flexibility in design and manufacturing.

... Rapid turnaround for specials and prototype parts.

... Eliminating the need for dies.

... Better material utilization.

The process described above could be applied to metals other than aluminum.

CONCLUSION

The use of laser energy to separate materials is being applied to a wide diversity of applications. As business pressures force the automation of currently labor-intensive manufacturing processes for cutting, the use of lasercutting techniques will continue to grow. This growth will be accelerated as the savings being realized by the Lasercutter "pioneers" become well known.

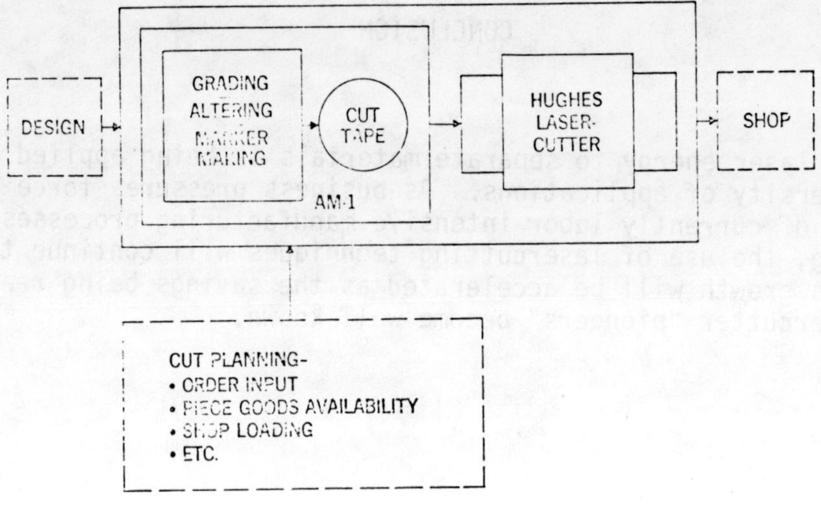

APPAREL LASERCUTTER FLOW DIAGRAM

FIGURE 1.

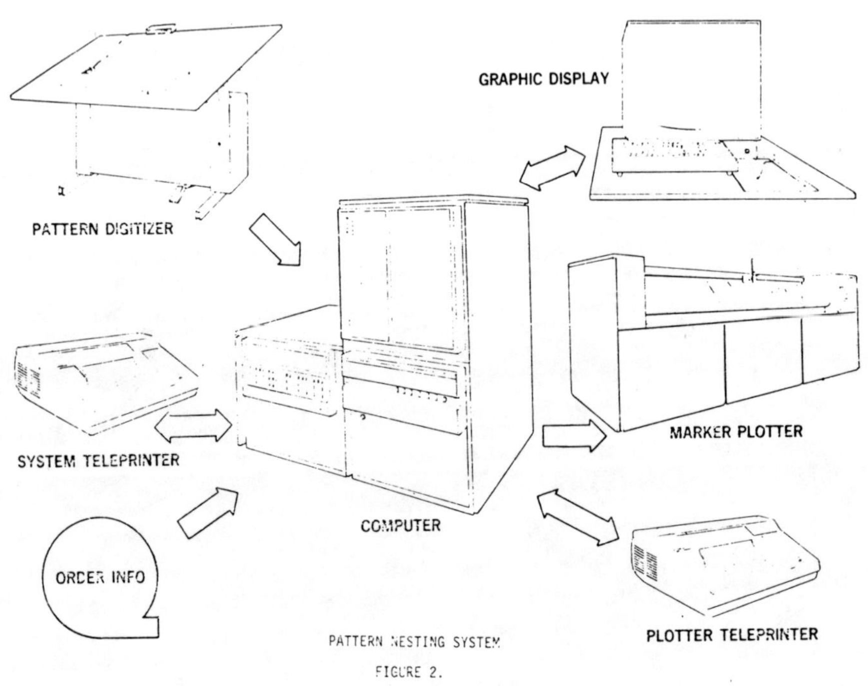

PATTERN NESTING SYSTEM

FIGURE 2.

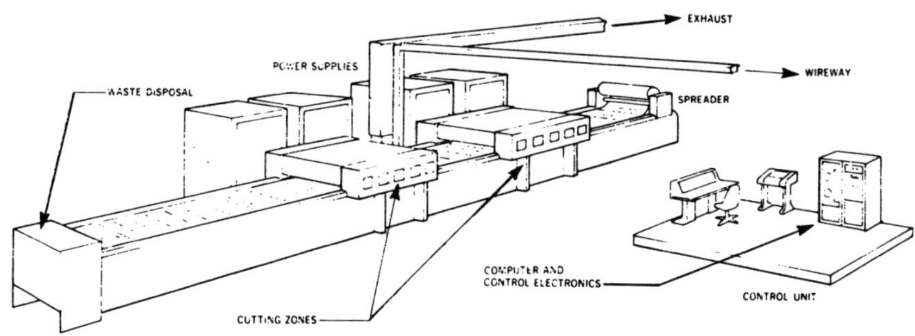

LC-2 LASERCUTTING SYSTEM

FIGURE 3.

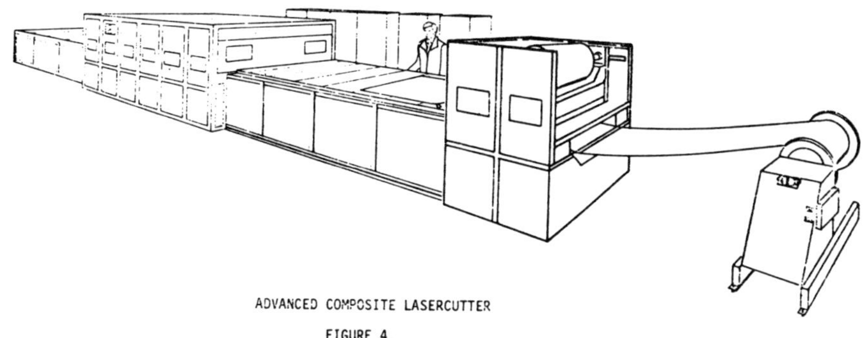

ADVANCED COMPOSITE LASERCUTTER

FIGURE 4.

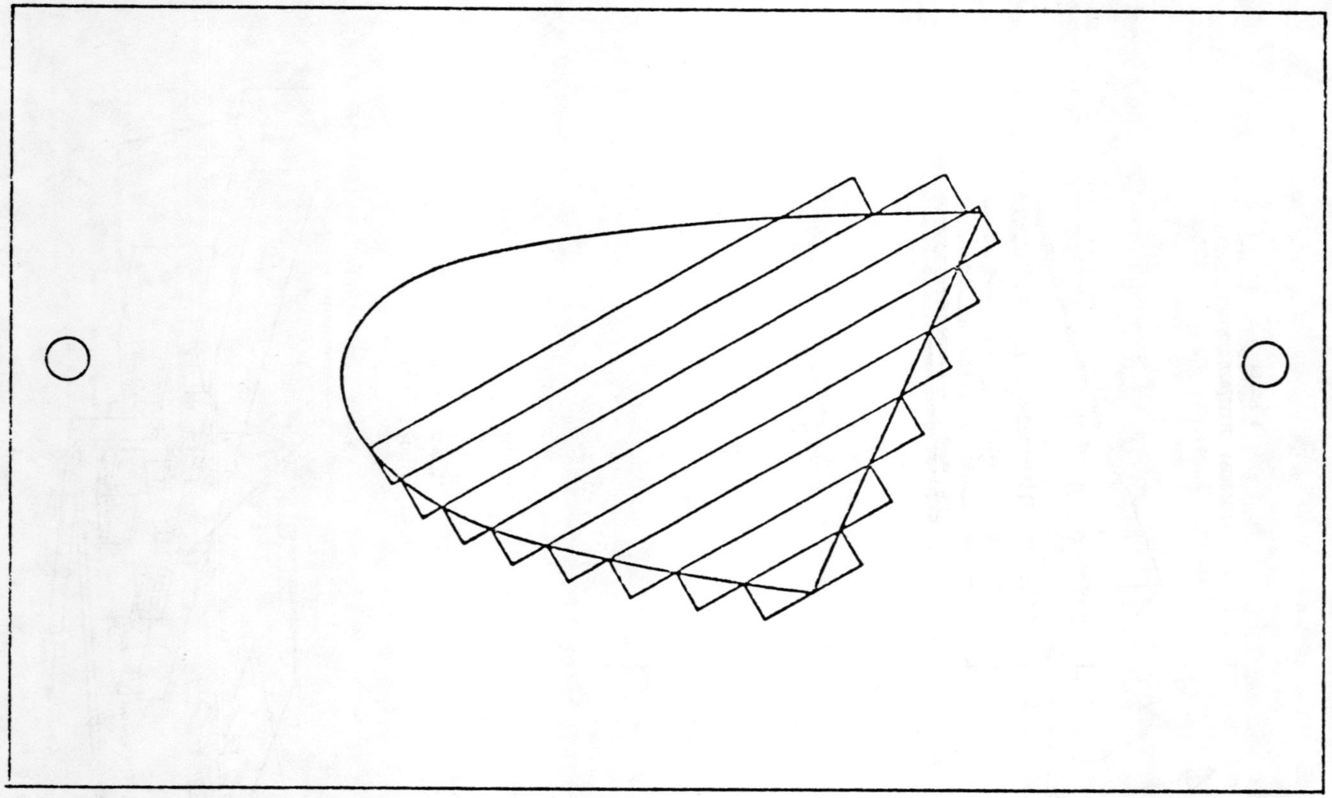

ADVANCED COMPOSITE HAND LAY-UP
USING 3-INCH STRIPS

FIGURE 5.

Material	Thick	Cut Speeds (IPM)	Power (KW)
Aluminum	0.040	250	3.0
	0.125	100	4.0
	0.250	40	3.8
	0.500	30	5.7
Nickel Alloy	0.125	120	4.0
Stainless Steel (300 or 400):	0.125	100	3.0
Steel (low carbon):	0.125	100	4.0
	0.660	45	4.0
Titanium:	0.250	140	3.0
	1.250	50	3.0
	2.000	20	3.0
Titanium Honeycomb Core:	0.250	100	3.0
Waspalloy:	0.125	140	3.0
Wood:			
Spruce:	1.620	780	3.0
Plywood:	0.750	200	2.0
Oak or Maple:	0.750	200	3.0
Advanced Composites:			
Boron/Aluminum Tape:	0.010	450	3.0
Boron Epoxy (cured):	0.120	60	3.0
	0.750	12	4.0

FIGURE 6

LASERCUTTING SPEEDS ON VARIOUS MATERIALS

LASER CUT HIGH STRENGTH ALUMINUM

3 kW CO_2 laser power
Material : 2024-T3, 0.040 in. thick
Rate : 200 ipm
Jet assist : air

Cut edge

100 X

Cut edge

2000X

LASER CUT IN 7075-T6 ALUMINUM

3 kW 200 ipm A.V. jet 0.040 in. thick

Laser cut edge —

Edge showing no HAZ

1.75 mils

UNITED TECHNOLOGIES RESEARCH CENTER

FIGURE 7.

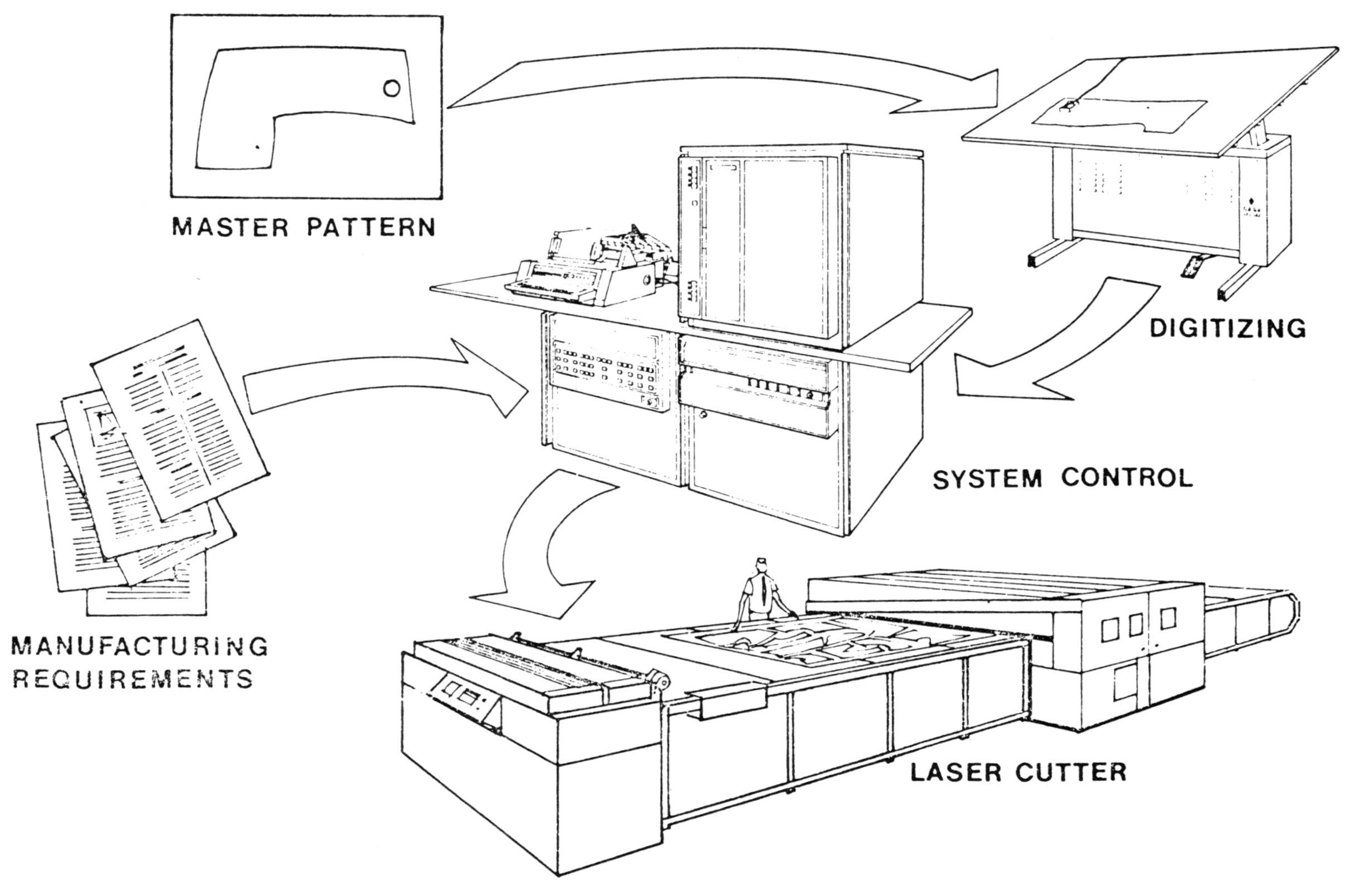

COMPUTER CONTROLLED LASERCUTTING
SYSTEM FOR METALS

FIGURE 8.

Presented at SME's NC Machining Conference, April 1975

NC/EDM—"Cutting" Costs With NC And A Wire Electrode

By C. Michael Mercincavage
IBM Corporation

The introduction of Numerically Controlled Electro Discharge Machining (N/C EDM) was about as well received as a draft notice during wartime. The idea of buying a piece of shop equipment incorporating the N/C EDM concept would have raised critical questions about ones' professional competence and state of sanity.

Fortunately, the early attitudes have changed. N/C EDM is finding its way into progressive thinking shops and paying off in significant savings when used for fabrication of tooling components, pressworking tools, miniature parts of complex profile, erosion of parts from exotic materials and, in many ways, resulting in a competitive advantage over other methods. The versatility and flexibility of the process will undoubtedly make N/C EDM a necessary addition to any shop expecting to remain competitive in the tool and die, job shop and production operation arena.

Figure 1 is a graphic representation of the basic N/C EDM process. The strength of N/C EDM lies in the basic makeup of the process. To the advantages of EDM, add the advantages of Numerical Control.

First, consider the general advantages of EDM. Some of the advantages include: The ability to erode hardened-tool steels, tungsten carbide, and some of the exotic materials with good finish and tolerance control. Also, consider the minimal fixturing requirements resulting from the fact that cutting forces are negligible. Further consider the in-cycle free time, available to an EDM operator, while a job is running. There are obvious advantages in an efficient arrangement that permits running of another machine tool in-cycle to an EDM operation. This situation is particularly true with N/C EDM.

Secondly, consider the advantages of N/C. At the top of the list is the inherent flexibility resulting from programmed sequences of instructions. New jobs can be set up and run with only minor changes (usually only tooling changes). Job runs may be stopped and started at specific program locations as desired. After a first piece is approved, repeatability and accuracy are consistent unless the operator intervenes or program changes are made. Because the machine tool functions are controlled by the Numerical Unit, a job may be run with only minimal operator attention.

The combination of features from N/C and EDM result in a process that will machine almost any material which is electrically conductive at room temperature. The process will produce two-dimensional shapes that are limited in profile complexity only by programming capabilities, and certain restrictions imposed by workpiece size. The former limitations are minimal if state-of-the-art programming languages are employed. Once set up, a job may be left unattended during the erosion

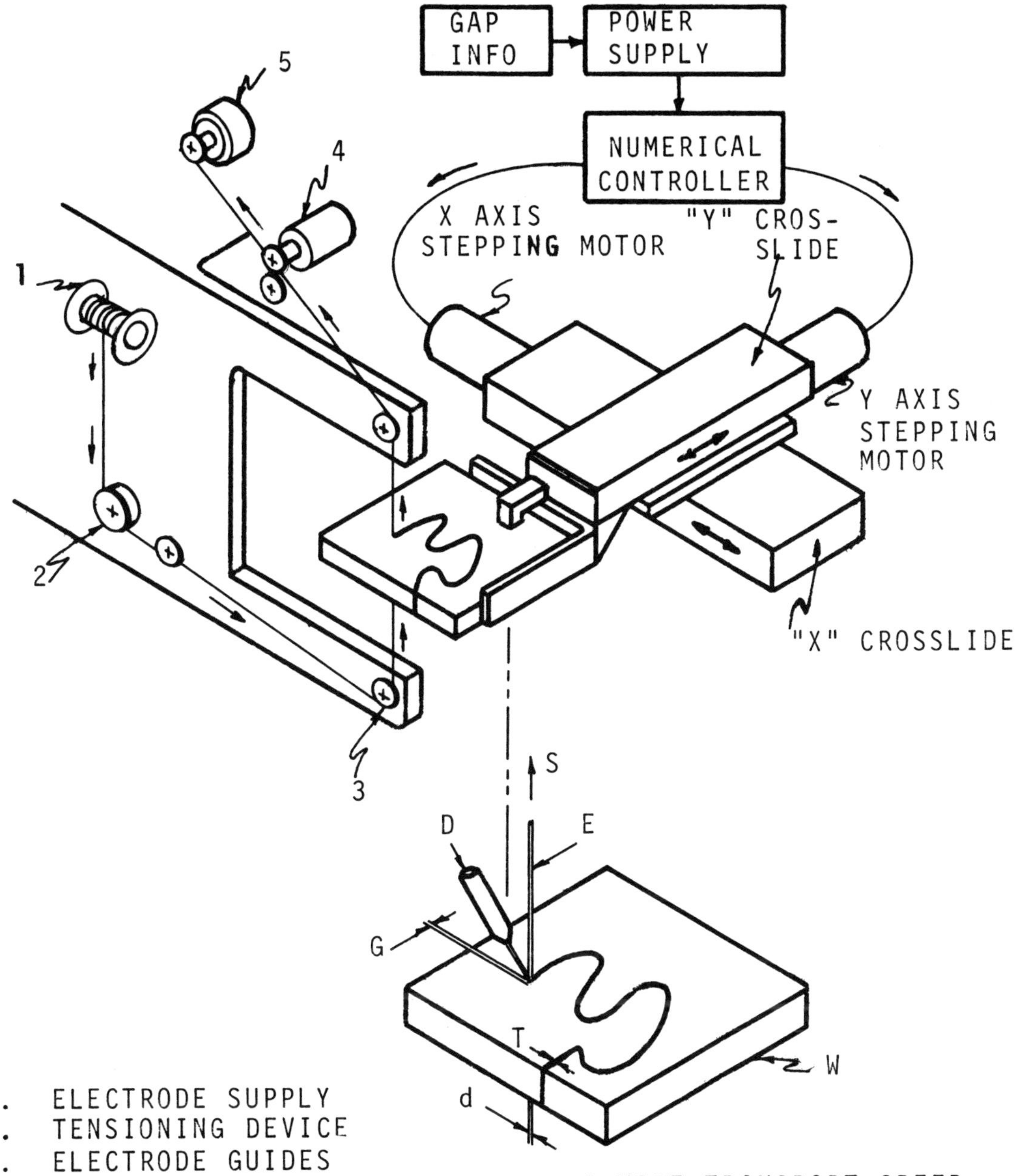

GAP INFO

POWER SUPPLY

NUMERICAL CONTROLLER

X AXIS STEPPING MOTOR

"Y" CROS-SLIDE

Y AXIS STEPPING MOTOR

"X" CROSSLIDE

1 2 3 4 5

S D E G T W d

1. ELECTRODE SUPPLY
2. TENSIONING DEVICE
3. ELECTRODE GUIDES
4. ELECTRODE DRIVE
5. TAKE-UP SPOOL

S=WIRE TRANSPORT SPEED
E=ELECTRODE WIRE
G=SPARK GAP
T=TRACK WIDTH
d=ELECTRODE DIAMETER
D=DIELECTRIC
W=WORK PIECE

FIG. 1
THE BASIC N/C EDM PROCESS

cycle. This combination of advantages offered by N/C EDM are best illustrated by the following example:

> Assume that you are required to produce a die plate, and two punches for a repress die with a complex contour, (with several tangent radii, points of inflection and fillets of 0.01-inch radius in the profile as shown in Figure 2.) The die plate is to be of 2.0-inch thick tungsten carbide and made in one piece (no sectioning). It's Friday morning, and the job must be completed by Tuesday morning. Completion of this order on the due date could result in a substantial follow-up order.

This is one of those situations where you wish it were possible to share the enthusiasm and confidence of your sales staff. But experience dictates otherwise. You know that the only practical way to do the job is with conventional EDM. The first big snag is fabrication of the electrodes; several roughers and finishers would be required for both the die plate and punches. Finding the free time or making some time available is a problem in itself, especially since the situation will call for substantial direct-labor overtime. After the electrodes are fabricated, possibly by late Saturday afternoon, the next problem would be soliciting an operator to monitor the EDM roughing operation and to change the electrodes as the erosion progressed. When this portion is completed, the finishing operation will be required. To minimize taper, frequent electrode replacements will be necessary. Assuming that the die opening was cored by a cylindrical electrode to minimize erosion demands on the shaped electrodes; the cutting times would be in the range of 8-12 hours as a minimum. Next the punches must be fabricated. Despite the fact that we are not going into all the details of manufacture, some hard, cold, expensive facts are already beginning to surface. First, is the fabrication of electrodes and second, is the requirement for direct-labor support on both a regular and overtime basis. Both of these factors, the electrode fabrication and direct labor, could be eliminated or minimize by N/C EDM. Figure 3 shows the estimated costs, in terms of direct labor, for conventional fabrication.

Consider the same job again, only this time assume your shop has an N/C EDM machine available for use on the job. Three basic requirements must be met to begin the job:

> First, machine-tool time must be available.
>
> Second, and N/C tape must be punched. This single tape will contain the information to describe the punch and die profile and the necessary functions for running or stopping.
>
> Third, an operator must be available to setup the work-

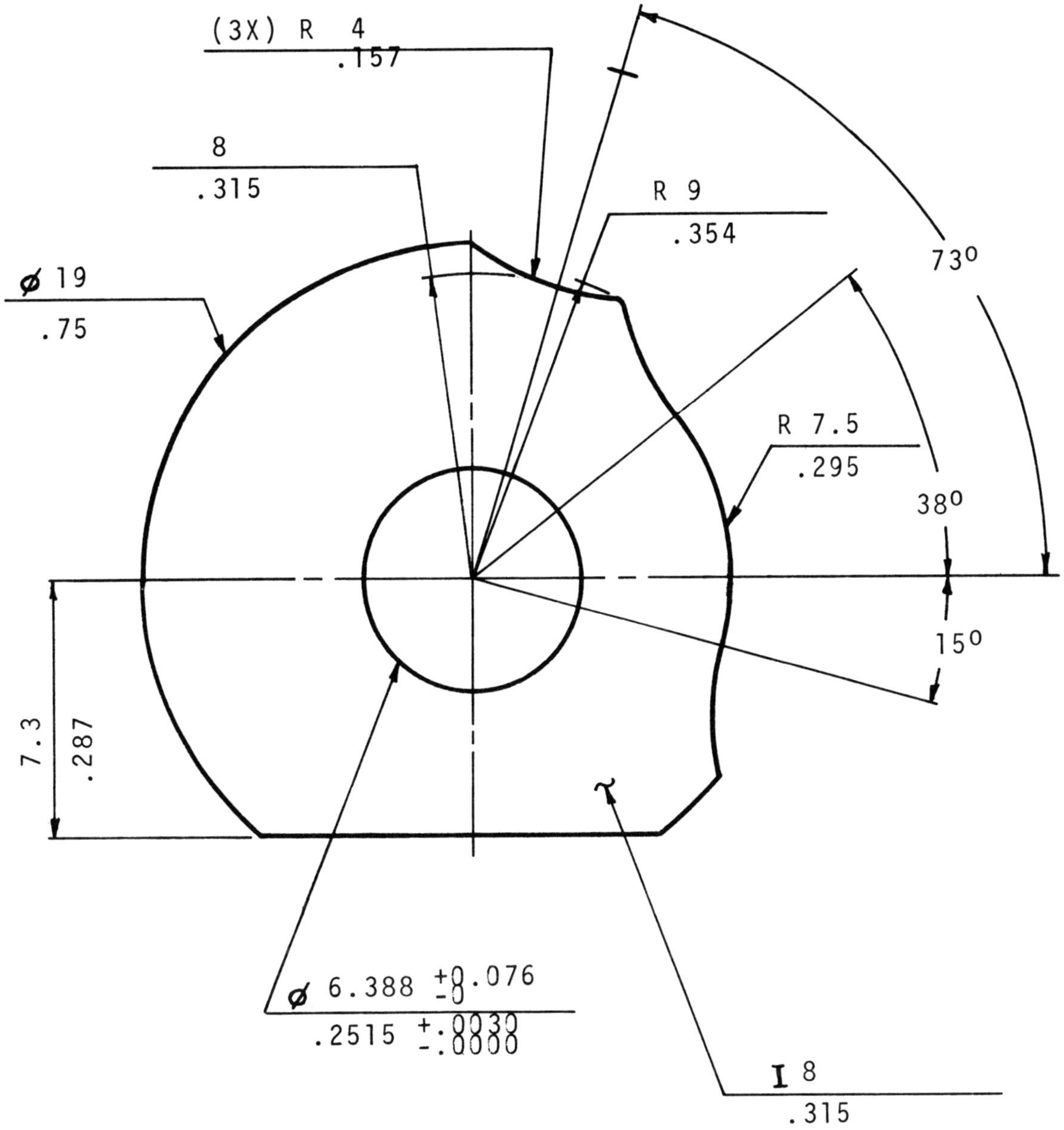

(3X) R 4
.157

8
.315

R 9
.354

73°

⌀ 19
.75

R 7.5
.295

38°

15°

7.3
.287

⌀ 6.388 $^{+0.076}_{-0}$
.2515 $^{+.0030}_{-.0000}$

I 8
.315

FIG. 2
REPRESS DIE CONTOUR

JOB DESCRIPTION/ OPERATION	HOURS CONVENTIONAL DIRECT	INDIRECT	HOURS N/C EDM DIRECT	INDIRECT	N/C EDM SAVINGS/(COST) DIRECT HOURS	INDIRECT HOURS
PREPARE DIE PLATE	-	-	-	-	-	-
FABRICATE ELECTRODES	16	-	0	-	16	
MOUNT ELECTRODES AND SET-UP	4	-	1	-	3	
PREPARE N/C TAPE	-	0	-	4	0	(4)
MONITOR EDM EROSION:						
ROUGH	4	-	0	-	4	
FINISH	4	-	0	-	4	
TOTALS (STRAIGHT TIME)	28	0	(1)	(4)	27	(4)

N/C EDM RUN TIME (HOURS)------------8

SYSTEM COST (APPROX.)-------------$90-110K

REPRESENTATIVE COST COMPARISION:

CONVENTIONAL FABRICATION VS N/C EDM

(DATA IS BASED ON FIG. 2 TUNGSTEN CARBIDE REPRESS DIE)

FIGURE 3

piece in a condition that will produce the required squareness, parallelism, and locations of the part features with respect to the machine crosslide-axes and wire transport axis.

Now, you should consider these three requirements with sufficient detail for clarity of the concepts that are involved. First, we must assume that the machine is available. This is a good assumption because existing setups can be easily broken and reestablished if a super hot requirement occurs. It is relatively easy to make the machine available, if the requirements occur.

Next, the method of tape generation for a complex contour of the type we are to erode will vary depending on tolerances on the part produced.

If part tolerances are in the ± 0.003-inch and greater range, data for describing the profile on tape could be obtained from a 10 or 20X size layout, a precision protractor with ± 5 minutes of arc resolution, and a precision scale graduated in 0.01-inch increments. It will be necessary to measure included angles of arc segments, arc center locations, lengths of straight-line segments, and angular positions of straight lines in space, etc. These data will then be punched into a control tape to run the machine tool. Time for our job should be less than 4 hours, if an experienced programmer handles the job.

If tolerances on the part produced from our repress die were less than ± 0.003-inch, computer programming techniques would probably be required to describe accurately the part profile and supply data for the N/C tape. Turn around times would vary widely; however, a time of less than 4 hours would be a realistic consideration. The job could be completed in much less time if the installation is up-to-date as far as state-of-the-art equipment.

Our third requirement, job setup, requires no more-or-less than a good precision milling setup. Wire electrode and critical work surfaces must be properly aligned (square and parallel) with respect to the work plane. Special tools that may be required are usually provided by the tool manufacturers to facilitate the setup operations.

After the tape is ready, the workpiece is setup (usually accomplished in less than 1 hour) and the power supply settings are calculated. Then the controller buttons are depressed and N/C and EDM, with all of their advantages, takes over. With the job started and running satisfactorily, the operator will return to some other task and probably will not bother with the N/C EDM job for several hours. Operator intervention is required only for such tasks as securing the die dropout piece, emptying wire electrode takeup spool, and other similar tasks. The EDM will take care of removing workpiece material while N/C does the driving and stopping operations.

Considering average capability in the shop, the job is begun and running in 5 hours or less. A punch or die plate erosion operation may run for 20-30 hours; the direct labor requirements for fabrication of the complex contour would usually be limited to 1 hour or less. You run N/C EDM in-cycle to other machining processes. Leave the machine alone and it will continue to run until a program stop or "end of tape" command is read into the controller.

When the die has been completed, the same tape (or a minor modification of it) may be used to erode punch, stripper plate, and other parts. Controller functions are used to provide proper offsetting of the wire electrode to generate precise relative clearances between the die set components. Tight process control can routinely produce clearance of 0.0001-inch or less per side.

Usually, the accuracy of the tool produced will be in the low tenths (0.0001-inch) range from the programmed values if the set-up is correct, if proper electrode offset values are entered correctly into the controller, and the process variables; held to the desired values.

Your total labor requirements for the N/C EDM method would be 4 hours indirect (programming) time and 1 hour direct (set-up) time. Conventional fabrication, using the most practical available method EDM) will require 24-30 hours of direct labor for fabricating only the die plate opening. One-half to two-thirds of this time would be used in making the EDM electrodes. The remainder would be required as dedicated direct labor for monitoring the EDM operation, electrode replacement, tool resetting, and other operations.

While secondary operations for producing the rough die plate may vary for conventional and N/C EDM methods, it is usually safe to assume that fabrication times that satisfy the N/C EDM set-up requirements would be less than time requirements for the conventional method. In addition, you might also save some costs related to preparation of the workpiece for the N/C EDM operation.

Pressworking tools are excellent examples of ideal workload for N/C EDM. The making of these tools is highly desirable because fabrication by alternate methods would require significant labor costs.

But, the flexibility of the N/C EDM process makes it adaptable to other machining areas as well with similar outstanding savings potential.

Assume, for example, that you require 50-100 small parts (part profile must fit into available N/C work plane) of complex profile with a tolerance ± 0.003-inch on most dimensions. Figure 4 illustrates a good example. If, you decided that a punch and die set is impractical for this quantity, (the probability of follow-up orders might justify a die set) the parts could easily be contoured from sheet stock in batches of 2-20 pieces, depending on part thickness. Because each load would be run in-cycle to other machining operations, an operator will be doing other operations, (drilling, tapping, etc.) on the same part while N/C EDM accomplishes the profiling. Profiling operations are extremely accurate and repeatable and the eroded surfaces can be used for location reference in subsequent operations.

One-time labor costs would include set-up and tape preparation. Other operator time charged against the job will be usually limited to loading and unloading of parts and minor maintenance. Any loading and unloading sequence should be achieved in less than 0.1-hour per load. When a load is 10 parts, direct labor per part charged against the N/C EDM operation falls into the 0.01-hour or less range, a respectable charge when evaluated by any criteria. This point is worth remembering: Once a job is set up and running satisfactorily, the direct labor required to produce parts is generally insignificant. In most cases, the cost is insignificant to the point of permitting an N/C EDM operation to be run in-cycle to other N/C or conventional operations. Usually, more time is required to ring on and off a N/C EDM cycle than to perform the load/unload sequence!

In these two examples the requirements are completely different from each other. In the first case, a single copy of a special tool is to be produced for pressing P/M parts. The second example involves actual contouring of part profiles from sheet stock. In the first case, we require precise control on punch-and-die plate <u>clearances</u>; while in the second case, we need a capability to <u>contour</u> a relatively high precision (low tolerance) profile in a part or stack of parts simultaneously.

In both cases, we need sharp corners and fillets, good surface finish and relatively fast speed. We have all these capabilities with N/C EDM. Also, in both cases, most of the direct labor was eliminated by taking advantage of N/C and EDM capabilities. Precision contoured parts and tools are routine products of the process. Differences in the types of work which may be handled are not a problem. Flexibility is unparalleled.

Routine N/C EDM applications will include the two examples discussed above; pressworking tools and small job lot contouring, and a greatly diversified collection of other types of work. Figure 5 lists good work load candidates. Machining

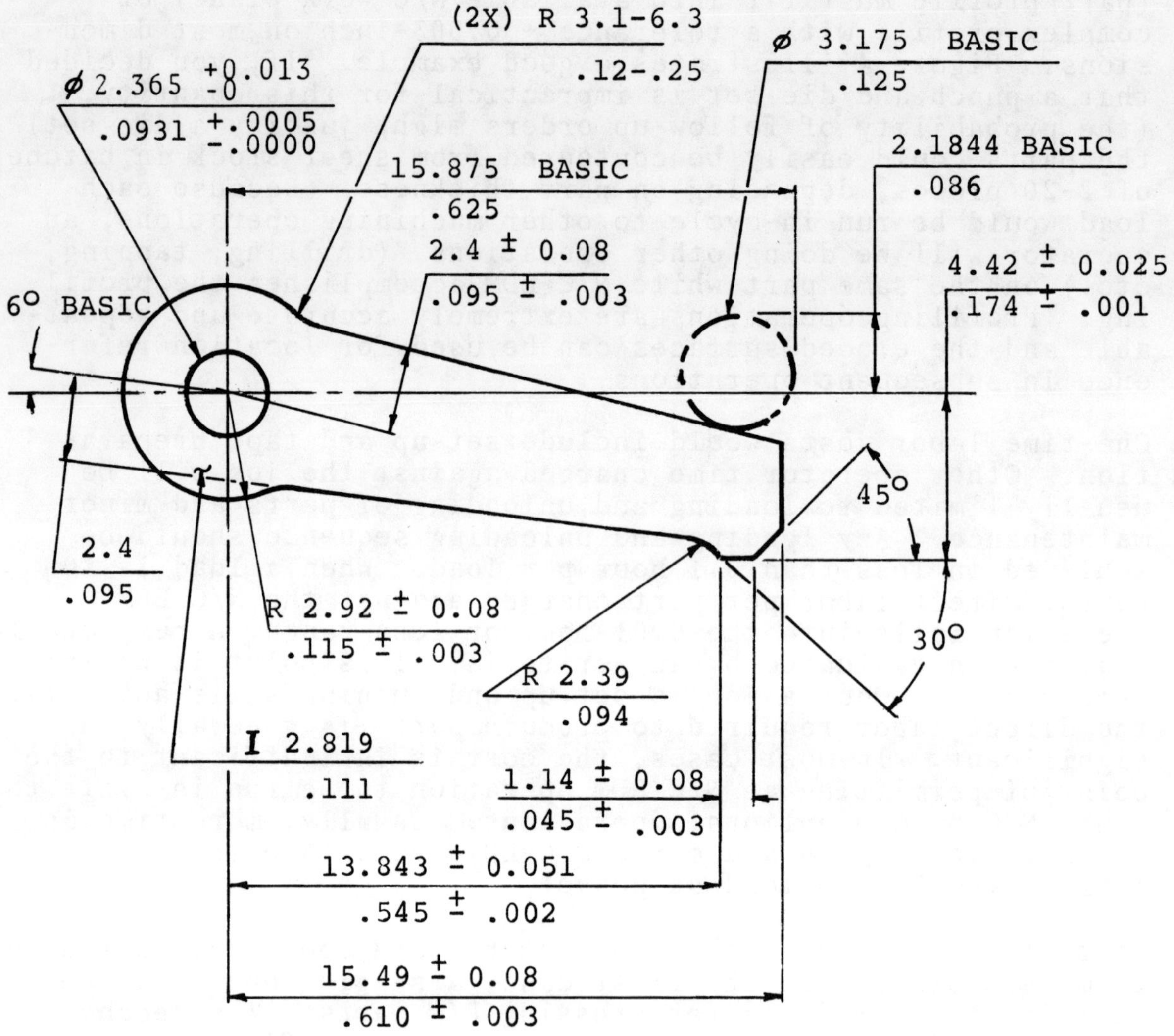

(2X) R 3.1-6.3

.12-.25

ϕ 3.175 BASIC

.125

ϕ 2.365 +0.013 -0

.0931 +.0005 -.0000

2.1844 BASIC

.086

15.875 BASIC

.625

2.4 ± 0.08

.095 ± .003

4.42 ± 0.025

.174 ± .001

6° BASIC

45°

2.4

.095

R 2.92 ± 0.08

.115 ± .003

30°

R 2.39

.094

I 2.819

.111

1.14 ± 0.08

.045 ± .003

13.843 ± 0.051

.545 ± .002

15.49 ± 0.08

.610 ± .003

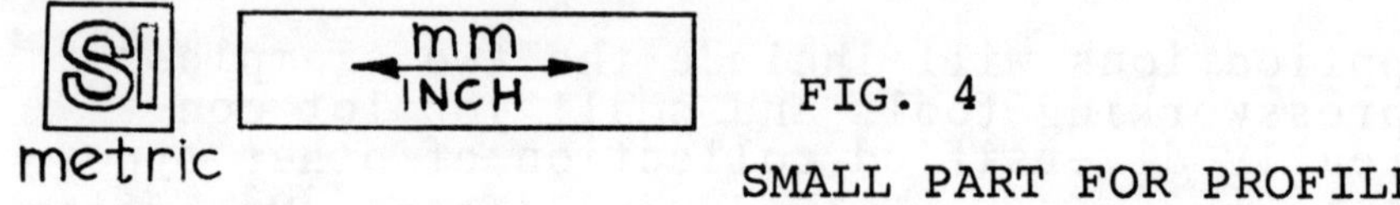

FIG. 4

SMALL PART FOR PROFILE CONTOURING

PRESS WORKING TOOLS

SMALL PART CONTOURING
- ENGINEERING PROTOTYPES
- LOW QUANTITY PRODUCTION RUNS

TOOL PROFILES
- TEMPLATES
- NESTS

NARROW SLOT MACHINING

SINGLE CRYSTAL ERODING
- SLICING
- CONTOURING

ROUTING OPERATIONS
- SQUARE INTERNAL CORNERS
- FRAGILE PART CONTOURING
- HIGH PRECISION PROFILING

COMPLEX RADIAL SHAPE PROFILING
- GEARS
- PULLEYS
- RATCHET WHEELS

'HARD' MATERIAL PROFILING
- TOOL COMPONENTS ERODED IN HARDENED CONDITION
- TUNGSTEN CARBIDE

FIGURE 5

GOOD N/C EDM APPLICATIONS

of hardened steel and tungsten-carbide guide bushings is a good application. Contours in tools or parts calling for square internal corners, (less than 0.005-inch R) are readily machined.

Also, consider machining complex profiles in small parts as specified in a routing operation. The parts may be small with thin sections.

"Impossible" jobs such as square internal corners, narrow slot machining (less than 0.005-inch) in tungsten carbide or hardened steel, single crystal rod machining, erosion of conventional EDM electrodes, with allowance for roughing and finishing generated from the same N/C tape, are typical applications. Square internal fillets can be cut in female electrode material for conventional EDM. As with other N/C operations, one tape may be used to produce as many copies as needed. This advantage makes it feasible to produce spares (punches, die blocks, conventional EDM electrodes, templates, etc.) at costs that will allow you to compile a significant return on your machine tool investment. When a job quote is based on a specific order size, any reruns of the same job should be highly profitable because the only actual costs associated with the job would be material, machine time, minor secondary operations, and set-up.

The case for N/C EDM thus far has shown significant savings in direct labor for production of parts and tools. So what is left for an encore? Plenty!

With negligible machining forces, fixturing requirements do not have to address cutting forces. Locating, ease of use and set-up, load-and-unload speed, and other factors become the design criteria. Reduced fixture size and complexity result in low-cost tooling, free of frills and design expense. In many cases, you can make tooling from the same control tape used previously to contour part profiles. A good example of this application is a part "nest". Another of the tooling advantages is the fact that a wire electrode may be used to pick up part features (bores, edges, etc.), and much of the necessary "tooling" capability is built directly into the machine tool. The wire electrode may be located within ± 0.0001-inch of the features.

Quoting N/C EDM jobs for profitable return is somewhat of a personal decision. Some shops are preparing quotations by computing costs at conventional rates and labor content and deducting a fixed percentage (5-10%) to make the bid more attractive. Based on your punch-and-die example, savings to the customer may be 5-to-10 hours on a 100-hour job. Profit will be based on the ability of the shop to program the N/C EDM system efficiently; however, a 50-hour profit (dollar equivalent on a job as described in the example will be typical.

As more shops use N/C EDM, this method of quoting jobs will probably change to the extent that the percentage deducted from the estimated hours--based on the conventional approach--will increase. That remaining percentage, covering costs and profit, will be determined by programming efficiency and other costs associated with N/C tape production. This assumes equal competence in the capabilities of their direct-labor competition. The shop with superior tape-generating procedures will be in the best competitive position.

Attention to the problem of programming should be a major effort before N/C EDM is installed. The best approach is a reasonable study of the programming support requirements before a machine tool purchase order is placed. The study should accomplish two purposes. First, the parameters for the required precision must be established for the type of workload to be handled. Second, as a result of these parameters it will be possible to make an accurate economic justification, based on the required range of precision of work to be handled, because these factors will directly affect programming costs.

The cost of programming <u>must</u> be considered in the justification. If pressworking tools for processing parts in your plant are to be your primary product, with a typical ± 0.01 - 0.02-inch tolerance, most programming data will be obtained satisfactorily from a multisize layout of the part profile. Precision punch and die-and-stripper plate clearances are obtained by varying electrode offsets on the machine tool control. However, if your workload mix is mainly profiles of tools or parts with tolerances in the low tenths range (0.0001-inch), manual data recovery will not be practical. Some level of computer-aided programming will be probably necessary for realizing competitive programming cost within reasonable turn-a-round time.

The scope of computer-assisted capability available today ranges from simple trigonometric and arithmetic processing using relatively simple stored algorithims, through sophisticated interactive techniques. In the more sophisticated systems, you may actually describe a part geometrically, store the geometric profile and interrogate the computer for necessary program data. Once these data are developed, execution of an output sequence will generate a tape and a hard copy printout. Some computer installations use error reduction or minimization techniques; others use programs to index spatially a single gear tooth to produce a computer-stored description of a complete gear or timing pully. The possibilities are unlimited; however, you obtain only what you are willing to put into the system and that fact should be recognized when you are considering your cost of programming and your actual savings.

With the great latitude and depth of computing that is available in the software and hardware market today, you will not have a problem in obtaining the competence level that is required in software and hardware support. If you have access to a time-sharing or onsite computing system, special computer programs may be developed by your own personnel. The ideal profit condition will be attained only if your requirements match your available programming and hardware levels. Do not overbuy or overestimate your requirements for computer support. If you exceed your requirements, it will initially present a weaker case for justification of N/C EDM and ultimately reduce your apparent return on investment and increase the payback period. You can always add to your capability at a later date if the workload justifies the added capability.

A manufacturing process without inherent faults or limitations is unbelievable and unavailable. N/C EDM comes close to being faultless; however, there are some facts of life which must be remembered.

As with conventional EDM, the effective size of the electrode--the finished electrode size plus overcut allowance--is dependent on the material being eroded, eletrode material, flushing, power levels, and other process variables. Therefore, effective electrode size must be determined before committing an electrode wire to the workpiece. For precision-tolerance work, optimum results may be obtained by running a "test cut" in the workpiece material to determine overcut, surface finish, and other conditions which are process-variable dependent. Running the actual job with the required electrode offset and the same process parameters will produce the desired results and precision. Because the test cutting will require machine time which is nonproductive, this testing should be classified as a process disadvantage.

For those cases where tolerances on parts are not as precise, the test-cut route should be used only until enough experience is gained to provide a data base for overcut prediction. A prediction of overcut, based on past experience, will usually fall into the 0.0005-inch range from the actual overcut. Here again, workload precision will determine the required method.

Another of the potential problems inherent in the N/C EDM process is the apparently slow rate of material removal. However, a comparison of the actual time required to move a job out the door, or into stock, will usually show N/C EDM on top over higher-labor conventional methods of fabrication. For example, considering precision profiling work in a hardened steel or tungsten carbide die plate, an experienced operator using grinding techniques will certainly remove more material per unit time if the period of measurement is confined to the actual grinding operation. If, the total time charged against the job is examined, the nonproductive aspects of conventional fabrication will appear. For example, in any contour grinding operation, wheel redressing, resetting, and changing are costly requirements for producing high precision work. If the workpiece is made of tungsten carbide, and diamond grinding wheels are required, obviously wheel cost is also a factor to be considered for conventional fabrication.

The cost of electrode wire for a week of N/C eroding of carbide is approximately $15.00. Compare this expense with the cost of the diamond wheel consumed for the same quantity of material removed by jig or surface grinding. The savings are substantial.

A representative completion time for running a job with conventional techniques should include redressing, resetting and wheel changes. N/C EDM run time will usually require less actual time, in shop hours, to complete the same basic profiling. The cost of conventional fabrication must include labor for redressing, resetting and wheel change plus wheel and other tool costs. Add to this cost the normal nonproductive factors inherent in any operation requiring operator attention. It is very likely that total job cost will far exceed N/C EDM costs.

If we stretch a point, the need for comprehensive operator training might be considered as a disadvantage. N/C EDM will not be producing at its maximum cost-saving potential unless the operator is proficient in handling a wide variety of workload situations without excessive "hand-holding" from indirect manufacturing support areas. It does not make good economic sense to assign just any "warm body" to run an N/C EDM system. The operator must be highly experienced in making precision setups, preferably of the light, vertical or two-axis N/C milling equipment and a capability to understand the implications of changes to power supply parameters, controller functions, and effects of electrode offset. It will require time, perhaps a man-month or more and a couple of scrap jobs to attain the desired level of competence.

If your shop cannot provide this type of skilled operator or the time to train a capable individual, then you will definitely be trying to run N/C EDM at a disadvantage. Both your profit and your regard for the process will be lowered.

There may be other disadvantages of the process, which are peculiar to a specific shop environment; however, these will be uncovered while you are examining your individual requirements.

Looking at the other side of the situation, another important advantage of N/C EDM, which is especially significant when a high pressure job situation arises, is the minimization of operator error. Once the job is set up and running, the pressure is off the operator. If the N/C tape, setup and operating parameters are acceptable, results will be acceptable and insensitive to the constant job queries and other error-producing factors which might influence a machine tool operator resulting in a costly and untimely error.

A discussion of the disadvantages of N/C EDM usually produces some additional advantages as outlined previously. Obviously, the multitude of advantages offered by the process will vary with respect to individual shop capabilities and requirements. The strong point to consider is the fact that overall costs for moving a job through the shop and out the door, are reduced with the use of N/C EDM which offers a clear cost advantage for high labor content tasks.

While N/C EDM has had a relatively unheralded beginning, the many advantages of the process will surely catapult its popularity to highly respectable levels in the immediate future. The early suspicions and caution have been pushed aside by user success. Process versatility will allow adaptation to many shop requirements, some of which are yet to be developed. A shop which lacks N/C EDM capability will not be in a position to compete with a shop possessing the capability if all other shop abilities are equal.

One final point deserves mention. At a time when skill shortages of tool, die-and-model makers is approaching a critical point; the advent of N/C EDM could not be more timely. While the process will not replace highly-skilled machining specialists, it will enhance their capabilities and increase productive output. Fewer men will be able to handle a larger workload.

Numerical Control has had a significant impact on production costs and philosophy. The latest brain-child of the N/C revolution, N/C EDM, combines the necessary control features of N/C with the remarkable machining capability of EDM to produce a process which yields substantial savings in areas where high skill and great cost were previously required. When you can remove high skill and high cost out of a job during routine operations, that's progress. N/C EDM offers this progress.

Reprinted from Production Engineering, March 1977

Electronic bandsaw sparks a revolution

Wire EDM—electrical discharge machining with a thin, copper wire electrode—is moving out of the toolroom and onto the shop floor. It is proving to be an economical way to carve out prototypes and small-volume runs of hard-to-make parts.

By DONALD E. HEGLAND
Associate Editor

Machining with wire is a special variation of electrical discharge machining. As in conventional EDM, the cutting mechanism is spark erosion. Material is vaporized from the workpiece by an electrical discharge. The big difference is in the electrode—the "cutting tool." Familiar, plunge-cut EDM employs a complex electrode, shaped to either match or mirror the shape of the finished part. The electrode is sunk into the workpiece, gradually eroding it to produce the required form.

Wire EDM employs an expendable wire electrode that is guided by a numerical control system to generate the required form in the workpiece. Think of it as an N/C, omnidirectional, electronic bandsaw with one exception—the wire electrode never touches the workpiece. High-resolution servos maintain the necessary spark gap between wire and workpiece.

The wire unreels continuously from a supply spool, passes through the kerf in the workpiece, and is taken up on a discard spool; hence the other common name for the process—traveling-wire EDM. The wire only makes one trip through the workpiece and is discarded, therefore electrode erosion is insignificant.

Tool and diemakers have embraced wire EDM with a passion; some users say it's the best application for N/C in the toolroom and one says the marriage of N/C with wire EDM is the greatest single advance ever seen in toolmaking. But the benefits don't stop at the toolroom door. Clever production engineers are finding more and more uses for wire EDM on the production floor.

A few limitations—a lot of advantages

Wire EDM has a lot going for it but like anything else it isn't a panacea. Although few, the limitations need to be considered. First, your workpiece has to be electrically conductive—no form of EDM works on an insulator. Second, you can't cut blind cavities—you can only do through-hole jobs. Third, wire EDM machines cost from $80,000-200,000, depending on size, capabilities, accuracy, and system sophistication. So, to get the best return from your investment, you have to be a little cagey about how you use wire EDM. As a general rule-of-thumb, it is for the tricky and complex jobs, especially on intractable materials. Experts advise that many garden-variety jobs still belong on conventional EDM machines.

Having defined the gross boundaries of applicability, what benefits lie within? Tooling economy is a big plus. Making the complex electrodes for conventional EDM usually represents about 40-60% of the total cost of the EDM work. And the electrodes can often be used only once because they are eroded as they cut the workpiece. There are no electrodes to make for wire EDM. The only tooling cost is for the wire itself, a mere

10-20¢ per hour of operation.

Wire EDM leads to simpler die designs and slashes toolmaking costs from 50-70%. Split and sectionalized dies that require subsequent form grinding are things of the past; wire EDM cuts them in one piece. Punches, die blocks and shoes, ejectors, and stripper plates can often be cut from the same tape. And it is also being used increasingly as the most economical method of making the complex electrodes used in conventional EDM.

Even if wire EDM is only used in the toolroom, the benefits spill over onto the production floor. Steel alloy tooling can be made after the blanks are heat-treated and stress-relieved (as in conventional EDM), resulting in close-tolerance tools that don't require finish grinding to remove distortions introduced by heat-treating. Long-wearing carbide tools that might be impractical or impossible to make by conventional methods because of their complexity are easy with wire EDM. The result is simpler, more robust tools that stay in the presses longer, making better parts faster.

Quick turnaround on die maintenance or repair is another plus. The workpieces are "stored" on tape and are available on demand, so replacing a broken or damaged tool with an exact duplicate is simply a matter of loading a blank into the machine and starting the tape through the tape reader.

Cams, gears, master gages, and any hard-to-define geometric shapes—especially in the exotic, space-age materials—are practically a piece of cake for wire EDM. Workpiece hardness is not a limitation; Alnico, Inconel, Waspaloy, and moly-rhenium are just a few of the "nasty" materials that wire EDM cuts easily. And there are even advantages to using wire EDM on softer materials like brass; zero-force machining leaves no burrs and no residual stress or distortion. Accuracy and surface finish are usually poor on coarse-grain graphite cut with wire EDM and the process is not generally recommended for this material.

Prototypes and short-run parts are also ideal candidates for wire EDM. Difficult blanks can be cut in comparatively short times, and layers of sheet metal can be stacked to cut a number of parts in one pass. Making samples in the development stage means design savings too. Prototypes can often be made for half, or less, the cost of production tooling. And the pro-

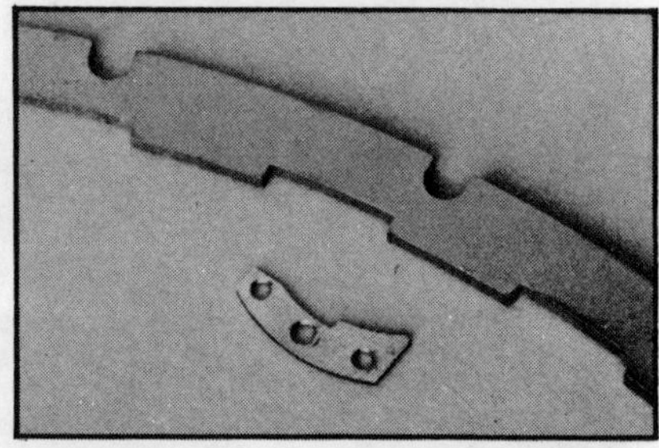

Stacking blanks makes it easy to cut quantities of thin prototype parts with wire EDM. Fifty of the small parts (0.02-in. thick) are cut in one pass in 1.5 hr. The large parts (0.08-in. thick rings) are stacked 12-high and cut in 50 hr.

Courtesy, Elox Div., Colt Industries

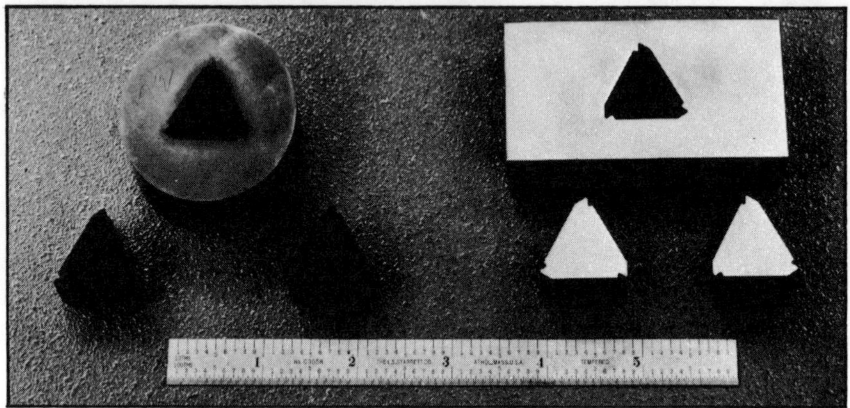

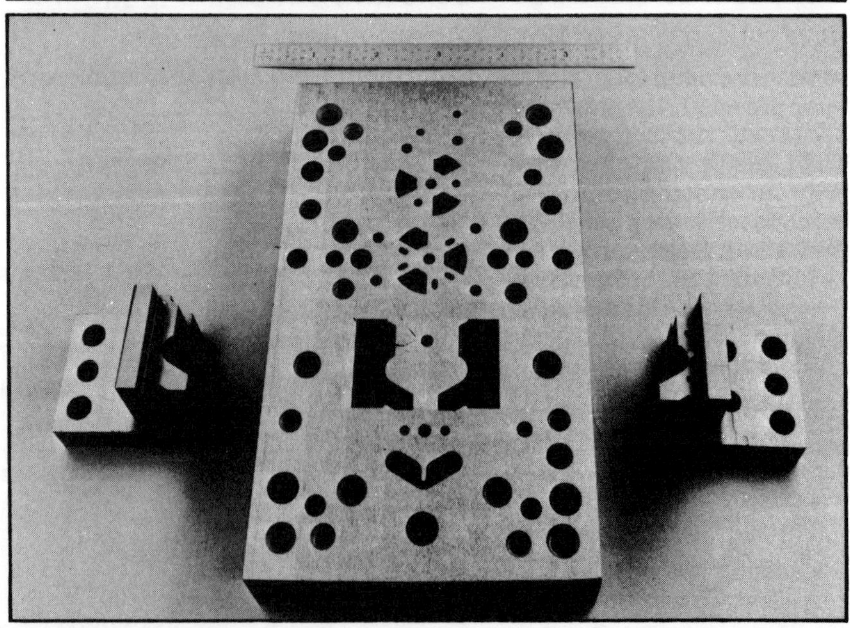

Total time to make the punch and die set (below) was 33.9 hr, including 10.5 hr for programming and 23.4 hr for cutting. Material is D2 hardened tool steel. Compacting tool set (above) was made in 31.75 hr, including 25.5 hr to cut the carbide parts, 5.25 hr to cut the brass laps and one hour for programming. *Courtesy, Andrew Engineering*

Courtesy, Elox Div., Colt Industries

Courtesy, Andrew Engineering

Two examples of linking programs together to cut a number of punches in a continuous run. The finished punches are separated from the parent material with a single, straight cut.

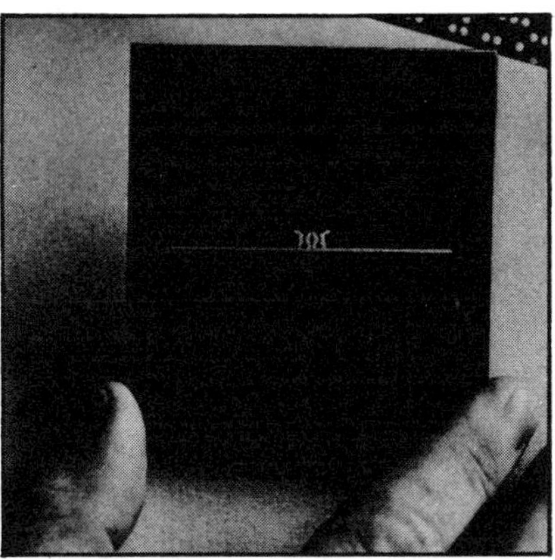

Tricky jobs are wire EDM's forte. The piece on the left has 12 dovetail slots, 0.1-in. wide by 2.25-in. long, with a radius on the bottom. Programming time—one hour. The extrusion die on the right includes 0.02-in. wide slots and 0.01-in. thick ribs with blended radii and angular straight lines. Manufacturing time—less than ten hours. Courtesy, Elox Div., Colt Industries

gramming effort isn't wasted—after proveout, the same tapes are used to cut the production tooling. Thin parts that require subsequent forming can be cut economically with wire EDM and used to establish forming station requirements at the same time that production stamping dies are being made.

A look at the process

By comparison with conventional metal-removal processes, the cutting speed of wire EDM is very slow. This is, however, deceptive. Zero lead-time for tooling, economical programming (zero time for an existing tape), and the ability to run continuously and unattended add up to make wire EDM competitive in terms of floor-to-floor time. And the linear cutting speed is independent of contour geometry, which simplifies estimating a job.

Typical cutting speeds in hardened D2 tool steel are 1-1.5 in./hr in 1-in. stock and 0.8 in./hr in 3-in. stock. The speed is not linear with thickness because thicker stock exposes more electrode area for better utilization of the available energy. Cutting speed in steel is virtually independent of alloy or hardness. Aluminum is less dense and has a lower melting point, so it cuts faster. Although carbides can be cut nearly as fast as steels, the surface finish is usually degraded because of the power required. Hence, recommended cutting speeds are from one-half to three-quarters that of steel.

The upper bound on workpiece thickness with present machines is about 3-4 in. There seems to be no practical lower bound. Material as thin as 0.002 in. can be cut with ease, either single-ply or stacked. Surface finish on parts cut with wire EDM looks like 10 rms, say users, but measures about 40 rms. The texture is matte, which is often advantageous for dies because it holds lubricant. As in conventional EDM, faster cuts—that require more input power—produce rougher finishes.

There are two schools of thought regarding the relative motion of workpiece and wire-guide system. One holds that the wire-guide system should remain stationary while the workpiece moves, as in an ordinary bandsaw. Advocates of this method claim that the wire-guide system is sturdier, providing superior accuracy. The other school favors holding the workpiece stationary and moving the wire-guide system. Those who favor this method say that controlling a constant mass—the wire-guide system—eliminates errors caused by differences in workpiece mass, and that this configuration simplifies loading and unloading. They also point out that it permits mounting several wire-guide systems on a single base. Accuracy is excellent regardless, with absolute positioning to ±0.0002 in. and repeatability to ±0.0001 in.

Water—filtered, deionized, and demineralized in a continuous closed-loop system—is the dielectric used with wire EDM; conventional EDM uses oil. Using water eliminates a potential fire hazard, hence the machine can safely run unattended. Water does a better job of flushing the debris out of the kerf than oil, and because of its higher heat capacity provides better cooling. The combination nearly eliminates the recast surface often found on parts cut with EDM.

Some wire EDM machines operate with the cutting zone—and often the entire workpiece—

completely submerged in a tank of dielectric as in conventional EDM. Other systems employ as many as three jets to direct dielectric into the cutting zone, with the runoff collected in an open tank that surrounds the workpiece. Those who favor submerging the workpiece in dielectric feel that this method provides better cooling of the electrode without agitating the dielectric. They also point out that workpiece temperature is more stable and homogeneous, and that submerging the workpiece protects it from atmospheric corrosion during machining. Proponents of the jet method of delivering dielectric to the cutting zone say that the jets keep the debris flushed out of the kerf better than submerging the workpiece does.

Kerf width is determined by wire diameter and applied power. Wire diameter typically ranges from 0.002-0.012 in., with copper used most frequently. Manufacturers generally recommend using molybdenum or tungsten to provide added mechanical strength if the job calls for wire smaller than 0.004 in. The overcut is generally 0.0015-0.002 in. per side. This can be reduced to 0.001 in. in thin materials by reducing the applied power, but cutting will be slower. Wire diameter is also affected by the minimum inside corner radius to be cut, and by the nature of the workpiece—harder and thicker materials generally call for heavier wire.

The cutting edge

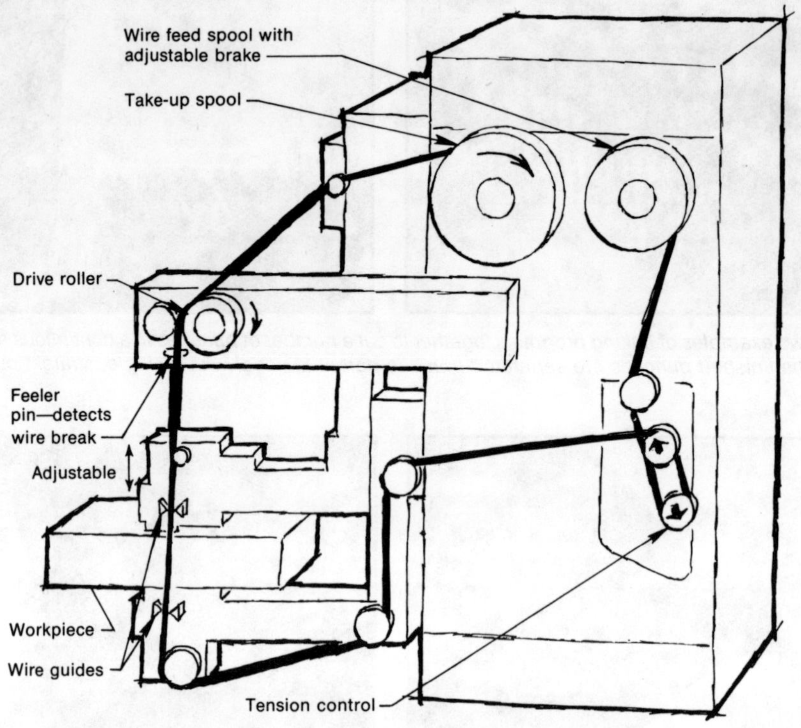

Typical wire EDM machine with stationary wire-guide system and moving workpiece. Heart of the system is the wire electrode that chews its way through the workpiece by spark erosion. Feeler pin shuts the system down if the wire breaks—a fail-safe feature for unattended operation. *Courtesy, Agietron Corp.*

N/C does it

Numerical control is the key that enables the user to avail himself of all that wire EDM has to offer; without N/C, the usefulness of the basic machine would be severely restricted. Consequently, all wire EDM machines incorporate some form of N/C. Some machines use simple, two-axis, hardwired N/C and respond either to direct keyboard input or punched tape prepared elsewhere. The more sophisticated machines are CNC, with a micro- or minicomputer built into the control system. These often include three-axis capability, useful for jobs like cutting small tapers for die relief in a single pass.

Other control systems interface with tabletop programmable calculators or in-house mainframe computers. In most control systems that include some form of computer, the computer can run the machine through one job and prepare or debug a tape for another job simultaneously.

Simple shapes can be programmed directly, using a calculator to determine coordinates from the part print. Complex jobs need computer assistance either from the built-in units (CNC) or separate computing facilities.

A series of programs can be stored in the computer and played back for continuous overnight or over-the-weekend unattended operation. The only restriction here is that all of the cuts have to be connected. Each new cut—unless it starts at the edge of the workpiece—requires manual rethreading of the wire through a starter hole. Rows of parts are often cut on a continuous basis and then separated with a single straight cut. Some users report trouble-free unattended operation for as long as 65 hr.

Information and illustrations for this article were provided by: Agietron Corp., Woodside, L.I., N.Y.; Andrew Engineering, Hopkins, Minn.; Charmilles Corp. of America, Plainview, N.Y.; and Elox Div., Colt Industries, Davidson, N.C.

Presented at SME's Nontraditional Machining Seminar, January 1970.
Updated for Machining Hard Materials

Rotary Ultrasonic Machining

By William R. Tyrrell
Branson Sonic Power Company

Ultrasonic energy is a well-established industrial aid, and widely used in such fields as cleaning, nondestructive testing, welding of thermoplastics and, to a lesser degree, fine wires in the semi-conductor industry.

There are, of course, several other application fields, including medical diagnostic and biological equipment.

The use of ultrasonics as an aid to machining has, in the main, been applied only to an impact grinding process. Although most effective in cutting irregular shaped holes or impressions, the materials that can be machined by this method are limited to glass, ceramics, and similar hard brittle materials.

Ultrasonic impact grinding, which will be described later in this paper, utilizes a reciprocating motion, without rotary action.

However, if a rotating diamond tool is vibrated at an ultrasonic frequency when machining hard brittle materials, a significant improvement in cutting rate and tool life will result.

One of the early exploiters of this principle was the Ceramics Division of the United Kingdom Atomic Energy Authority at Harwell, England. This establishment required small quantities of special ceramic components fabricated from fired material. The limitations of some machining operations with conventional machinery stimulated their interest in investigating new techniques.

A commercially available high intensity ultrasonic probe designed for producing intense cavitation in a liquid was purchased and modified to excite a diamond core drill attached to its tip. The assembly was engineered to rotate, providing a rotating diamond tool that was simultaneously vibrating up and down 20,000 times a second.

This principle enabled the people at Harwell to drill holes from .040 to .500 inch diameter and up to one inch deep in glass and ceramics quicker and more accurately. The machine was used successfully in trepanning glass discs, thread-forming glass and ceramic workpieces internally and externally.

The ultrasonic equipment used at Harwell in that development was of American design, and the principle has since been extensively developed in this country.

Rotary ultrasonic machines utilizing diamond tools have been available in the United States for about one year.

Before describing this new ultrasonic machining method in detail, a brief explanation of the more familiar and well-established ultrasonic impact grinding system may help to avoid confusion. Ultrasonic impact grinding does not generally utilize diamond tools and has no rotary motion. A tool shaped to correspond to the required hole or impression, and made of a relatively soft metal, is axially vibrated at an ultrasonic frequency. The vibrating tool tip is brought to bear upon the material to be machined with an abrasive slurry flooded or forced between the tool tip and the material. Boron carbide suspended in a liquid medium is a common abrasive slurry. It is the boron carbide particles that are hammered into the work piece by the reciprocating tool tip that cut or abrade the hole.

Since the tool has no rotary motion, it is able to grind any shaped impression and is generally applied for cutting irregularly shaped holes or several holes at a single application. The principle suffers from quite rapid tool wear and from tapering of the hole being cut. Tapering is the result of the loose abrasive slurry cutting on the side of the tool as well as the tip. The deeper the hole, the more pronounced is the taper. Ultrasonic impact grinding, however, is likely to be more effective in cutting complex-shaped holes, particularly when corners of minimum radii are required, than end milling with the rotary method.

The major components that make a Rotary Ultrasonic Machine Tool are:

1. POWER SUPPLY that converts 110V, 60 Hz to 20,000 KHz electrical energy.
2. ROTARY HEAD ASSEMBLY consisting of a motor driven, bearing mounted spindle or sonic converter. The heart of the spindle is two lead zirconate titanate discs which convert the electrical energy from the power supply to a mechanical vibration. These discs possess piezoelectric properties and physically expand and contract in accordance with the polarity of the voltage applied to them. The resulting vibration is amplified by a titanium alloy "horn" or impedance transformer, accurately machined to have a longitudinal natural frequency of 20 KHz and to deliver the maximum amplitude of vibration at the tip. In practice, the assembly is tuned to a slightly higher frequency to enable it to accommodate the diamond tool which is attached to the tip.
3. CUTTING TOOL. With few exceptions, the application of this new rotary ultrasonic principle has been in the glass and ceramic industry which has necessitated its use with diamond tools. Standard commercially available diamond core drills and mandrels, both plated and impregnated types, are used. It is necessary that the diamond tool be attached to the horn tip tightly enough to ensure good acoustic coupling; i.e., it effectively becomes part of the horn, and the assembly will tune at 20 KHz.

Although good mechanical coupling can be achieved with conventional chucks and collets, acoustically they are inefficient, providing only limited energy transfer. The energy loss is dissipated as heat, which has undesirable effects. To achieve the desired coupling, the diamond tool is brazed or epoxied to a metal threaded adaptor and screwed firmly into the tip of the horn.

Since the horn and tool assembly must have a resonant or natural frequency of 20 KHz, the size of tool that can be used is limited. The natural frequency of the assembly will change with the weight or mass of tool used. Too heavy a tool will throw the frequency of the horn assembly beyond the resonant frequency of the transducer and power supply, and the tool will not vibrate. At the present stage of development, this factor limits the weight of the tool to about 35 grams. Tool length is also critical. Within the length of the horn-tool assembly there are nodal and antinodal points, i.e., points of minimum and maximum vibration. To ensure maximum amplitude of vibration at the tool tip, certain critical lengths must be avoided.

Tool weight limitation has been helped by the use of titanium threaded adaptors and epoxy for attaching the tools to the adaptor. Diamond core drills up to one inch in diameter have been successfully employed.

These three components, Power Supply, Rotary Transducer, and Cutting Tool, comprise a basic system and can be applied in the most effective way to the particular requirement. They can be used in a conventional milling machine or drill press, (Fig. 1) or attached to more sophisticated machinery. If required, the rotary ultrasonic spindle (Fig. 2) can be used remotely on special or automatic equipment.

DRILLING

When ultrasonic vibrations are applied to a rotating diamond core drill, the reduced friction between the tool and the material being drilled has an advantageous effect.

First, it produces an acceleration in cut for a given tool RPM and pressure. For example, a ½" diameter, 120 grit, impregnated core drill rotating at 3,000 RPM at a constant pressure of 12 lbs./sq.in. with ultrasonics will drill through 99.5 aluminum oxide at a rate of 1¼" per minute. If the same drilling operation is attempted under identical conditions but without ultrasonics, a much reduced cutting rate will initially be achieved, but it will rapidly decay and cease cutting after a short time. Drilling tests with ultrasonics enabled the drilling of ½" diameter holes through ½" thick aluminum oxide in an average time of 27 seconds. Without ultrasonics, but with a constant 12 lbs./sq.in. pressure, the drill reached a depth of ¼" in 80 seconds and then stopped cutting. If, however, the ultrasonics were switched on at this point, the tool would again commence to cut and rapidly reach the normal ultrasonic cutting rate.

The accelerated cutting that results from the application of ultrasonics to the diamond tool is greater at low than at high tool pressures.

Being able to cut at low pressures is particularly advantageous when drilling or machining delicate components. It enables the drilling of holes in thin material without causing breakage. Holes can be drilled with only a few thousandths of an inch separation. They can be drilled close to the edge or corner without breakdown of the edge of the piece. (Fig. 3)

The problem of edge-cracking of hard brittle material at the point of breakout,when drilling a through hole, is helped by the ultrasonic method. Light pressure on the tool enables the cutting action to be maintained to a greater depth before the push-out occurs. Of course, rigid clamping or backing of the piece being drilled will minimize or eliminate breakout when drilling conventionally. Ultrasonic vibration of the tool will not eliminate breakout without backing or clamping, but it does minimize it.

The drilling of small diameter deep holes with conventional tooling can present problems with respect to wander of the tool or inability to drill a perfectly vertical hole. An ultrasonically vibrated rotating diamond tool is able to significantly improve the degree of accuracy for such an operation. Fig. 4 shows two .040" holes drilled into the edge of a piece of glass. The holes are about 2½" deep and separated by only .020". They were drilled with a diamond core drill and the average drilling time per hole was 14 minutes.

The technique has also proved advantageous when the surface of the work piece is not horizontal to the drill. Even with the work piece at a steep angle, the drill will commence cutting upon contact and negligible sliding of the tool tip occurs.

Another example of the deep hole drilling capabilities of the rotary ultrasonic technique was the drilling of a .040" diameter hole through a 3½" length of beryllium oxide. A .040"OD core drill was used, and on completion of the drilling operation a .017" diameter core piece, a full 3½" long, was extracted from the bore of the resulting hole.

Drilling deep holes with diamond tools but without ultrasonics often necessitates the frequent backoff of the tool after having drilled a relatively short distance. This allows the coolant, if fed through the center of the tool, to flush away the debris and loosen the core that may have begun to jam in the bore. The backoff or pumping action lengthens the time required to drill.

When ultrasonics are applied to the rotating tool, the periodic backoff becomes unnecessary, and the required pressure can be applied and maintained until the hole is complete.

Ultrasonic vibration of the tool reduces friction both at the cutting face and in the bore. When a core drill is used with coolant fed through the bore, it allows a greater flow of coolant, providing more effective flushing. It is thought that some cavitation of the coolant liquid occurs at the tip which has a desirable ultrasonic cleaning action upon the diamond particles. In practice, negligible loading of the diamonds occurs when the tool is ultrasonically vibrated.

The prevention of tool loading, as well as lower working tool pressure, undoubtedly contributes to minimizing tool wear. One company engaged in machining beryllium oxide has reported longer tool life with the Rotary Ultrasonic Machine Tool than that experienced with similar tools used without ultrasonics.

Core drilling of deep holes without ultrasonics is often complicated by the core becoming jammed in the bore of the tool. It frequently necessitates the removal of the tool from the machine so that the core piece can be extracted. Core hang-up is much less of a problem when the tool is ultrasonically vibrated. Even on the deepest hole, the core is generally left loose in the hole after the drill is withdrawn.

Although the principle offers advantages, both in speed and precision when drilling with diamond core drills, drilling smaller holes which necessitate solid mandrels is limited with respect to drill life. At the current state of development, rotary ultrasonic machining is not the answer to the electronic industry's prayer for a method of drilling .020" diameter holes fast and by the thousand.

The rotary ultrasonic drilling with diamond tools has proved particularly advantageous in the following applications:

Drilling glass printed circuit boards. More than 300 .040" diameter holes were required in "Chemcor" glass. Up to three pieces having a total thickness of .255" were drilled simultaneously in an average time of 35 seconds per hole. Because of the subsequent printing of the conductive circuit on the glass, no breakout or cracking could be tolerated.

Deep hole drilling. A research project required the drilling of four .095" diameter holes through the sides of a 7½" square block of glass. Accurate intersection of the holes at a depth of 6½" was required and was achieved within .005". In preliminary tests with a 1/8" diameter core drill through a 4½" square of similar material, a misalignment of only .0008" resulted at a depth of 3½".

MILLING

When ultrasonics are applied to a diamond mandrel for milling, the operation benefits in the same way as when drilling. The pressure required to cut is much less, progress through the material is more rapid with ultrasonics than without, and the operation becomes smoother.

Surface grinding can be accomplished with small diameter, 1¼", diamond plated wheels when the bottom of the tool is the cutting surface.

Slotting can also be achieved with the same tool when the edge or wall of the diamond wheel would be the cutting surface.

The stresses induced in the material by the ultrasonic vibration make wheels thinner than .080" impractical at this time. Because of this .080" minimum wheel thickness, the machine is not applicable to slicing.

When end milling, it is possible to mill parallel grooves and maintain a dividing wall as thin as .030". Surface finish at the bottom of the cut is improved and edge cracking is reduced.

THREADING

External and internal threads can be ground with a diamond wheel and a rotary chuck attachment. The component to be threaded is held in the rotary chuck which is motor driven and incorporates a lead screw which raises or lowers the chuck one thread pitch per revolution. A full depth of thread can be cut at one pass and cutting rates of up to four threads a minute achieved.

Threads finer than 16 TPI may prove difficult in some softer materials when the thread peaks will break down.

The smallest diameter of internal thread that is considered practical by this method is ¼", the limitation being the small shank diameter of the diamond tool.

The method of cutting external and internal threads is shown diagrammatically in Fig. 5.

One very promising application being investigated is the cutting of sheet "pyroceram". The particular requirement is the cutting of large rectangular and other shaped holes in 8' x 4' x 1/4" and 3/8" thick sheets for laboratory bench tops. An impregnated 120 grit diamond mandrel, shaped like a spool, is used, which cuts through the material and bevels both top and bottom edges. In preliminary tests, cutting rates of 7½" per minute were achieved.

Another application where the addition of ultrasonics to a diamond tool has proved advantageous is drilling boron-epoxy composites. Not only does it enable faster drilling but there is much less wear on the tool, tool wear being a problem with this very abrasive material. For example, ¼" diameter holes were drilled through a 3/4" thickness in 70 seconds. After ten holes, the loss of length from the impregnated diamond section of the tool was barely measurable. Very encouraging results have also been achieved when drilling boron-epoxy composite, laminated both with sheet stainless steel and titanium. Drill wear is of course much more severe, but an effort to determine an optimum type of diamond tool is being made. The quality of hole drilled by this method was much superior to that drilled by more conventional methods.

Another user of a Rotary Ultrasonic Machine had a requirement to ream the bore of thousands of components, where the bore surface finish was critical. They determined that a diamond reamer plus ultrasonics was able to ream faster and provide a better surface finish than other machinery available to them.

A nuclear fuel test center in California uses the machine for machining uranium oxide cermets and benefits greatly from the smoother, more gentle cutting action. Prior to acquiring the machine, the breakage rate of these expensive components during machining with diamonds was high. The introduction of the Rotary Ultrasonic Machine using the same diamond tool as before greatly reduced the number of breakages. The same establishment also uses the machine for fabricating test apparatus from alumina, which they are able to do quickly and accurately.

The applications described in this paper have in the main related to non-metallic, hard, brittle materials necessitating the use of diamond tools.

Most Rotary Ultrasonic Machines are concerned with these materials and until recently have not been applied to ductile metals. However, some interesting results have been obtained by the Grumman Aerospace Corporation when ultrasonic vibrations were applied to a cobalt twist for drilling 1/8" thick titanium. Preliminary tests have indicated that a more rapid rate of cutting is achieved with less tool wear. The prospects of reducing the time for each hole and obtaining more holes per drill has stimulated further work.

One set of results is reproduced on the graph shown in Fig. 6.

Another metal application has been the grinding of plasma sprayed coatings. One use of a Rotary Ultrasonic Machine Tool is the grinding of a plasma sprayed titanium dioxide coating on an aerospace component. The coating is very hard and abrasive, and conventional metal cutting tools have a very limited life. Diamond tools are more durable but an ultrasonically vibrated rotary diamond tool has proved to be the most effective grinding method for this particular application.

When this new machining method was introduced, much of the initial interest expressed was in solving existing machining problems, often the impossibles. Although ultrasonic machining with diamonds has been able to do some work not previously considered practical, in the main it is applied to the more routine job, which it does faster, with less breakage, and with less tool wear.

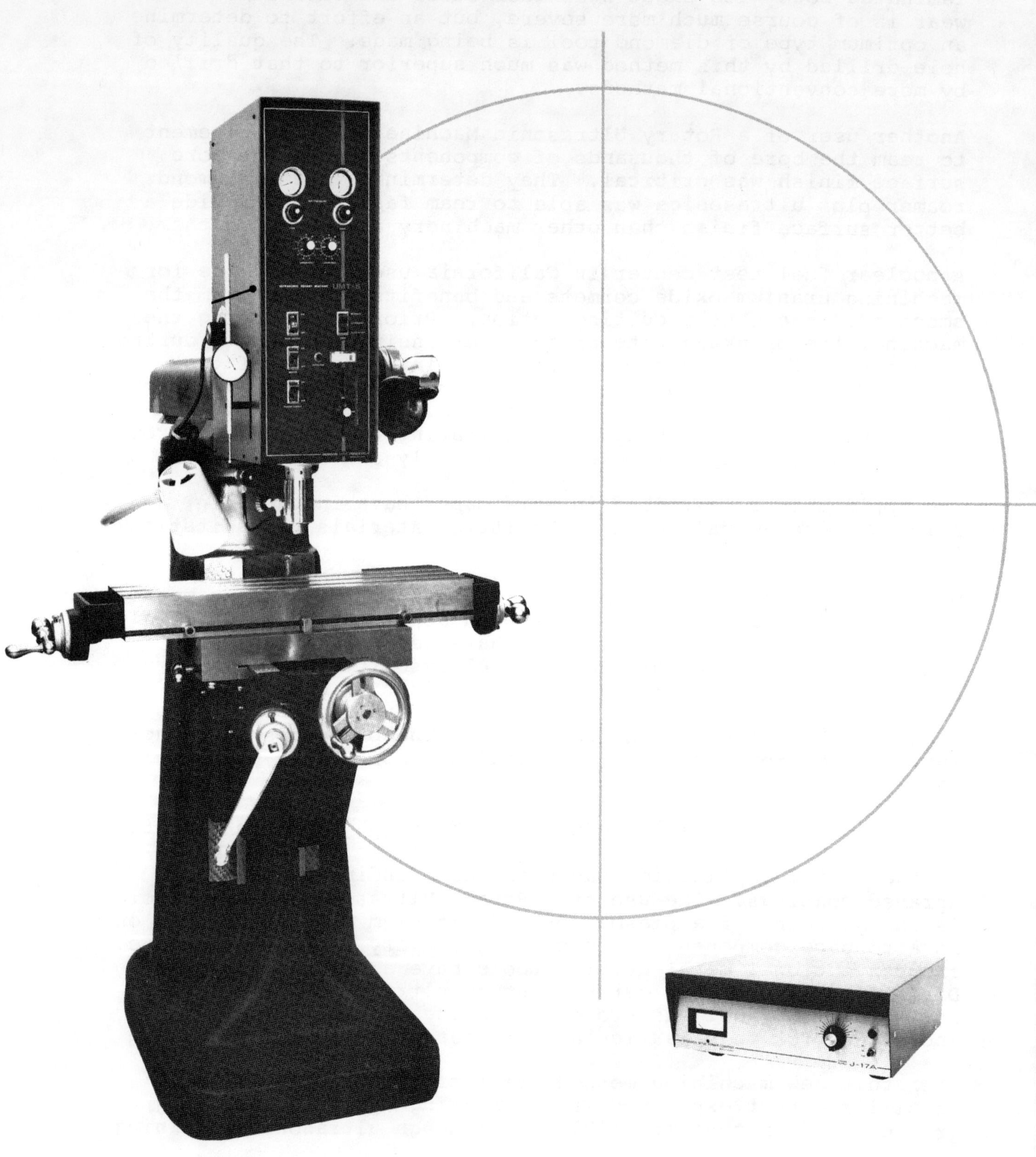

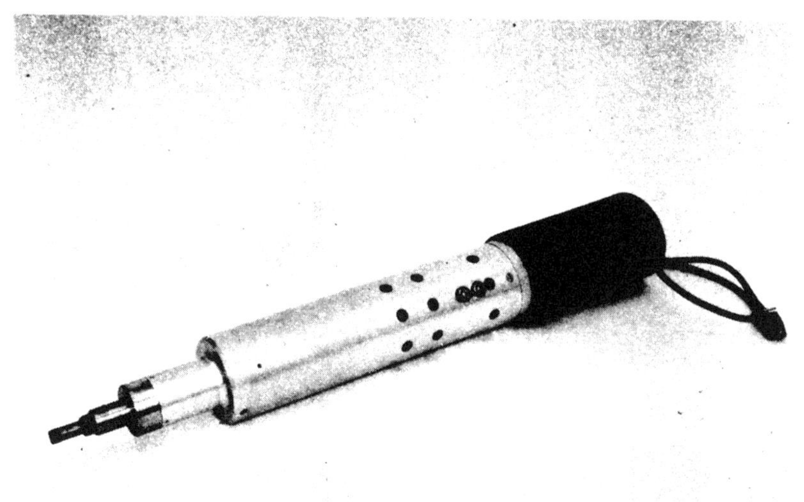

Figure 2.

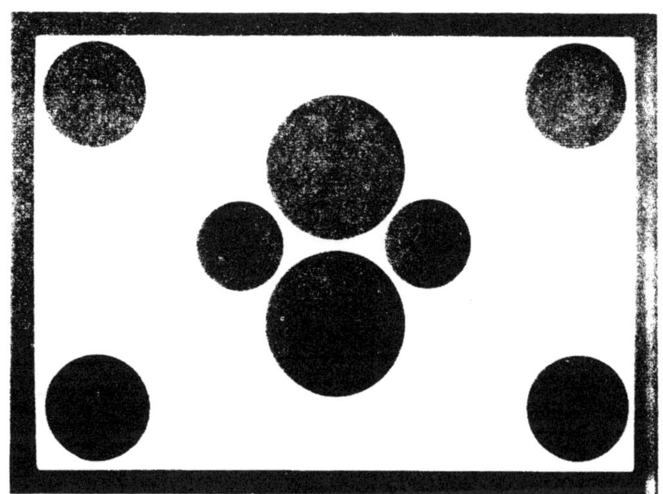

Figure 3.

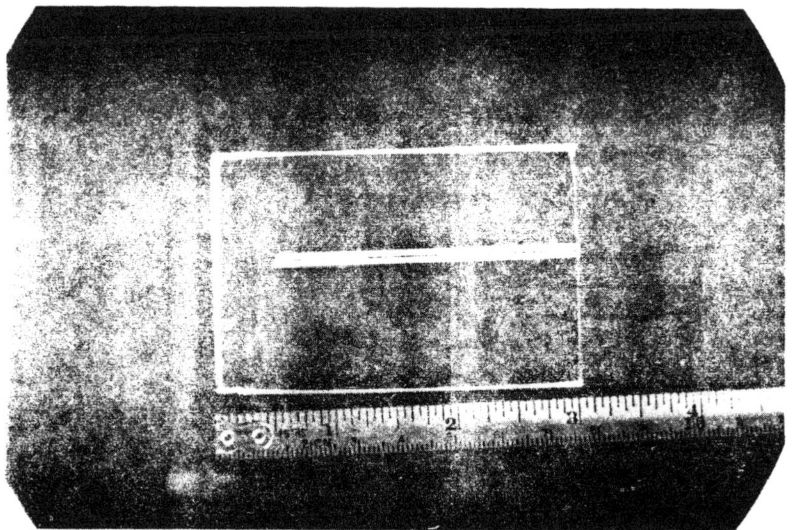

Figure 4.

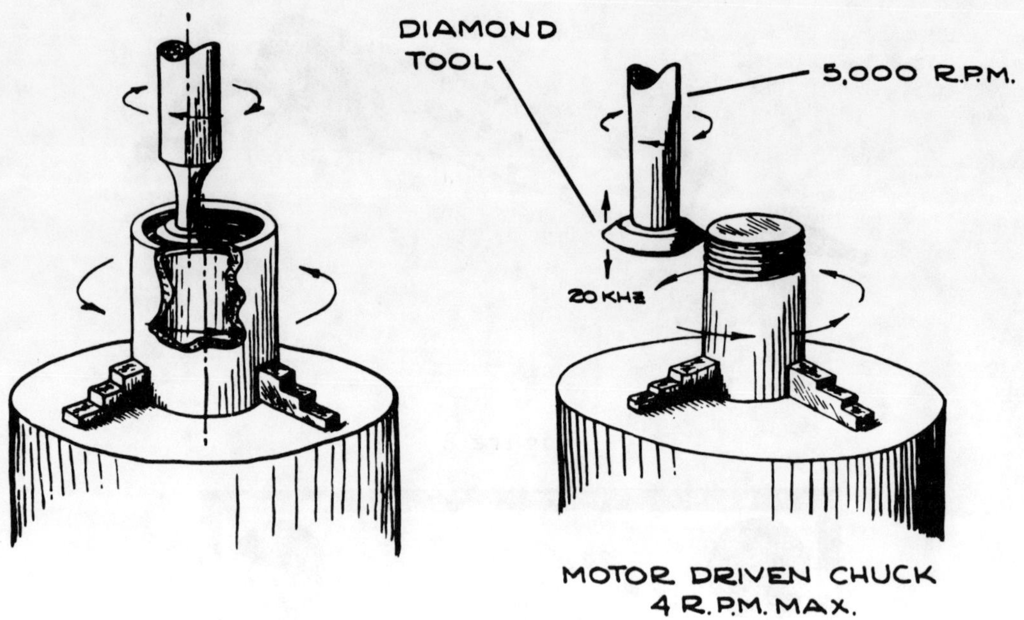

THREAD CUTTING FIGURE - 5

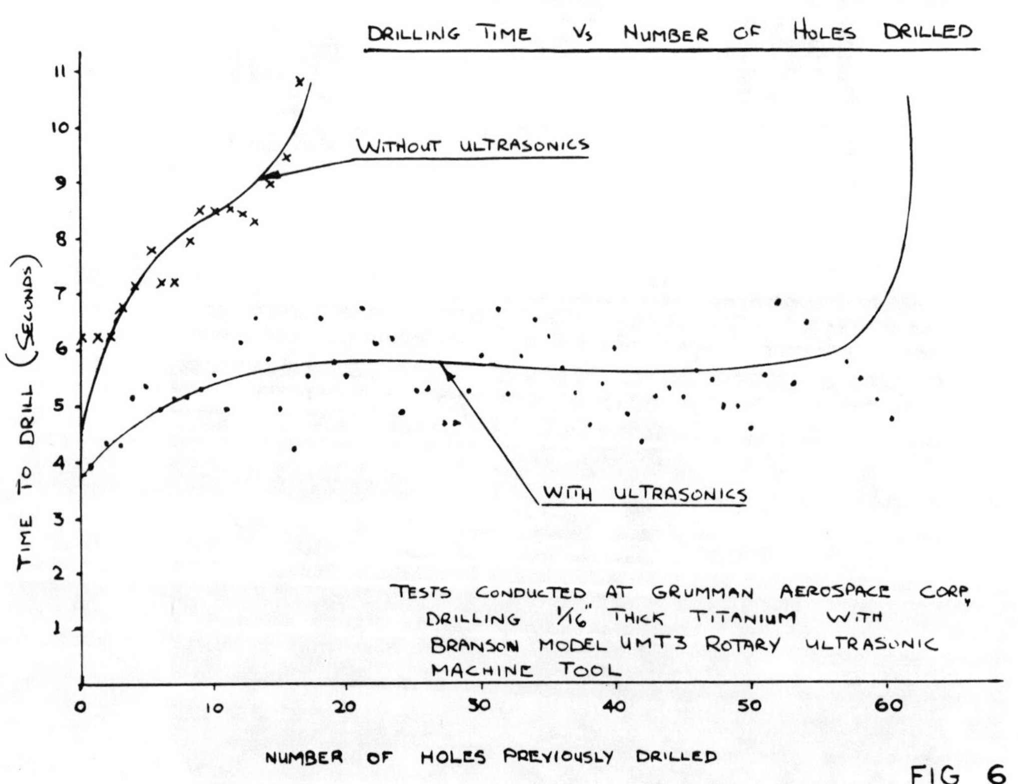

FIG 6

CHAPTER 2

CUTTING

Presented at the Industrial Diamond Association of Japan's 30th Anniversary Meeting and Seminar, May 1978.

The Application of Polycrystalline Tooling

By S.K. Bhattacharyya and D. Aspinwall

ABSTRACT *The machining of silicon aluminum alloys and hardened steels are examined in the light of new developments in cutting tool materials. Results from an industrial case study undertaken in a leading UK manufacturing company, highlight the many advantages to be obtained from using COMPAX® polycrystalline diamond blank tooling, not least in terms of extended tool life, better component quality and overall tool cost reduction. Machinability studies using BZN® Compacts relative to a fully hardened, Rc60, AISI D3 cold work die steel, indicate a material highly sensitive to changes in both cutting speed and feed rate. The study also shows that the product is capable of increasing metal removal rate by up to 4 to 1 when compared with optimal grades of carbide tooling.*

1.0 Introduction

Developments in the single point machining sphere over the last 15 years have centered primarily on machine tools, control aspects being the key area. While the early 60's saw the growth of NC systems, the 70's have witnessed the extension to CNC and DNC with all their inherent complexities.

On the tooling front, however, the situation is comparatively unchanged, sintered carbide being the state-of-the-art for the vast majority of manufacturers. While the introduction of coated carbides, predominantly TiC or TiN, has eased the machining situation for common ferrous materials, the machining of silicon aluminum alloys and hardened steels has received little or no specific attention.

2.0 The Problem Area

The efficient machining of a growing band of silicon aluminum alloys poses a complex problem in terms of tool material and in relation to machine condition.

To be effective, the use of tool materials such as WC-Co or single crystal natural diamond requires relatively high metal removal rates with cutting speeds in the region of 305-914 m/min for carbide and 213-457 m/min for natural diamond. Machine stability and rigidity play a vital role at these speeds, particularly in relation to natural diamond. At low cutting speeds built-up-edge (b.u.e.) outlaws the use of WC-Co (primarily as a result of poor surface finish), while the susceptibility of chipping increases for diamond.

Tied in with all these aspects is the nature of the work material, the abrasive effects of silicon and fracture tendencies of inclusions. For the tooling engineer, faced with out-of-date, well used autos, capstan and center lathes, efficient machining can mean no more than minimum output requirements. Choice of tooling is dictated primarily by machine condition rather than as a direct result of component material, design, or production requirement.

The machining of hardened steels presents a simpler but equally important picture. Typical machining speeds for AISI, D^2 and D^3 steels having a mean hardness of around Rc60 are currently in the region of 15-25 m/min, using optimal grades of Carbide tooling.

Rapid tool wear, both progressive and catastrophic, effectively limit metal removal rates to a level where grinding becomes the only feasible production alternative.

This paper outlines the results of an industrial case study using COMPAX Polycrystalline Diamond tooling and data obtained from machinability studies using General Electric BZN Compacts.

3.0 Industrial Case Study

3.1 Background

No single tool material provides the complete solution to all machining application problems. In general, a compromise situation exists where properties such as high abrasive resistance are offset against low impact resistance. COMPAX blanks represent an amalgamation of tooling in order to minimize this "play off" approach. The intrinsic properties of the polycrystalline diamond layer provide both wear resistance and immunity against built up edge, while the base carbide layer of the laminate is responsible for its toughness and resilience.

Tests using COMPAX blank tooling were performed under actual production conditions at a leading UK, Midlands manufacturing company. A high

® ® Trademarks of General Electric Company, USA

percentage of the company's basic component manufacture is being absorbed into the non-ferrous area, with silicon aluminum alloys playing a key role.

Machining of this material is currently done with WC-Co in both insert and shank form, however due to moderately high silicon content of the castings used, problems of rapid tool breakdown occur on a fairly frequent basis. The situation is further worsened by both the age and fixed speed/feed capabilities of many of the machines used, a feature that is highly relevant considering the affinity of WC-Co to the formation of built-up-edge at low cutting speeds.

TABLE 1
Component Material—Silicon Aluminum Alloy Casting—LM24M BS 1490 1963
Silicon Content 7.5 to 9.5%

		OP1	OP2	OP3
Bore Dia	(mm)	289.94	39.64	26.94
Machine Sped	(rev/min)	540	486	486
Cutting Speed	(m/min)	476	60	41
Feed Rate	(mm/rev)	0.13	0.15	0.15
Depth of Cut (nominal)	(mm)	0.33	0.381	0.381
Triangular Insert		x		
Shank Tool			x	x
WC Tooling—Rake Angles		0°	7°	15°
—Relief Angles		11°	7°	15°
—Nose Radii	(mm)	0.812	—	1.219
COMPAX —Rake Angles		0°	7°	7°
Blank —Relief Angles		11°	7°	7°
Tooling —Nose Radii		0.812	0.406	1.219

3.2 Tool Evaluation

Performance evaluations using diamond polycrystalline tooling with a silicon aluminum alloy were based on the results of a ten week study, where a direct exchange was made between COMPAX blank tools and those currently in use on three different bore diameters. No special machine features were provided and the tools were run under normal line operating conditions.

Component material specification and general machining details common to both WC-Co and COMPAX blank tooling used during the trials are given in Table 1, along with comparisons of tool geometry and form. The existing WC-Co geometry on all but one of the tools was judged compatible. Recommended rake and relief angles for the COMPAX blank tools were in the order of 0° to +10° and +5° to +10° respectively.

The cutting speeds used for operations 2 and 3 were well below defined optimum levels for WC-Co tools, and even more so for COMPAX blank tooling. The realistic values were in the region of 500-1000 m/min.

A comparison of the operating characters of both the WC-Co and COMPAX blank tools for the ten week period is given in Table 2, however specific points arising from the tests are detailed below:—

a) Unlike their conventional counterparts the COMPAX blank tools ran continuously for the duration of the test period without the need for additional setting or adjustment after initial tool installation.

b) Surface finish using the COXPAX blank tools was at least 50% better, with Ra values of 1-1.27 μm (40-50μin) compared with 2-2.8μm (80-90 μin) for the WC tooling.

c) The COMPAX blank tools showed no sign of built-up-edge even at cutting speeds as low as 41m/min.

d) Depth of cut variations and intermittent loading of the COMPAX blank tools had no appreciable effect.

TABLE 2

	WC-Co Tooling			COMPAX Blank Tooling		
	OP1	OP2	OP3	OP1	OP2	OP3
CapitalCost/ Tool £	0.56	1.90	1.90	54.00	41.65	41.65
Total Number of Regrinds Possible	Nil	6	6	3	4	4
Component Output/Cutting edge (10 wk basis)	500	600	600	40,000	40,000	40,000
Setting Time Hrs (10 wk basis)	2.66	33.3	33.3	0.033	0.50	0.50
% Output In-increase due to Reduced Setting Time (10 wk basis)	—	—	—	0.30	3.7	3.7
Machine Down-time Costs due to Setting Time (10 wk basis) £	14.18	188.25	188.25	0.175	2.82	2.82
Operator Down-time Costs due to Setting Time (10 wk basis) £	4.12	51.66	51.66	0.051	0.775	0.775

3.3 Conclusions

A major factor to be considered in the purchase of any new tool material is the initial outlay. At first glance the polycrystalline diamond tools appear expensive and uneconomical for machining operations, particularly when compared with carbide. All too often the assessment of both cost and performance is based primarily on equivalent numbers of cheaper tools, with little or no regard for parameters such as setting time, machine cost and utilization, operator goodwill or surface finish.

The test results show the advantages to be gained from polycrystalline diamond tools, particularly in terms of reduced setting time and higher machine utilization. Indeed, when viewed in the full context of cutting speeds well below optimum, intermittent loading, depth of cut variation and poor machine conditions, the performance of the tools is even more startling.

As a direct result of the study the company has purchased polycrystalline tooling for a complete line. Both roughing and finishing operations are to be performed using COMPAX blank tooling in single- and multipoint form.

4.0 Machining of Hardened Steel Using BZN Compact Indexible Inserts

4.1 Background

Cubic Boron Nitride (CBN) is finding increasing use not only in the abrasive machining sphere, but also in single point cutting. Its application as a cutting tool has, up to now, been restricted primarily to finishing applications. BORAZON CBN polycrystalline blanks form the basis of the BZN Compact — Kravechenko (2).

General Electric BZN® COMPACTS offer considerable advantage over the direct use of plain CBN polycrystal blanks in allowing semi-finish and finish machining operations to be performed.

Visually resembling COMPAX blank tooling, the BZN COMPACT has a layer of polycrystalline BORAZON CBN bonded to a WC-Co substrate. In addition to providing strength and resistance to shock, the laminate allows flexibility in handling.

4.2 Experimental Equipment and Test Plan

All cutting tests were undertaken on a Dean Smith & Grace Lathe, incorporating variable speed control, 5-2000 rpm, and having a 15 kW drive motor. Standard BZN COMPACT indexible inserts, type B-SNG-433, were used for the majority of tests together with reground and lapped 9.53 mm squares, having the same 1.19 mm nose radii.

A standard Wimit Econdex 'FC' insert tool holder was used to provide the following geometry:— BR = 0°, SR = –6°, SCEA or Plan Approach angle = 45°, ECEA = 45°, SR = 6°, ER approximately = 5°. A similar geometry was used to accommodate the 9.525 mm square inserts. In both cases the clamping action of the tool did not act as a chipbreaker. Composition details of the A.I.S.I D3 cold work die steel are given below. Bar hardness was within the range Rc 58 to 62.

C	Si	Cr	Mn
2.2	0.30	12.0	0.20%

Initial bar size was 228 mm diameter x 914 mm long. For tests using cutting lubricant, Cincinnati Cimperial 20 water soluble emulsion type fluid was used. Flood application of fluid to the cutting zone was maintained for all tests at 4*l*/min.

Both cutting speed (v) and feed rate (f) were varied for the given tool geometry. Table 3 gives details of the test plan.

TABLE 3

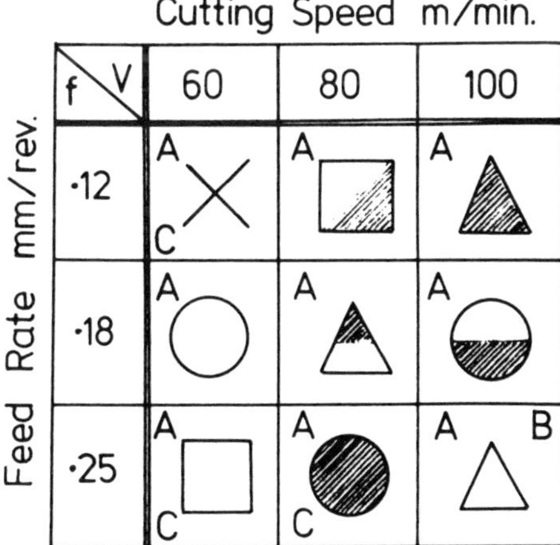

Cutting Speed m/min.

Feed Rate mm/rev. f \ V	60	80	100
·12	A, C	A	A
·18	A	A	A
·25	A, C	A, C	A, B

KEY

A - Speed Feed Tests Using Cutting Fluid Having A Mean Oil Content Level Of Approximately 8 %

B - Tests Involving Oil Concentration Levels Of 5-6%, 11-12% & 18-19%

C - Dry Cutting Tests Involving Thermal Investigation

Symbols Correspond To Graph Plots

Test Plan

Depth of cut throughout all tests was maintained at 2 mm. Replications were undertaken on a purely practical basis, the results at each intersection being analyzed relative to adjacent blocks. Edge finish on the majority of inserts were such that no chipping was visible when viewed at 15x magnification. A new cutting edge was used for each of the block arrangements.

4.3 Parameters Measured

Figure 1 gives details of tool flank and crater wear measurements taken at each time interval, plotted flank wear being the average of three readings. Stylus probe examination using Rank Taylor Hobson Talysurf 4 Machine, was employed to give an indication of crater topography, primarily midband wear forms.

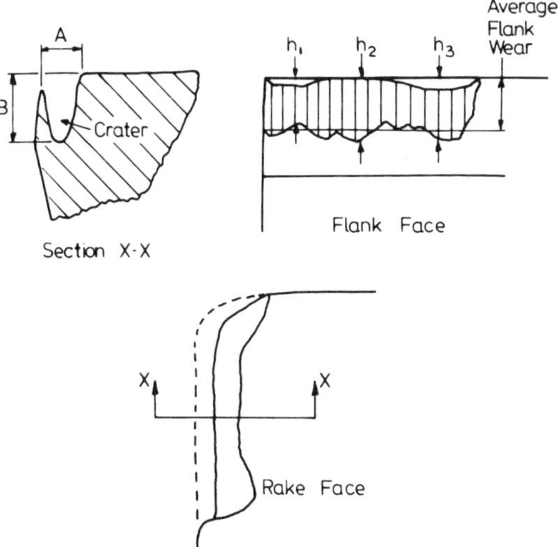

FIGURE 1
Details of Tool Wear Measurement

Force data was obtained using a Kistler Piezo Electric Three Component Force Dynamotor, Type 9263. Measurements were made in both the static and dynamic mode. Thermal analysis was undertaken using an AGA Infra Red Scanning Thermovision Camera mounted directly onto the lathe saddle. In addition, surface finish measurements were made after each pass, using the Rank Taylor Hobson Surtronic Machine on a (sample) cut-off length of .25 mm.

In order to obtain a permanent visual record of the wear mechanism, in-process photography was used showing both flank and crater growth profiles.

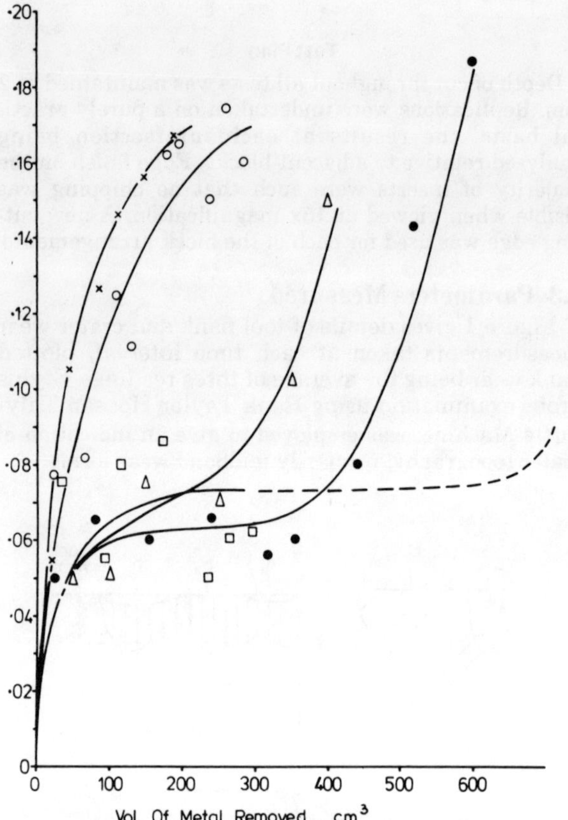

FIGURE 2
Crater Width Response During Fluid Cutting

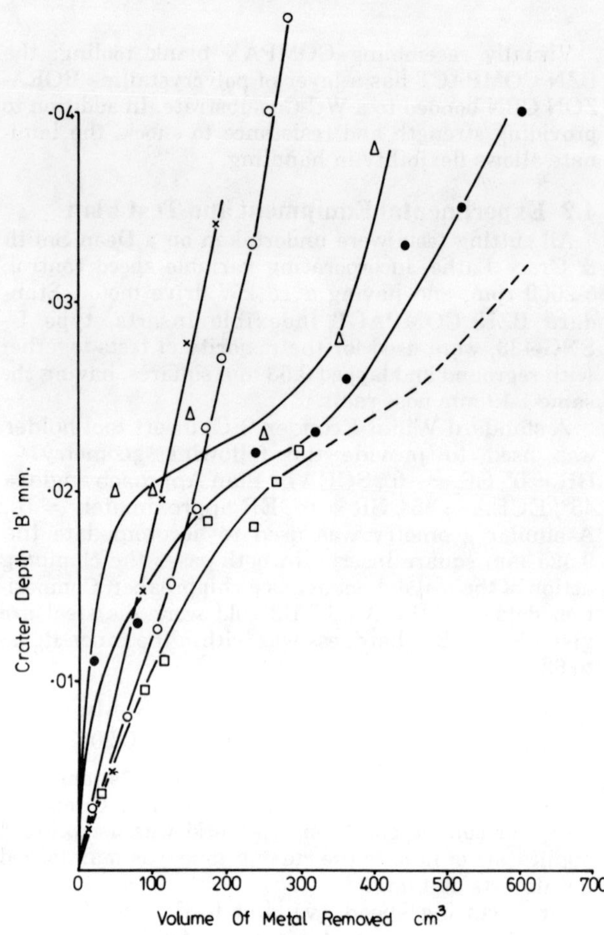

FIGURE 3
Crater Depth Response During Fluid Cutting

5.0 Results and Discussion

5.1 Crater Wear

Crater wear measurements plotted in Figures 2, 3 and 4 represent wear formation at the mid point of the cutting edge. Figures 2 and 3 show crater width and depth for changes in both cutting speed and feed rates during fluid cutting.

The prime feature exhibited by both sets of graphs is the pseudo steady state secondary stage, occurring at high feed rate, particularly in relation to crater width, 'A'. Though not as exaggerated, a similar trend is visible for crater depth 'B'.

Variation of crater width shows a dependence only on feed rate, as well as an increase in feed value which produces a decrease in mid-band width and in the rate of growth. In contrast, crater depth is responsive to changes in both feed and speed. An increase in feed value produces a decrease in depth, while increasing cutting speed results in increased crater depth.

The apparent stability of crater width at high feed rate results from the longitudinal growth pattern of the rake face crater. Photographs 1 to 8 show the non uniform crater wear produced at f = 0.25 mm, v = 80 m/min, for specific time intervals. Increased localized wear occurring initially at the nose radii and depth of cut locations spreads along the cutting edge during machining. The extent of crater width formation along the whole of the cutting edge at test measure corresponds to the crater width at the nose radius and depth of cut, after approximately nine minutes of cutting.

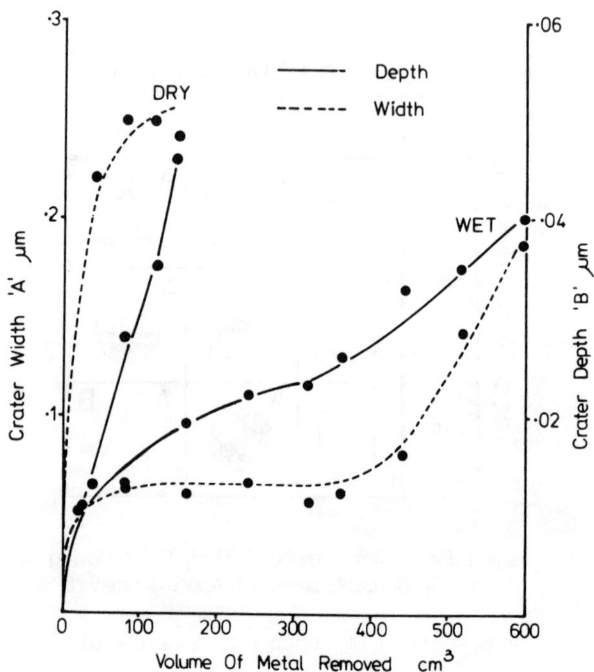

FIGURE 4
Crater Topography During Dry Cutting

Characteristic of the crater formed at high feed rate was its dual identity (i.e., both a primary and secondary crater forming).

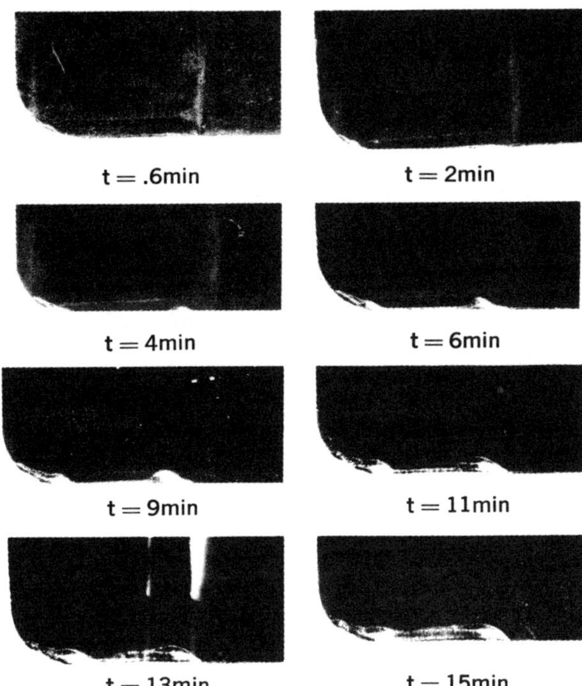

Photographs 1-8, v= 80 m/min, f = 0.25 mm/rev, Fluid Cutting 10x Mag.

In contrast photographs 9-13 show the rake face crater development at f = .12 mm/rev, v = 60 m/min. The width of the crater shows uniformity over the entire cutting edge.

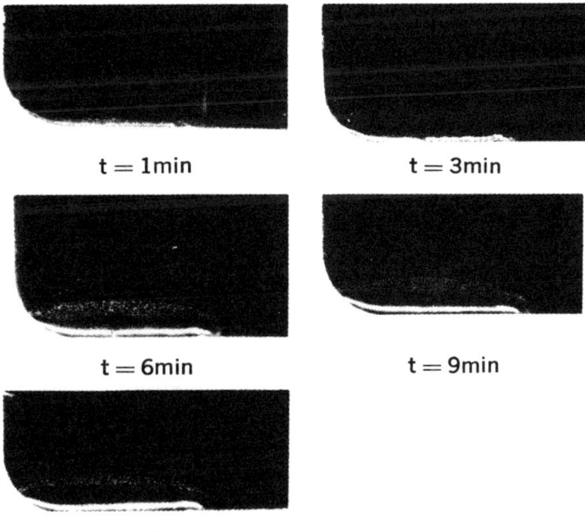

Photographs 9-13, v = 60 m/min, f = 0.12 mm/rev, Fluid Cutting 10x Mag.

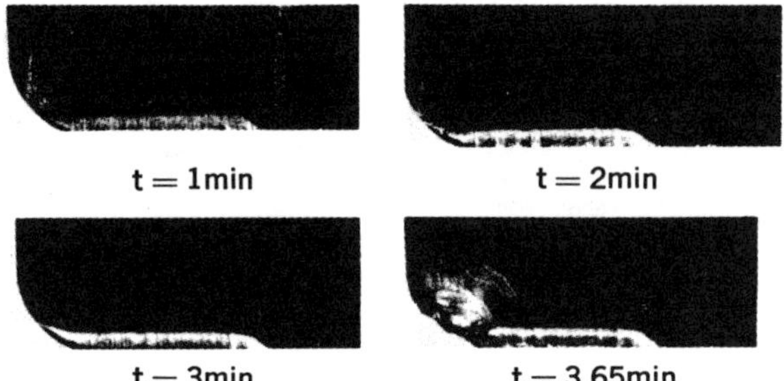

Photographs 14-17, v = 80 m/min f = 25 mm/rev, Dry Cutting 10x Mag.

Details of comparative wet/dry cutting results for crater topography are shown in Figure 4, photographs 14-17 indicating the development of the 'dry' crater.

Rapid crater growth is evident during dry cutting with complete absence of notching at either the depth of cut or nose radii locations. The form of the crater resembles closely that achieved after extensive fluid cutting.

Wear commonly associated with notching is caused by conventional sliding of both abrasive and material transfer. Chemical interaction with the surrounding atmosphere also plays a key role. Notching of the BORAZON CBN layer at high feed rates points to chip pressure as an additional factor. SEM analysis of the notching sites shows very smooth surfaces with ridging characteristic of diffusion wear—Trent (11).

Chip sliding over the rake face of the BZN Compact did not occur in the majority of fluid tests other than during the initial machining period. At high feed rate, f = .25 mm/rev, chips were predominantly continuous with greater visible resemblance toward a medium soft steel than to a fully hardened Rc60 die steel. Chip flow followed an almost vertical trajectory with minimal rake face contact. Under dry cutting conditions at this feed rate immediate chip production was of the small intermittent type, with the chips leaving the cutting zone in a vertical spray manner. To some extent the lack of conventional sliding provides an answer to the non-representative notching produced, particularly with regard to lack of extensive score and groove marks.

The dependency on feed rate of crater position is shown in Figure 5. The tool profiles are from Talysurf 4 Machine tracings taken at the mid point of the cutting edge for different feed rates.

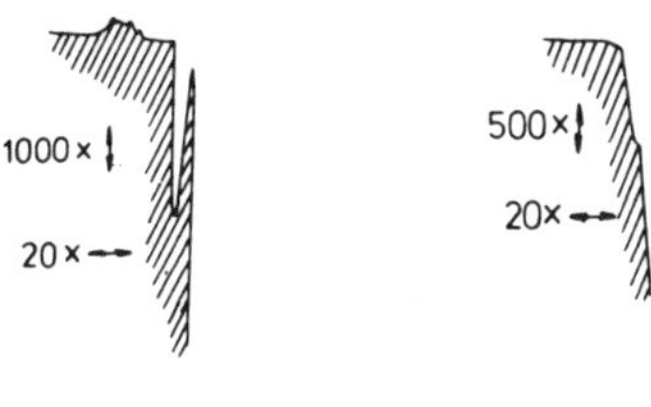

FIGURE 5
Tool Profile Traces

At f = .12 mm/rev, the position of the bottom of the crater occurs at some distance behind the cutting edge, resulting in a fairly non stable cutting configuration. At f = .25 mm/rev, the crater has moved toward the cutting edge, its base occurring on the flank face, thus providing the self sharpening effect noted by Focke et al (1) and greater tool edge resilience during machining.

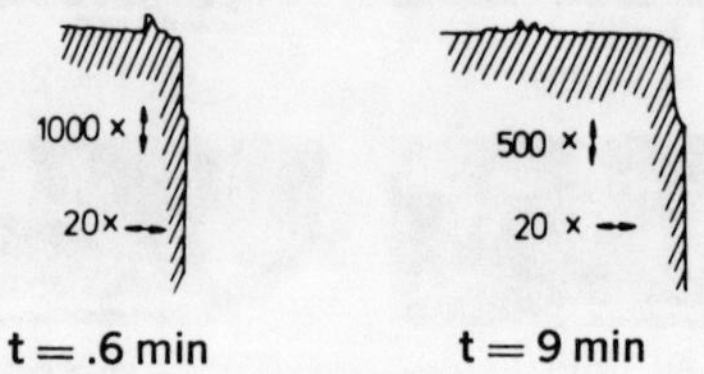

FIGURE 6

Tool Profile Traces v = 80 m/min, f = 25 mm/rev, Fluid Cutting

Ridging on the rake face of BZN COMPACT tooling (shown in photographs 1-8 as a line running parallel and in close proximity to the cutting edge, and in profile on Talysurf Machine tracings shown in Figure 6) occurred during all tests. Though apparent on photographs taken in situ during the machining cycle, the ridge showed little physical trace on Talysurf Machine readings taken after 2-3 minutes of cutting. Coincident with ridge decline was the growth of a broad spatter band positioned at some distance behind the cutting edge. Location of the ridge varied according to feed rate, moving away from the cutting edge with increased rate of feed. In addition, its initial size showed feed dependency, increased feed rate producing a larger ridge profile. The variation of cutting speed had minimal effect on either ridge size or location.

Identification of its composition using X-ray microanalysis techniques showed Fe, Mn and Cr as the primary deposits, suggesting a process of oxide layering during machining similar to that observed by Opitz and Konig (12).

Transformation of nose radii geometry during cutting noted by Kravchenko et al (2) occurred primarily at the high feed rates during both wet and dry cutting. Collapse of the flank face at this location effectively provided a double lead cutting edge having a primary SCEA of 75°-78° and active length of .85 mm. Photographs 18-21 and Talysurf Machine tracing (Figure 7) show details of this effect prior to and after formation, v = 100 m/min, f = .25 mm/rev.

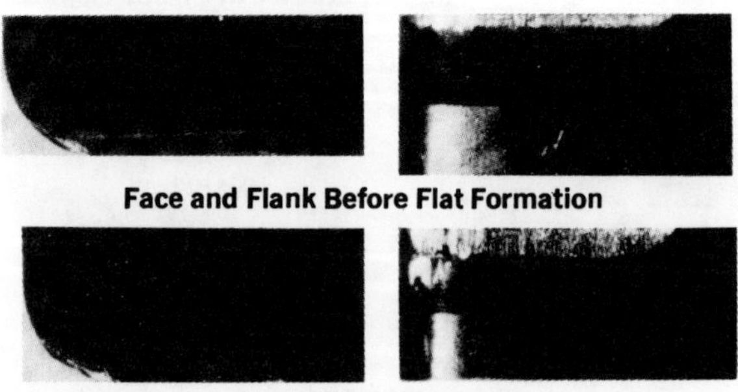

Face and Flank Before Flat Formation

Face and Flank After Flat Formation

Photographs 18-21, v = 100 m/min, f = .25 mm/rev, Fluid Cutting 20 x Mag.

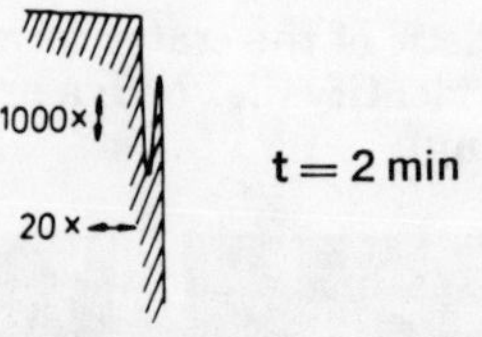

FIGURE 7

Tool Profile Trace v = 100 m/min, f = .25 mm/rev, Fluid Cutting

Crater growth at the nose radius is shown with the base of the crater at some distance behind the cutting edge. The mean value of crater depth prior to fracture is 0.021 mm. The fracture corresponds to a metal removal value of between 150-250 cm³ on all tests undertaken at v = 60, 80 and 100 m/min for f = .25 mm/rev.

The dependency of the BORAZON CBN layer on the catalytic effects of temperature for crater growth either through dry cutting or as a result of increased cutting speed, suggests a "diffusion" process similar to that observed in carbide tooling as the primary wear mechanism.

5.2 Tool Temperature

Measurements of tool rake face temperature during dry cutting are shown in Figure 8. The data relate to the effects of variation in feed rate, cutting speed remaining constant at 60 m/min. Thermal tests involving variation of cutting speed provided little quantitative data, because fracture of the BORAZON CBN layer occurred almost on commencement of cutting.

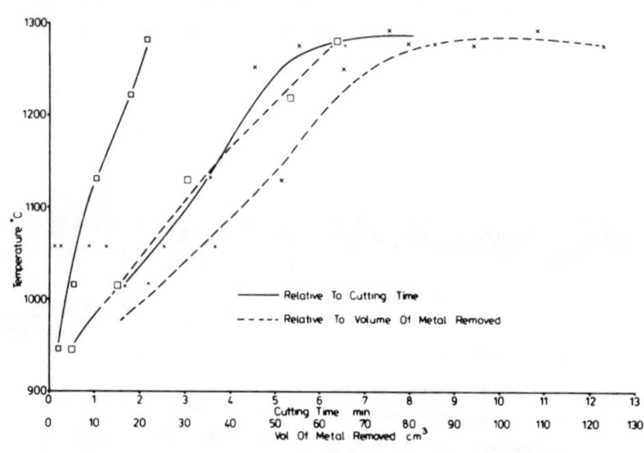

FIGURE 8

Rake Face Temperature During Dry Cutting

Plotted values represent the maximum recorded rake face temperatures at specific time intervals. Both visual observation and infra-red scanning of the rake face during cutting pinpointed the region near the nose of the tool as the hottest area during cutting—Trent (13). Measurement of tool temperature during fluid cutting was not possible due to absorption of the infra-red radiation by the coolant liquid.

The rate of thermal input relative to cutting time limits tool performance. The reason for this is fracturing occurs at the nose region of the BORAZON CBN layer extending onto the flank face after only 2.1 minutes of cutting for f = .25 mm/rev. At a feed value of f = .12 mm/rev, a balanced thermal situation exists which allows the tool to operate effectively with rake face temperatures in the region 1275-1300°C.

Relatively poor thermal conductivity is characteristic of ceramic tooling. Excessive heat results in large thermal gradients, which puts stress on the tool. Cutting tests undertaken at f = .25 mm/rev, v = 80 m/min results in fracture of the BORAZON CBN Layer at the nose region. This occurrence exemplifies the sensitivity of the BZN compact to large temperature gradients.

5.3 Flank Wear

The relationship between flank wear and volume of metal removed during fluid cutting is shown in Figure 9. Plotted values correspond to both speed and feed effects for the whole of the test plan. A linear trend in all graphs is evident. Increased cutting speed results in an increase in the rate of flank wear. The effects of feed rate are shown superimposed on speed effects and indicate minimal change in flank wear for an increase in feed value.

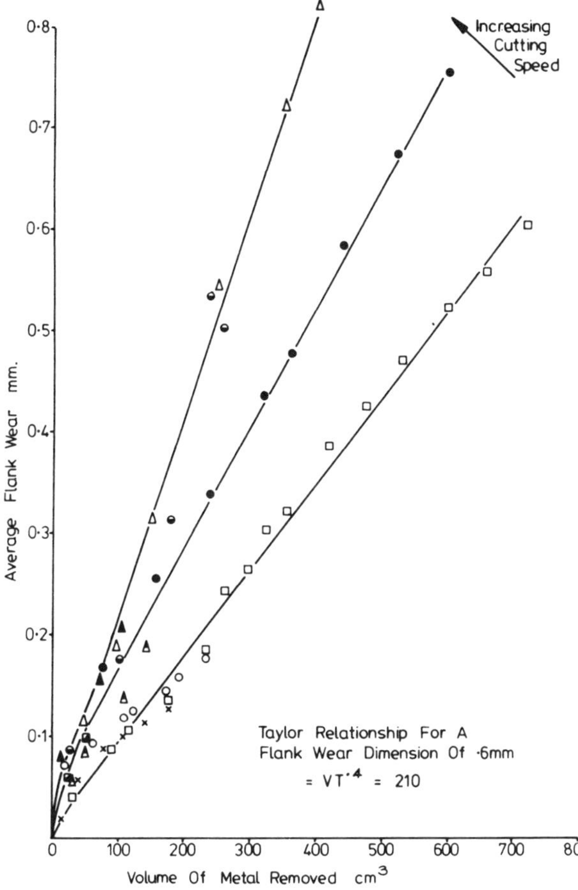

FIGURE 9

Flank Wear To Volume Of Metal Removed During Fluid Cutting

The BZN Compacts were found to be highly sensitive to variation in both speed and feed rate in relation to the uniformity of flank wear growth. Photographs 22-29 indicate the flank growth pattern at v = 80 m/min, f = .25 mm/rev. Uniform wear patterns occurred in general under combined elevated cutting speed and feed conditions.

Tests involving variation of oil content of the cutting fluid showed little change in either flank or crater wear rate, and in the relationships obtained for static force development or surface finish. The effects of dry cutting on flank wear rate produced a similar result, where the 't' statistical value of .397 indicated a non-significant effect at all levels. Actual penetration of the cutting fluid to the flank face appears doubtful and in general the effective action of flood application to the cutting zone is associated primarily with chip and rake face cooling.

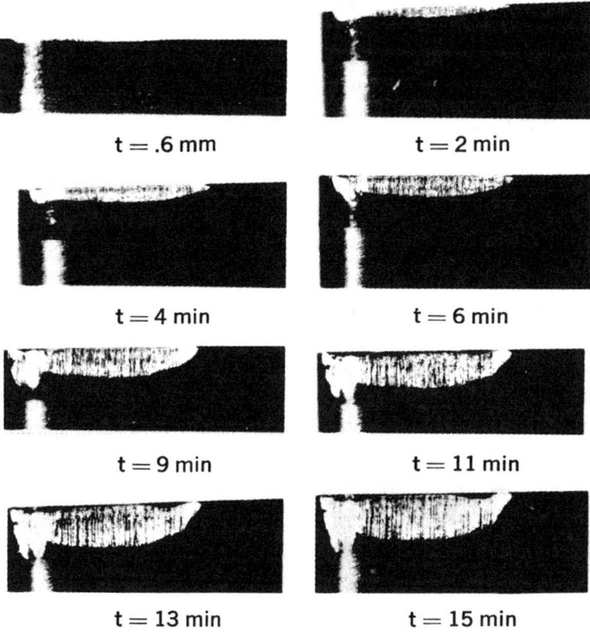

Photographs 22-29, v = 80 m/min, f = .25 mm/rev, Fluid Cutting 10x Mag.

As with crater formation the mechanism of flank wear appears temperature dependent. SEM examination of both wear forms at v = 80 m/min, f = .255 mm/rev shows a strong similarity in surface topography with an almost featureless narrow band of BORAZON CBN running along either side of the cutting edge. During all tests built-up-edge was not apparent. The X-ray microanalysis showed minimal adhering to the metal. The Taylor relationship obtained for a flank wear dimension of .6 mm is: $VT^{.4} = 210$

5.5 Brittle Failure

Fracture of the BORAZON CBN layer as a result of either thermal stressing or excessive loading, showed a characteristic brittle failure mode. At v = 100 m/min, f = .25 mm/rev, fracture of the tool, as shown in photographs 30-31, occurred on initial work contact in four out of seven tests undertaken. All four tools show severe fracture of the BORAZON CBN layer in the regions associated with high pressure during cutting. The fractured surfaces show a granular appearance with abrupt contouring.

t = 0 min
(Initial work contact)

Photographs 30-31, v = 100 m/min f = .25 mm/rev, Fluid Cutting 10x Mag.

The adoption of a large SCEA for all tests was made in order to reduce the effects of chip loading on the rake face of the BZN Compact. By increasing the SCEA to 45°, the effective length of the cutting edge is increased, resulting in a reduction in chip thickness for a given feed rate. The increase in cutting force associated with this move, due to the greater contact length, is negligible. In real terms the thinning out of the chip effectively maintains a constant chip thickness for increases in feed rate, thereby allowing higher rates of feed to be used. The effect is important especially when viewed in relation to observed feed rate/flank wear growth.

The spasmodic nature of tool fracture at high feed/cutting speed points to edge quality as a major feature in tool life. The 15x magnification chipping code proves inadequate at this level. Recent work has shown that additional edge preparation in the form of 45°, 30% feed rate chamfer allows machining at v = 80 m/min, f = .31 mm/rev to be undertaken satisfactorily with minimal notching and uniform flank wear.

Figure 10 shows Talysurf Machine profile tracings of the cutting edge taken within the nose region of the tool at v = 80 m/min, f = .25 mm/rev. The BORAZON CBN layer shows a deformation pattern occurring under the high feed rate conditions, similar to that observed by Trent (13).

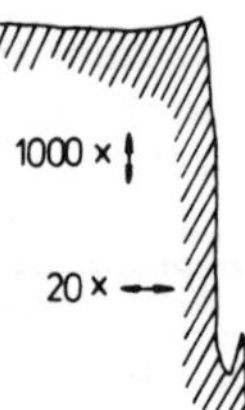

FIGURE 10

Tool Profile Trace v = 80 m/min, f = .25 mm/rev, Fluid Cutting

5.4 Static Force and Surface Finish

A comparison of both the static cutting (Tangential) and Normal Forces experienced at f = .25 mm/rev, v = 60, 80 and 100 m/min during fluid cutting is shown in Figure 11.

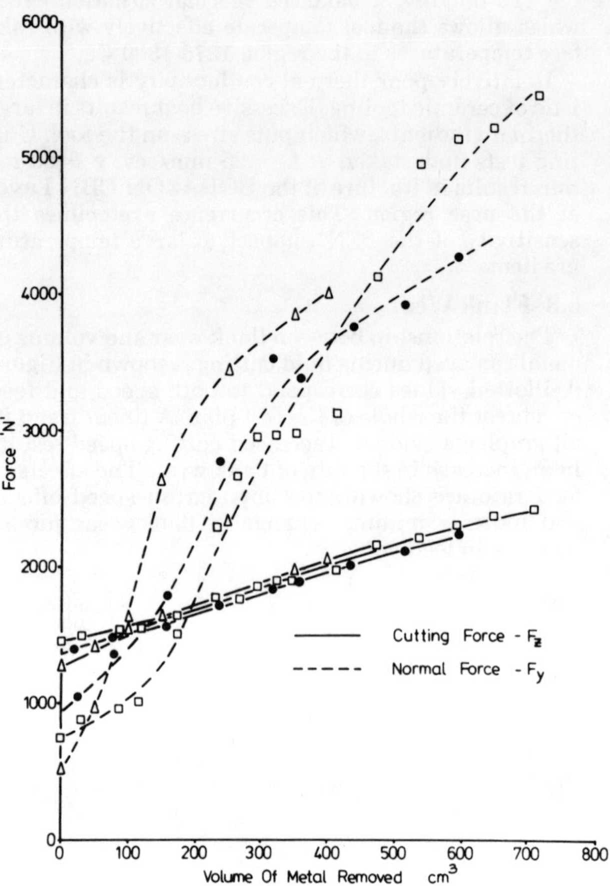

FIGURE 11

Force Data Under Fluid Cutting

In purely practical terms the data imply the use of a reasonably robust lathe. A high degree of rigidity is required in the cross slide and tool mountings in order to counter the large normal forces which occur as a result of both material condition and tool geometry. Power requirements based on cutting forces indicate a moderately powered machine and indeed the 15Kw motor used proved adequate (2 mm/side depth of cut).

Surface finish values relative to volume of metal removed are plotted in Figure 12, the graphs showing the combined effects of feed rate and cutting speed during fluid cutting. The effects of tool nose fracture, which occur at f = .25 mm/rev, are evidence that a rapid rise in Ra value occurs within the 150-250cm³ metal removal tolerance band on all high feed rate plots.

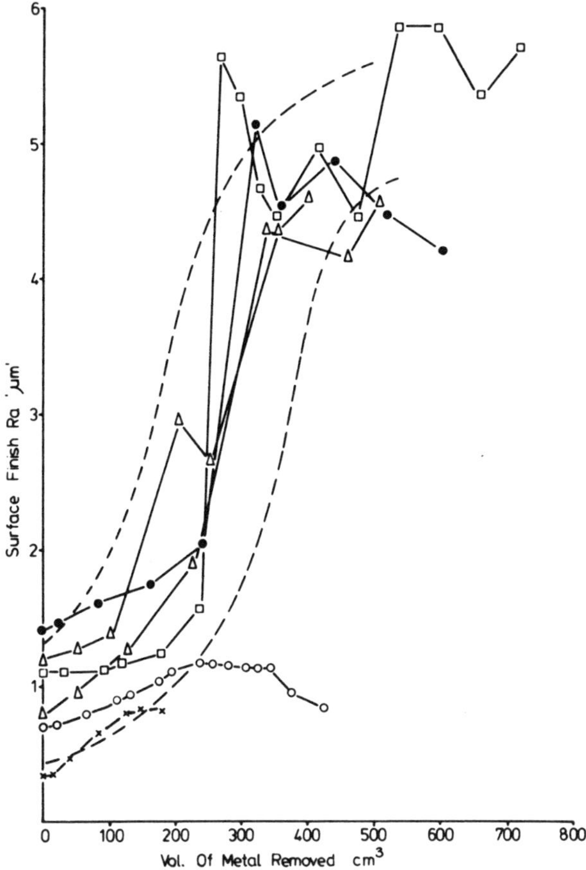

FIGURE 12

Surface Finish Produced During Fluid Cutting

A similar response, is evident in the static force data. Surface finish produced at v = 60 m/min, f = .12 mm/rev and .18 mm/rev shows a complete contrast, minimal rise in Ra value occurring due to the retention of initial nose geometry.

6.0 Conclusions

The machining of hardened die steel using BZN Compacts, has been shown to be effective, particularly under high loading semi-finish conditions. In addition to laboratory feed rate and cutting speed evaluations, close working links with a large number of tool fabricators and users, has highlighted a number of practical operational considerations. These considerations are detailed below and presented as machining guidelines for the effective use of BZN Compact tooling.

a) Work Material Hardness should be be ⩾ Rc 45. Initial machinability studies undertaken on Rc 18-20 and Rc 30 steels showed machining to be possible though not economic in comparison with carbide tooling.

b) Use a robust lathe having a high degree of cross slide/tool post rigidity. Power requirements are dependent on depth of cut, etc.

c) Use negative tool geometries with the largest SCEA possible (45°). For medium—heavy machining use a large nose radius.

d) To maximize tool life, cutting speed and feed rates should be in the region of v = 80 m/min, f = .25 mm/rev.

e) Ensure an adequate edge finish, using the 15x magnification chipping code as a minimum requirement. At high rates of feed, special edge geometry and finish will be required.

f) Use a plentiful supply of cutting fluid. Flow of the coolant to the cutting zone should be constant. Intermittent flow causes thermal stress resulting in spalling of the BORAZON CBN layer. A water soluble emulsion type fluid, (a mixture of 20:1,water:oil) is adequate.

g) Chamfer the work material at 45° prior to cutting.

h) Use a chip breaker.

i) Avoid intermittent forms of cutting.

References

1. Failure Mechanisms of Superhard Materials when Cutting Superalloys. A. E. Focke et al. Proceedings of the 4th International Conference of Material Technology—Caracas, Venezuela, 1975.
2. CBN Machining of Hardened Chrome—Alloy Steel. B. A. Kravchenko et al. Machines and Tooling Pages 38-42 Volume (43), Number 6.
3. BORAZON® Compact Cutting Tools. R. E. Hanneman and Hibbs. SME Technical Paper MR73 143.
4. Some experiments on the Finish Produced by Diamond Turning Tools. A. G. King and J. Wilks. International Journal of Machine Tool Design Research Volume 16.
5. Surface Morphology of Synthetic Diamonds and Cubic Boron Nitride. R. Komanduri and McShaw. International Journal of Machine Tool Design Research Volume 14, 1974.
6. Synthesis of Polycrystalline Cubic BN. M. Wakatsuki et al. International Research Bulletin Volume 7 , 1972.
7. Edge Prepraration: Its Importance and How. J. Agnew. Tooling and Production Volume 38, 1973.
8. Nature of Brittle Failure of Cutting Tools. J. N. Loladze. CIRP Volume 24/1/1975.
9. Use of CBN Lathe Tools in Engineering. A. S. Kamenkovich and Y. A. A. Muzykant. Machines and Tooling, Page 35-38, Volume (43), Number 6.
10. Wear Mechanisms in Coated Tools. A. S. Raju et al. Proceedings of Third International Conference on Production Engineering—Kyoto, July, 1977.
11. Metal Cutting. E. M. Trent—Butterworth and Co. (Publishers) Limited, 1977.
12. On the Wear of Cutting Tools. H. Opitz and W. Konig. Eighth International MTDR, 1967.
13. Hot Compressive Strength of Cemented Carbides. E. M. Trent. Eighth International MTDR, 1967.

Presented at the International Conference on Manufacturing Engineering, August, 1980. Copyright General Electric

Machining Sintered Carbides With Polycrystalline Diamond Tools

By G.F. Wilson
R. Alworth
General Electric
And S. Ramalingam
Georgia Institute of Technology

ABSTRACT

Die grade tungsten carbide-cobalt alloys are machined with commercial polycrystalline diamond tools made from COMPAX blanks and the results obtained are presented. It is shown that it is commercially feasible to machine hard die materials with diamond tools. The chips produced are examined with a scanning electron microscope. An attempt is made to analyze the mechanism of chip formation. It is suggested that chip formation may involve an indentation-shear process suggested previously by Doyle. The worn diamond tools are also examined with the scanning electron microscope. It is shown that abrasive tool wear occurs in which micro-crystalline cleavage apparently has a significant role.*

INTRODUCTION

Diamond, with a Knoop hardness of about 7,000 Kg mm^{-2}, has the highest hardness among all known materials. Its hardness and wear resistance make it a good candidate for use as a cutting tool material. Under machining conditions where surface chemical reactions and decomposition or reversion reactions are not severe, diamond can be and is used as a cutting tool material. Use of diamond to machine hyper eutectic aluminum-silicon alloys in the automotive industry is now common and is made possible by recent developments in high pressure diamond synthesis capability.

The high hardness of diamond tools may permit their use to machine other hard materials such as tungsten carbide-cobalt alloys which are usually classified as "hard tool materials". Ability to machine these alloys is an advantage, since the only other commercial techniques to shape these alloys are grinding, ELG, ECM and EDM. While grinding, ECM and EDM can be used to shape these alloys, the metal removal rate is limited with these processes and they can only be considered as "finishing" processes.

The WC-Co alloys used for tooling typically possess a Vickers hardness between 1,100 and 1,900 Kg mm^{-2} with the hardness dependent on the cobalt content. Since the diamond hardness is several times the hardness of these materials, it should be feasible to machine these alloys with diamond tools. To test this concept and to document the chip formation while machining very hard tool materials, an experimental study was undertaken. The results obtained are presented here.

Since there apparently has been no previous study of the machining of WC-Co hard materials, the aims of the present study were (a) to demonstrate that such materials can be machined with commercial diamond tools, and (b) to obtain data regarding tool forces that prevail during machining. Unlike most engineering alloys, the WC-Co hard materials contain a hard compound, WC, as the principal phase. The mean free path in the soft, binder phase is also very small. Very little is known regarding chip formation in alloys possessing such microstructures. Documentation and analysis of chip formation in such alloys is hence also an aim of this work. A machining study in which one hard material is used to machine another can cast some light on the prevailing wear mechanisms. Worn diamond tools were therefore examined and the results obtained are presented.

EXPERIMENTAL MATERIALS AND TEST CONDITIONS

Since it is known that abrasive wear can be a severe problem in sliding wear when the hardness ratio (hard to soft) is less than three, a higher cobalt alloy containing tungsten carbide with a bulk hardness less than 1,400 kg mm^{-2} was chosen for the machining study. Commercially available

*Trademark of General Electric Company, U.S.A.

forming tool type WC-Co alloy (Cobalt is the binder phase) is reported[1] to possess a Vickers hardness of 1204±12 kg mm^{-2}. The bulk hardness ratio in this instance is greater than five and was hence considered to be suitable work material. The typical mechanical properties and structural parameters characterizing it, taken from Reference 1, are shown in Table 1.

TABLE I
Properties of Tungsten Carbide—Cobalt Work Materials Machined

Composition:	**Forming tool grade WC-Co material**
Vickers Hardness	1204 ± 12 kg mm^{-2}
Young's Modulus	530 GPa
Proportional Limit	1.04 GPa (0.02% strain)
Transverse Rupture Strength	2.69 GPa
Mean Grain Size (WC)	1.50 micrometers
Mean Binder Film Thickness	0.51 micrometers
Nominal Mean Free Path	0.404 micrometers

A tool made from a commercial diamond COMPAX blank, 12.5 mm in diameter, was used to machine a block of WC-Cobalt work material, larger than 50 mm in diameter. The circular diamond insert was held in an ISO type tool holder (no. MRGNR-16-4) to provide −5° back rake and −5° side rake. The clearance angle was 5°. A heavy duty Sheldon lathe was used to machine the work piece. During cutting a 10% water-base emulsion containing EP additives was used for flood cooling.

Typical cutting runs were made at a cutting speed of 28 m min^{-1} with a feed per revolution between 0.25 mm and 0.55 mm. Additional higher speed cutting studies were also carried out at 57 m min^{-1}. In all test runs, a depth of cut of about 0.5 mm was used. Typical cutting runs lasted between 1 and 3 minutes as needed for force measurements. A piezo-electric, 3-component dynamometer was used to measure the cutting forces during machining.

EXPERIMENTAL RESULTS

The turning tests carried out showed that in every instance, it was possible to machine the WC-Co work material used with the COMPAX blank material.

Satisfactory surface finish, comparable to that usually obtained in rough machining, was obtained and the machined bodies produced were suitable for general engineering applications. The cutting forces which prevailed during continuous cutting with flood cooling are shown in Table II.

The chips produced during machining superficially resembled the type of chips produced during the machining of cast irons. They exhibited the commonly encountered lamellar structure[2] observed in machined chips. A SEM micrograph of the chips produced is shown in Figure 1 (inset shows a low magnification view). The similarity to commonly produced machined chips is superficial. This can be best appreciated by an examination of the sectional view (longitudinal) of the chips produced, shown in Figure 2. Typical serrated chip structure is shown in the left with intense shear localization in a macroscopic "shear plane" approximately 5 microns in thickness.

According to the thin shear plane model of cutting[3,4], the lamellar structure of the chip is due to repeated shear on parallel shear surfaces that become activated as the tool advances into the work material. The shear plane is "flat" or nearly so with slight curvature. In contrast to this, the lamellar structure of the chip produced in this instance originates from a more complex shear surface. The shear surface is chevron-shaped as shown in Figure 3. This can be discerned also by a closer examination of Figure 2.

FIGURE 1.
Lamellar structure of WC-Co chips produced viewed from the free surface of the chip. (400 X.) A low magnification view of the chip is shown in inset.

TABLE II
Tool Force Turning Test Data Tungsten Carbide-Cobalt Work Material

	Tool	Commercial Polycrystalline Diamond 12.5 mm Dia.				
	Back Rake	−5° Side Relief 5°				
	Side Rake	−5° End Relief 5°				
Run No.	**Depth of Cut mm**	**Feed mm Rev^{-1}**	**Speed m min^{-1}**	**F_C Newton**	**F_L Newton**	**F_R Newton**
1	0.50	0.254	28.34	1068	445	2893
2	0.50	0.305	28.34	1113	445	2670
3	0.50	0.406	28.34	1291	445	3026
4	0.50	0.467	28.34	1335	512	2893
5	0.50	0.508	28.34	1467	623	2893
6	0.50	0.554	28.34	1467	534	2626
7	0.50	0.305	56.99	757	223	1958

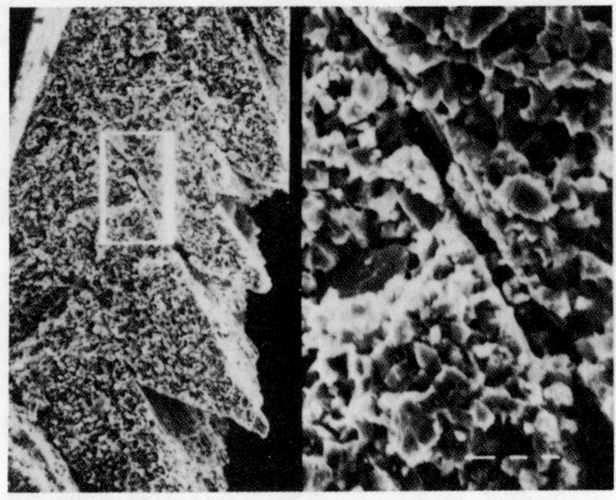

FIGURE 2.
Section of the chip parallel to its longitudinal direction demonstrating periodic shear. (500 X.) Windowed region is shown at left at 2500 X.

FIGURE 3.
A section of the WC-Co chip illustrating the chevron-shaped shear surface and periodic shear. Free surface of the chip is in the top. (400 X.)

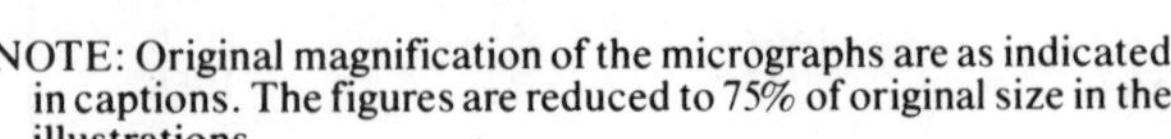
NOTE: Original magnification of the micrographs are as indicated in captions. The figures are reduced to 75% of original size in the illustrations.

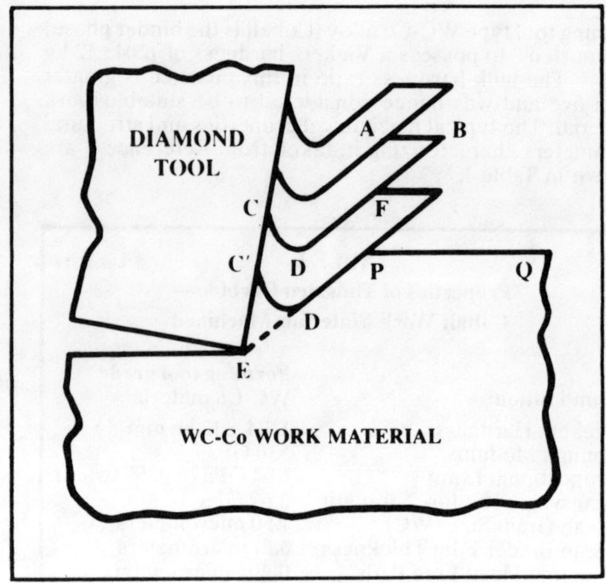

FIGURE 4.
Schematic illustration of the chip formation process in WC-Co hard materials consistent with the observed chip morphology.

The free surface of the chip lies towards the top of Figure 3, while the rubbed surface of the chip can be seen in the lower part of the same micrograph.

An examination of the chip sections using metallographic techniques[5] suggests that chip formation in the present instance must occur as shown schematically in Figure 4. The edge AB of the chip corresponds to the unmachined surface PQ and CDF corresponds to the shear surface seen in the chips. The chip surfaces corresponding to AB and CDF respectively are shown in greater detail in Figure 5. During continuous cutting, as in the longitudinal cutting in the present instance, instead of long chip ribbons, shorter segments are produced by fracture on the segment DF of

FIGURE 5.
Free Surface of the chip corresponding to edges AB and BF indicated in the previous figure illustrating chip formation process. (1750 X.)

FIGURE 6.
"Fractured surface" of chip segment viewed normal to CD at 600 X. Shear on CD produces "flat" region C'D' **Fracture shown in inset reveals carbide grains. (6500 X.)**

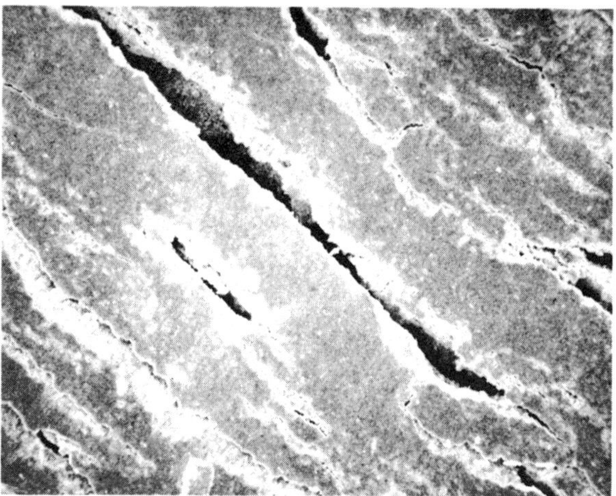

FIGURE 7.
Shear on CDF penetrates the full thickness of the chip. Rubbed surface of the chip that has been in contact with the tool shows "fissures." (500 X.)

FIGURE 8.
SEM micrograph of the polycrystalline diamond tool. Murakami's reagent was used to reveal the microstructure. (560 X.)

FIGURE 9.
Rake and flank faces of the worn diamond tool. Note the "cobalt" smear on the rake and flank surfaces and the edge sharpness. (500 X.)

the shear surface. Figure 6 shows that surface segment CD has apparently smoothed out the "flat" region C'D' of Figure 6. An examination of the rubbed surface of the chip, Figure 7, offers additional evidence to indicate that segmental shear on surface CDF penetrates the full thickness of the chip.

The tool used for the machining tests was examined in a SEM for initial structure and to elucidate the tool wear and tool-chip interfacial processes encountered during machining. The microstructure of the polycrystalline diamond tool is shown in Figure 8. The rake and flank surfaces of the worn tool are shown in Figure 9. It is seen that as a result of machining, the binder phase cobalt is smeared on the rake as well as the flank surfaces. While cobalt accumulation at the rear of the crater is heavy, the thin films of cobalt found on the tool-chip interface are usually electron-transparent at 30 kV.

Surprisingly well-bonded layers of cobalt, as much as five microns in thickness, can be seen on the flank surface of the tool. The smeared cobalt layer covers almost all of the flank surface as can be seen from Figure 10. While abrasion tracks can be seen on the rake surface, the mechanism responsible for its generation could not be identified by SEM examination of worn tools.

Selective etching techniques were hence used to identify the mechanisms responsible for the wear of diamond tools. The worn tools were etched with Murakami's reagent to reveal the presence of WC on the tool surfaces. The SEM examination following two minutes of etching showed that the smooth "cobalt" layers observed on the rake and flank surfaces were in fact cobalt-tungsten carbide alloys, rich in cobalt. The observed tungsten carbide grains were submicron in size and in many instances as small as 0.1 micron in size. Visual estimate in the SEM suggests that the "cobalt layers" actually contain between 25 and 50% of fine carbide fragments.

The etched tools, etched for WC, were again etched with a 33% solution of hydrochloric acid to partially dissolve cobalt layers in order to reveal the worn surfaces of the diamond tool. Figure 11 shows the fine structure of the "abrasion grooves" observed on the rake surface. One complete grain together with evidence in favor of micro-crystalline cleavage of the diamond grain can be seen.

Change in groove orientation across the grain boundary GB may also be seen. An examination of the structure of the edge of the cutting tool suggests small scale tool edge failure by grain cleavage in addition to micro-crystalline cleavage within each grain. This is illustrated in Figure 12.

DISCUSSION

The turning tests and the forces measured during machining suggest that hard materials such as WC-Co sintered products can be machined with polycrystalline diamond tools. The metal removal rates obtained ($1-5 \times 10^3$ mm^3 min^{-1}) by diamond turning are far higher than those feasible with other metal removal processes such as EDM, ECM and grinding. While the tool forces observed for the metal removal rates obtained are higher, they are well within the loads permissible in heavy duty tool room lathes. Since this class of work materials are usually used for the fabrication of metal forming tools, machining with COMPAX* diamond blank offers a new and economical means for the fabrication of WC-Co based forming tools.

The cutting conditions used were far removed from those normally employed (orthogonal cutting) for quantitative analytical studies of chip formation. It is nevertheless possible to make rough estimates of the shear and normal stresses on the "shear plane" using the force data obtained and the chip thicknesses measured. To facilitate this estimate, the chip thickness was measured using the SEM (approximate t_2 of 0.125 mm at a cutting speed of 59 m min^{-1}) and the chevron shaped shear surface was assumed to be symmetric with respect to the chip thickness. That is, point D of the shear surface CDF was assumed to lie in the center of the chip thickness with CD and DF inclined equally with the long axis of the chip. From chip metalographic section, the shear angle was estimated as 42°.

FIGURE 10.
View of the flank surface of the worn tool showing well-adhered layers of "cobalt" nearly coverning the full depth of flank wear region. (400 X.)

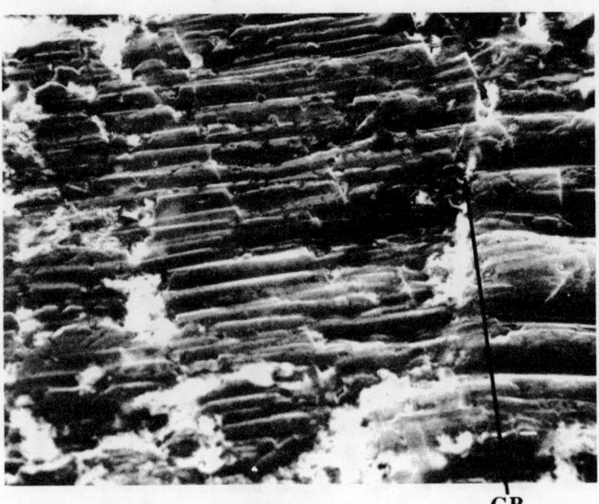

FIGURE 11.
Rake surface of worn tool etched to show micro-crystalline cleavage below observed "abrasion" tracks. Grain boundary is shown as GB. (2500 X.)

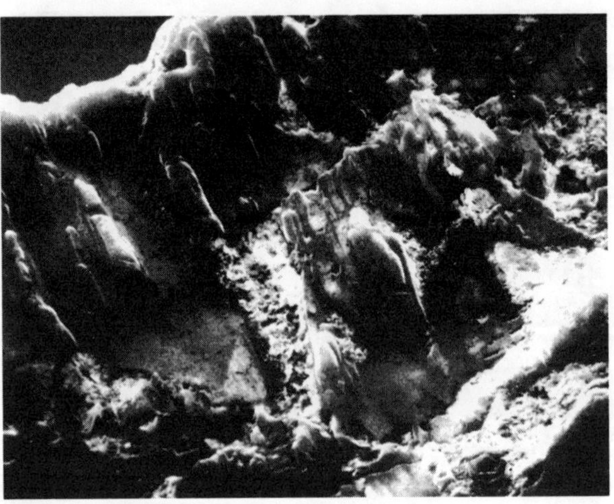

FIGURE 12.
Transcrystalline cleavage fracture of diamond grains in the vicinity of the cutting edge. Tool edge is seen at upper left. (2100 X.)

Using the measured tool forces and the values estimated above, the friction coefficient during machining is estimated to range between 1.65 and 2.28. The corresponding friction angles range between 58° and 66°. The shear and normal stresses on the "shear surface" with the approximations cited earlier are computed to be of the order of 1.83 GPa and 4.20 GPa respectively.

The work material used in the present study contains substantial WC with a hardness of 2,500 kg mm^{-2} and the remainder is cobalt. The mean free path in the cobalt phase is of the order of 0.5 μm[1]. The SEM and the metallographic analysis[5] show no evidence of plastic deformation within each lamella. Yet the measured cutting forces show large power dissipation during machining (0.14 Watts. mm^{-3}) at the speeds and feeds used. This rate of dissipation is some 2 to 3 times the specific energy requirement for metal removal in machining alloy steels[3].

Since the present WC-Co alloy has a low fracture toughness, about 15 $MNm^{-3/2}$ according to Pickens[1] the measured dissipation can not be accounted for by fracture along the shear surface. Hence, despite the small mean free path, bulk shear must occur between the lamella. Such shear, propagating through the carbide grains, must lead to their fragmentation by shear separation (estimated shear strain in the present instance is 2.18 for a shear angle of 42° and rake of −5°). Metallographic examination of chip sections[5] shows this to be the case. High normal stresses across the shear surface and local shear stress in excess of that for shear in compression, estimated earlier, facilitates carbide shear during machining.

As noted earlier, finely fragmented carbides are observed in the "cobalt smear" produced at the rear of the crater region. The "cobalt smears" on the flank surface also show the presence of fragmented carbides. It is hence reasonable to claim that chip formation in the machining of WC-Co alloys involves extensive carbide grain shear on the shear surface. The large friction coefficient observed may well be due to the presence of cobalt-fine carbide mixture on the tool-chip interface during machining.

If indentation-shear model of chip formation were to occur, it is possible that as the tool advances into the work material, the interfering material is initially indented and "piled-up" ahead of the tool (triangular region C′E′D′ of Figure 4). As the tool advances further, growth in size of the "pile-up" is retarded by the high tool-chip friction with concommitant rise in compressive stress within the "pile-up". When the stress build up is sufficient to induce shear on an appropriate plane, such as D′P of Figure 4, bulk shear in a thin shear surface is induced. Thus, indentation and shear, alternating in sequence, lead to chip formation and yield the observed lamellar structure. Such a model has been proposed earlier by Doyle[6]. Additional orthogonal cutting studies are however necessary to clearly demonstrate that this in fact is the chip formation mechanism in the present case.

Since the tool material work material hardness ratio is of the order of three, abrasive wear mechanism must be expected to operate during machining. The abrasion tracks observed on the rake and flank surfaces of the diamond tool tend to confirm the operation of the abrasive tool wear mechanism. The SEM micrographs suggest that abrasive wear, in the micro-scale, appears to be by microcrystalline cleavage within individual diamond grains. Although diamond tools appear to be abrasively worn by WC grains, tungsten carbide is anisotropic[7] with a basal plane hardness of 2500 kg mm^{-2} and a pyramidal plane hardness of 1400 kg mm^{-2}. Hence in the machining of polycrystalline WC-Co compacts, abrasive wear of diamond is not as severe a problem as it would appear at first sight.

The observed transcrystalline cleavage of diamond grains at the tool tip can be attributed to the high local normal stresses. Since the polycrystalline diamond tools are essentially binder-free, strain incompatibility at the grain boundaries may well be sufficiently severe to induce transcrystalline cleavage (initiated at the grain boundaries). Cleavage of this type is however found to be restricted to a depth of only two to three grains from the cutting edge (Figure 12). Moreover, a self sharpening action apparently operates during machining, as is evident from Figure 9. It would therefore appear that edge breakdown of the cutting tool is not as serious a problem as it would initially appear.

CONCLUSIONS

The results obtained in the present study show that die grade WC-Co in the machining of Cobalt cemented WC, abrasive can be machined successfully with polycrystalline diamond tools. Satisfactory surface finish and tool life can be obtained at metal removal rates much higher than those possible with ECM, EDM and grinding. Diamond tool machining thus offers a new means of shaping die grade carbide forming tools.

While the chips produced superficially resemble those produced in the machining of engineering alloys, a different mechanism of chip formation apparently operates. A preliminary analysis suggests that an indentation-shear mechanism of chip formation may be adequate to account for the chip morphology. Since the observed power requirements for metal removal can not be accounted for by fracture within the work material, bulk shear during chip formation must occur. Microscopic evidence in support of bulk shear is found.

While several mechanisms of tool wear may operate simultaneously, SEM analysis of the worn tool suggests that abrasive wear of diamond is an important component of tool wear while machining WC-Co alloys. In the microscopic scale, abrasive wear appears to occur by microcrystalline cleavage within individual grains of diamond in the polycrystalline diamond tool.

We close by noting that the average surface finish Ra obtained by turning with the polycrystalline diamond tool is 0.94 μm at the low cutting speed (29 m min^{-1}) and 0.64 μm at the higher cutting speed (59 m min^{-1}). The corresponding peak-to-valley measurements are (Rt/Rmax) 4.4 μm and 3.5 μm respectively. At the cutting conditions used (feed rate of 0.305 mm rev^{-1} and tool radius of 6.25 mm) the best theoretical center-line-average roughness (h_{CLA}) that can be expected is only 0.47 μm. The surface finish obtained is thus only some two to three times the best that can be expected on theoretical grounds. It is hence considered to be quite good. Hard material tools turned with such surface finishes can be finished to smoothness needed for die applications with relative ease. Machining with polycrystalline diamond is hence a feasible proposition for the production of metal forming tools made of WC-Co materials.

ACKNOWLEDGEMENTS

The machining study reported here was conducted in the laboratories of General Electric Company and Georgia Institute of Technology. The authors thank Specialty Materials Department for the experimental facilities and for the tools fabricated from COMPAX* industrial diamond blanks. One of the authors (S.R.) in indebted to Dr. D. M. Turley of the Materials Research Laboratory, Department of Defence, Melbourne, Australia for the private communication regarding chip formation analysis. The authors acknowledge with thanks the expert manuscript preparation by Miss Melinda Wilson.

REFERENCES

1. PICKENS, J. R. (1977) Ph.D. Dissertation, Brown University, Providence, R.I., U.S.A.
2. BLACK, J. T. (1972) *J. Eng. Ind., Trans.* ASME., Vol. 94, pp. 307-315.
3. MERCHANT, M. E. (1945) *J. Appl. Phy.*, Vol. 16, pp. 267-274 and 318-324.
4. PISSPANEN, V. I. (1948) *J. Appl. Phy.*, Vol. 19, pp. 876-885.
5. TURLEY, D. M. (1979) *Private Communication* to S. Ramalingam.
6. DOYLE, E. D. (1974) *Proc. Int. Conf. Prod. Tech.*, Melbourne, pp. 161-167.
7. PONS, L. (1968) *Anisotropy in Single-Crystal Refractory Compounds*, Plenum Press New York, Vol. 2, pp. 393-444.

Presented at the Industrial Diamond Association of Japan's 30th Anniversary Meeting and Seminar, May 1978.

Machining Of Hard Ceramics With Sintered Diamond Tools

By Hidehiko Takeyama

1. Introduction

Machine parts have been encountered very often with physically, chemically and mechanically severe conditions as industrialization or civilization advances. From this point of view ceramics have been regarded as attractive materials for some special machine parts or components because of their excellent heat, erosion and wear resistance. The trend will increase more and more in the future, and then larger ceramic materials will be utilized as members of machine structures. In that case the processings of ceramics will be a bottleneck in view of productivity. If grinding with diamond wheels can be replaced by cutting with diamond tools a marked advance can be expected in processing ceramics.

The objective of this study is to verify the possibility of ceramic cutting through the challenge of cutting extremely hard aluminium oxide ceramics (1650 Hv) with sintered diamond tools, and also to expand the possibility by analysing the tool wear behavior.

2. Test procedure

Type of cut: Longitudinal turning, dry cutting

Material cut: Dense aluminium oxide ceramic, purity 99%, volumetric density 3.8, porosity 0, hardness 1650 Hv, elastic constant 3.5x10 Kg/cm², dimension 40mmϕ

Tool: Sintered diamond (GE COMPAX Diamond Blank Tool Material), hardness 6500-8000 Kg/mm²Hk, tool shape (0,0,5,5,45,45,0.8mm) and (0,0,5,5,70,20,0.1mm)

Cutting condition: Depth of cut 0.15mm, feed 0.0125mm/rev~variable, cutting speed of 6 m/min~variable.

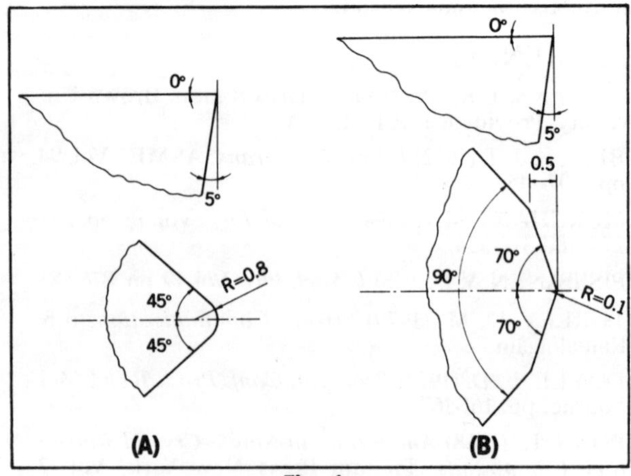

Fig. 1
Round tool (A) and pointed tool (B)

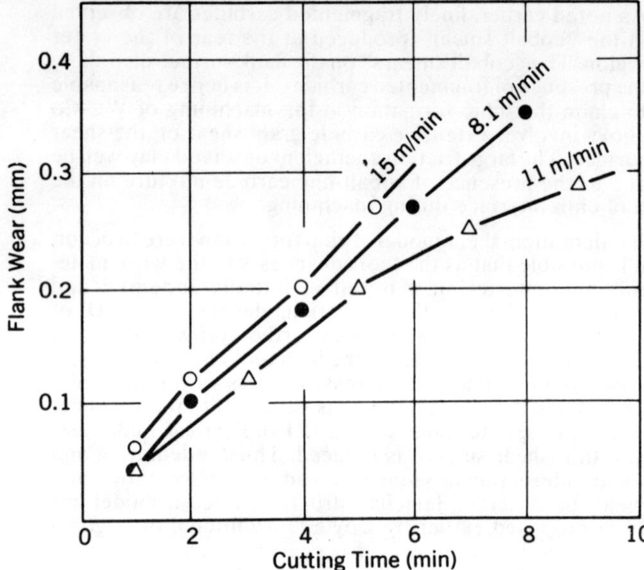

Depth of cut: 0.15mm, feed: 0.0125 mm/rev, tool shape: 0,0,5,5,70,20,0.1mm

Fig. 2
Tool wear process of pointed tool

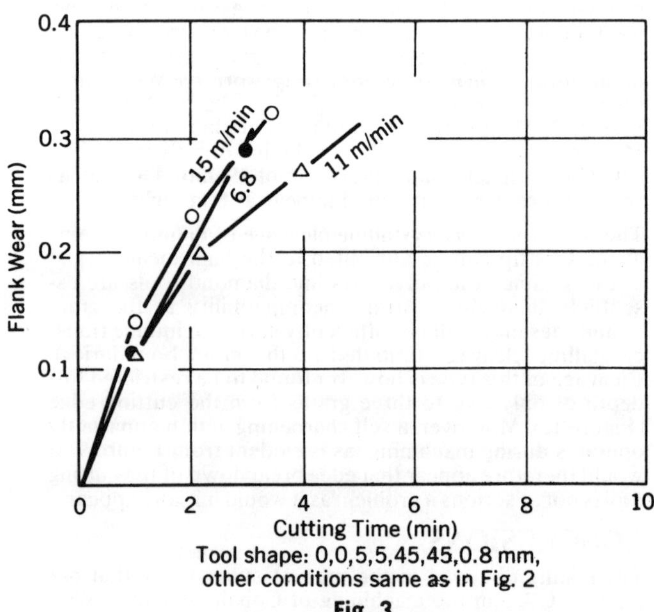

Tool shape: 0,0,5,5,45,45,0.8 mm, other conditions same as in Fig. 2

Fig. 3
Tool wear process of round tool

3. Experimental results

The experimental tool wear process when turning the aluminium oxide ceramic with round and pointed tools, which are illustrated in Figure 1, at 0.0125mm/rev feed and 0.15mm depth of cut, is shown in Figures 2 and 3, respectively. Apparently the wear land of the round tool is not only excessive but also irregular compared with the case of the pointed tool. This naturally leads to the shorter tool life as seen in Figure 4.

A characteristic feature revealed in Figure 4 is that the tool life curve is greatly different from that in metal cutting and does not follow so-called Taylor's tool life curve. In other words, the tool wear of sintered diamond, when machining ceramics, shows up neither temperature, cutting speed nor cutting distance dependence, but rather cutting time dependence. Furthermore, as far as this experiment is concerned, there exists an optimal cutting speed (approximately 11m/min.)

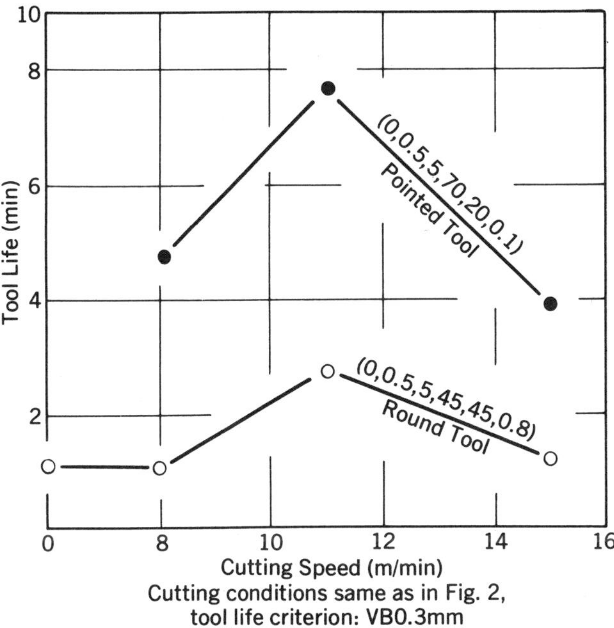

Cutting conditions same as in Fig. 2,
tool life criterion: VB0.3mm

Fig. 4
Relation of cutting speed and tool life

4. Investigation of results

4.1 Relationship between hardnesses of tool material and material cut

With advancements of tool material, the utilization field of cutting process has been expanded, and materials subjected to machining have become harder. Since sintered diamond has been verified to be able to cut hard aluminium oxide ceramic (as mentioned above) the relation between hardnesses of tool material and material to be cut at ambient temperature can be plotted over the wide range by referring to the data obtained so far by the author as demonstrated in Figure 5. As seen in the figure, the highest hardness of work material possible to be cut with the tool material is approximately linear with the hardness of the tool material, and it indicates that the hardness of tool material must be approximately 4.5 times harder than the work material in order to realize satisfactory cutting.

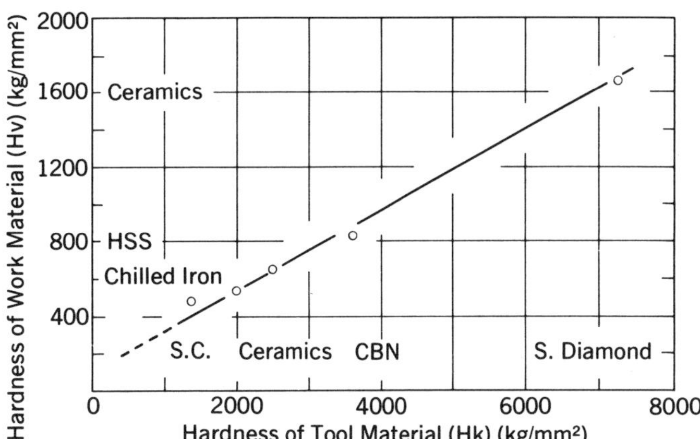

Fig. 5
Relation of hardnesses of work and tool materials to realize cutting

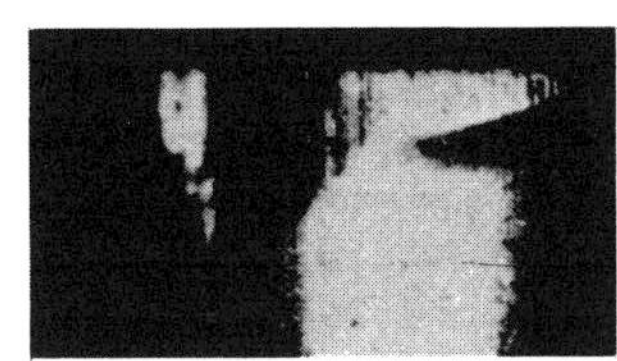

(A) Round Tool (B) Pointed Tool

Cutting conditions same as in Figs. 2 and 3

Fig. 6
Typical tool wear patterns of round tool (A) and pointed tool (B) at life time

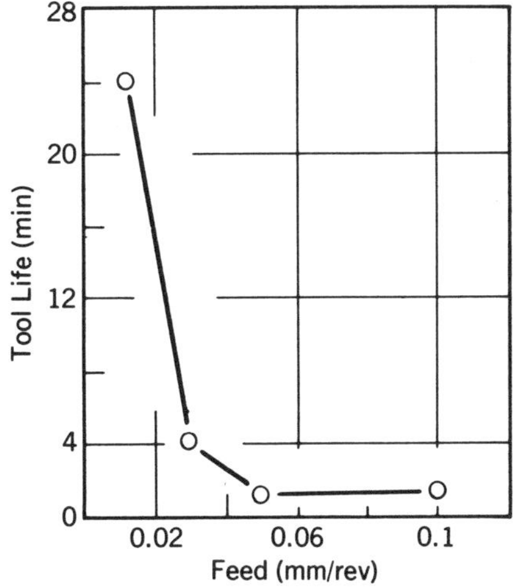

Depth of cut: 0.05 mm,
cutting speed: 8m/min

Depth of cut: 0.05 mm, feed: 0.0125 mm/rev

Fig. 7
Relation of feed and tool life (round tool)

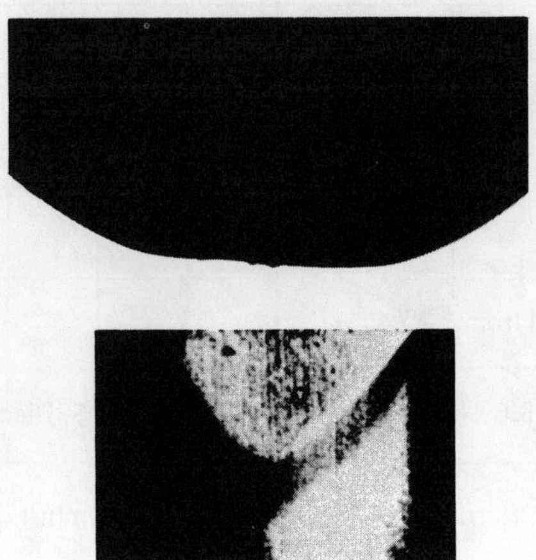

Fig. 8
Wear pattern of round tool at small uncut chip thickness

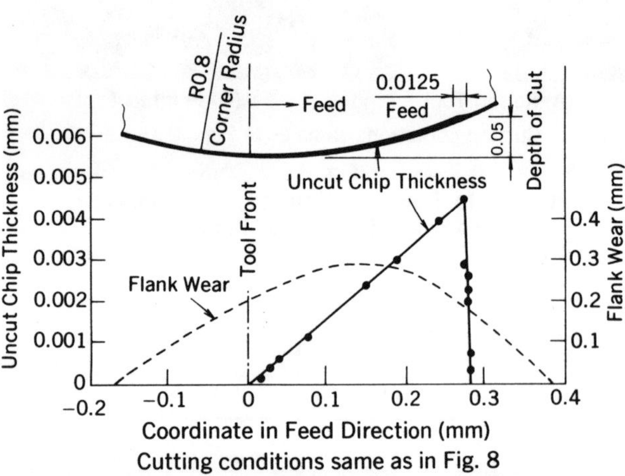

Cutting conditions same as in Fig. 8

Fig. 9
Correlation between uncut chip thickness and flank wear

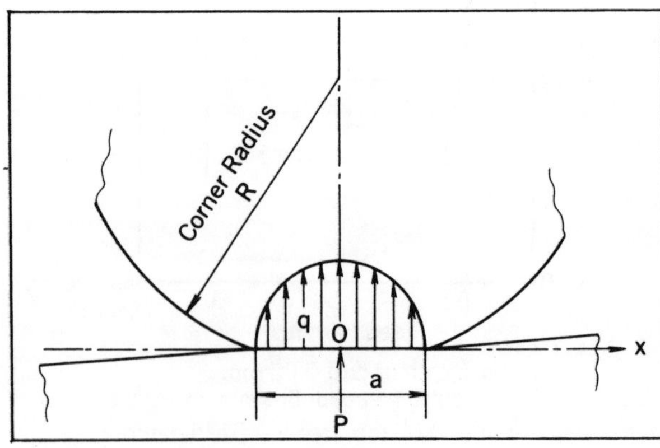

Fig. 10
Analogy of elastic contact

3.2 Characteristic feature of tool wear

Typical wear patterns of round and pointed tools at the life time when turning the ceramic at 0.15mm depth of cut, 0.0125mm/rev feed rate and 11m/min cutting speed are shown in Figure 6. The wear patterns clearly indicate a close correlation between the width of flank wear and the uncut chip thickness.

The flank wear seems to be primarily governed by the cutting pressure at the tool-chip interface. In order to uniformly distribute the cutting pressure over the cutting edge, a pointed tool is desirable for a longer tool life. Since the tool wear is greatly affected by the uncut chip thickness or the cutting pressure, a larger side cutting edge angle with a comparatively small corner radius which prevents chipping is desirable.

For the same reason the tool life will be greatly affected by the feed rate. This has been experimentally verified as shown in Figure 7, which is the relationship between tool life and feed rate with the round tool. Larger feed rates than approximately 0.02mm/rev are detrimental because of excessive wear or chipping.

What is worthy of attention is that Taylor's tool life curve does not hold with the combination of ceramics and sintered diamond tool. This seems to suggest that the mechanism of tool wear is appreciably different from that in metal cutting. Although the mechanism of tool wear in this case has not been clarified yet, it looks like a crushing type of wear as far as the observation in the experiment is concerned.

As a too large uncut chip thickness is harmful to the tool life, one that is too small is also detrimental. This is apparent from the wear pattern shown in Figures 8 and 9, in which, even at the cutting edge area, geometrically never contacts with the work material (i.e., on the trailing side of the cutting edge in the feed direction, appreciable wear land can be observed, and also the flank wear increases towards the middle of the tool-work contact area). Under this condition appreciable residual depth of cut or elastic deformation of work material is likely to occur.

The situation could be analogous to the elastic contact between a cylinder with a radius identical to the corner radius of the tool and a plane as shown in Figure 10. Computing the contact pressure q by means of Hertz's theory it is expressed as follows:

$$q = \left\{\frac{1}{\pi^3} \cdot \frac{P}{bR\,(k_1 + k_2)}\right\}^{1/2} \sqrt{1 - \frac{x^2}{a^2}}$$

$$k_1 = \frac{1 - \nu_1^2}{\pi E_1}$$

$$k_2 = \frac{1 - \nu_2^2}{\pi E_2}$$

where **b** is the contact length parallel to the cylinder axis, **a** the contact length in the contact area, E_1 and E_2 the elastic constants of cylinder and plane, respectively, and ν_1 and ν_2 the Poisson's ratios of cylinder and plane, respectively.

The contact pressure calculated by the foregoing equation increases towards the middle of the contact area as shown in Figure 10. The real wear pattern seems to interprete a severe elastic contact as mentioned. Under such the condition a tool wears in vain and expected dimensional accuracy cannot be obtained.

5. Conclusions

(1) A dense and hard aluminium oxide ceramic (1650 Hv) can be turned with the sintered diamond tool. Ceramics commonly utilized for machine components or structures will be softer than that, so that the possibility of diamond cutting of ceramics will be larger.

(2) The relation between hardness of work material and that of a tool able to cut the material is nearly linear, and the latter is to be approximately 4.5 times harder than the former.

(3) The cutting pressure greatly affects the tool wear, and consequently the permissible feed rate is limited.

(4) For the same reason the tool shape should be designed so that the cutting pressure or the uncut chip thickness may be uniform along the cutting edge. In this sense a pointed tool with a larger side cutting edge is desirable.

(5) A too small uncut chip thickness causes a severe elastic contact with the work material and hence rubbing wear.

(6) The relation of tool life and cutting speed in the combination does not follow so-called Taylor's tool life curve.

Acknowledgements: The author would like to express gratitude to Mr. N. Iijima for his cooperation in the experiments and Kyoto Ceramic Co. Ltd. who offered the material cut.

HIGH-SPEED MACHINING of a hardened steel roll with a standard large BORAZON CBN insert at Tool Steel Gear & Pinion. Despite its hardness, the roll is being turned at speeds and feeds typical for turning mild steels. Turning the roll takes 90 minutes; it was formerly ground in 360 minutes.

Reprinted from Manufacturing Engineering, October 1975

Now: Turn Hardened Steels and Tough Superalloys as Easily as Mild Steels

The extension of BORAZON CBN (cubic boron nitride) technology from grinding wheels to indexable inserts marks a significant advance in turning tough-to-machine materials. Here's how these new tools are being applied

DANIEL E. HERZOG
Specialty Materials Department, General Electric Company

HIGH-SPEED TURNING of hardened alloy steels and tough superalloys — practically impossible with conventional tools because of the rapid rate of edge wear — has been made economically feasible by the advent of a proven material in a new form. The case in point: indexable inserts made of BORAZON CBN.

Developed by General Electric's Specialty Materials Department, Worthington, Ohio, these inserts are capable of turning hardened alloy steels and tough superalloys as easily as mild steels are cut by conventional tools. Tool life is usually at least an order of magnitude longer than the lives of conventional tools used to turn the same materials under the same conditions.

In field applications, these inserts have proven out for rough and finish turning of hardened steels such as S5 and D2, tool steels in the M and T series, modified 52100 steels and chilled cast irons. The Rockwell C hardnesses of these materials range from about 50 to over 70.

Superalloys that are being successfully machined with BORAZON CBN inserts include Inconel 718, Rene′95 and others. As a class, the superalloys have somewhat lower hardnesses than the hardened steels — typically around $45R_C$ — but their combination of high strength, gumminess and other characteristics, such as their tendency to deform plastically under pressure from the cutting edge, places them in the difficult-to-machine category.

Superhard Cutting Edges. BORAZON CBN inserts are made by bonding a layer of polycrystalline cubic boron nitride to a cemented carbide substrate. Thickness of the CBN layer is 0.5 millimeter. Overall thickness of the inserts is 3/16 inch. The CBN layer does the cutting; the carbide substrate provides a strong, tough, shock-resistant base for the cutting edge, which consists of literally millions of tiny crystals of cubic boron nitride, tightly bonded to each other.

Cubic boron nitride is — next to diamond — the hardest material known to man. It has a Knoop indentation hardness of 3500 kilograms per square millimeter. This is nearly double the hardness of Grade C-2 carbide, which runs from 1800 to 2000 kilograms per square millimeter.

Polycrystalline BORAZON CBN has other properties that contribute to its performance as a tool material. It doesn't react chemically with metals or oxidize at temperatures below around 1000 deg C — a temperature that's unlikely to be reached in machining operations. And CBN has excellent hot hardness so inserts made of this material stay sharp under severe machining conditions that weaken and soften conventional tool materials.

Turning Hardened Rolls. Turning of hardened steel and chilled cast iron rolls may prove to be one of the most important applications of BORAZON CBN inserts, on the basis of production turning tests conducted by The Tool Steel Gear & Pinion

Company at its Plant 2, Sharonville, Ohio.

Initial tests to determine optimum machining parameters were conducted on a variety of steel rolls with hardnesses of 64 to over 70 R_C. Because of the short lives of conventional tools, the company regarded turning as impractical. All hardened rolls were ground.

The largest, hardest roll production-turned at Tool Steel Gear & Pinion to date is 168 inches long, with a ground length of 80 inches. Final ground diameter is 25 inches and about 0.090 inch is removed from the diameter in turning, with 0.010 inch of stock left for finish-grinding. Hardness of the material, a modified 52100 steel, is close to 71R_C.

These rolls are usually slightly out-of-round when they arrive at the lathe and are covered with a dense scale produced by heat treatment. Neither the out-of-roundness nor the scale appears to have adverse effects on the performance of the BORAZON CBN inserts. There's no need for sand-blasting the rolls, which was mandatory before the rolls were ground.

An American Tool Works lathe with a 36-inch swing is used for the turning operation. This 25-horsepower machine can handle rolls up to 234 inches long.

The BORAZON CBN insert is a standard 1/2-inch diameter round, which is clamped in a standard toolholder identical to those used for carbide inserts. Since the brittle steel chips automatically break into short lengths, no chipbreaker is required. The tool has a 5-deg negative rake angle and a 5-deg side clearance angle.

The cutting fluid is a 20:1 mixture of water and soluble oil, applied in copious amounts to the roll at a point where rotation of the roll carries it directly into the tool/workpiece interface. Chips are carried away by an under-the-floor conveyor.

Tool Steel Gear & Pinion's production tests established that a surface speed of 275 fpm resulted in an optimum combination of productivity and tool life for turning these rolls. Feed is 0.22 ipr; depth of cut is 0.35 inch.

Under these cutting conditions, rough turning takes 86.5 minutes. Removing the same amount of stock by grinding took 360 minutes — turning takes less than 1/4 as long as grinding.

The BORAZON CBN inserts produce consistently good finishes — typically in the 25-60 microinch (rms) range. Out-of-roundness (runout) is reduced from 0.012 inch before turning to 0.002 inch or less after turning. Because of the good finishes and geometric accuracy produced by rough turning, finish grinding can be done in less time.

On the basis of experience in machining hardened steel rolls at Tool Steel Gear & Pinion the edge wear of BORAZON CBN inserts is comparatively slow and uniform. The edge is considered to be expended when 0.030 inch of uniform flank wear is attained. After turning one of the 25-inch diameter rolls, the amount of edge wear is barely detectable. On this job, edge wear rates were accelerated when running at surface speeds in excess of the 275 fpm established as the optimum. It was observed during the tests that minor nicks in the edge — acquired while experience was being gained with the tool — had no adverse effects on its cutting performance.

Horsepower requirements for turning the 25-inch diameter rolls with CBN inserts have proved to be surprisingly low, ranging from 10 to 15 horsepower. Tool Steel Gear & Pinion tool engineers believe that the low horsepower requirement is due to the ability of the tool to stay sharp.

Significantly, the experience at Tool Steel Gear & Pinion indicates that the CBN inserts need not be limited to hardened steel rolls. Performance tests are planned for hardened crane wheels, brake wheels and, in fact, all hardened steel parts that currently are being ground.

Turning Small Precision Parts. Another pioneer in the application of CBN inserts is Precision Products Company, Grove City, Ohio. Precision Products is a job shop that specializes in the production of high-precision parts from difficult-to-machine materials. Stainless steels. Monel. Hardened steel alloys. Stellite. Superalloys. And composites.

In some cases, Precision Products has replaced carbide turning tools with CBN tools; in other cases, turning with CBN inserts has replaced grinding operations.

The first part machined was an insertable valve seat with a flame-sprayed area where the valve contacts the seat. This ring-like part is of Type 302 stainless steel. The flame-sprayed material is Colmonoy No 43 — a powder that fuses at 2200 C. Hardness of the stainless steel is 20-25R_C, which presents no machining problems. The hardness of the flame-sprayed material is 50R_C, and its toughness is comparable to that of high-temperature superalloys.

Triangular standard CBN inserts are used for two operations on the critical area — plunge cutting and finish cutting the face, which has a 45-deg chamfer, and turning the ID. In these operations, all excess flame-sprayed material is removed.

When Grade 883 triangular carbide inserts were used for these operations, tool life averaged three pieces per edge. With CBN inserts, tool life averages 1000 pieces per edge.

Geometry of both tools is the same — a 5-deg clearance (side relief) angle and a zero-deg side cutting edge angle. Nose radius for the carbide tool is 0.015. Nose radius for the CBN tool is 0.047 inch.

The machine is a 7.5-horsepower Clausing lathe. The cutting fluid is a 15:1 water-soluble oil mixture. Tool change time is five minutes.

Eliminating the need for frequent tool changes has, of course, improved productivity. In addition, surface speed has been increased to 295 fpm — more than

Rough Turning Parameters

Type of material	*Speed (fpm)*	*Feed (ipr)*	*Depth of cut (inch)*	*SCEA (deg)*
A-2, A-6 (R_C58)	250	0.005	0.060	45
D-2 (R_C58)	250	0.005	0.060	45
D6A6 (R_C54)	400	0.006	0.060	45
8620 (R_C63)	250	0.005	0.060	45
Iron, chilled (R_C60)	400	0.008	0.200	45
Iron, Meehanite (R_C56)	600	0.008	0.250	45
M-2 (R_C62)	250	0.005	0.060	45
Moly-Chrome (R_C68)	300	0.008	0.125	45
S5 (R_C60)	600	0.005	0.040	45
52100 (R_C70)	270	0.020	0.035	45
Colmonoy	600	0.006	0.125	15
Incoloy 901	800	0.006	0.125	45
Inconel 600	600	0.006	0.125	15
Inconel 718	600	0.006	0.125	45
K-Monel	600	0.006	0.125	15
René 41	600	0.006	0.125	45
René 77	500	0.006	0.015	71
René 95*	900	0.005	0.125	45
René 95**	450	0.005	0.125	45
Stellite	600	0.006	0.125	15
Waspaloy	600	0.003	0.060	45

**Hot isopressed　**Forged*

double the speed used with the carbide inserts. Similarly, feed has been increased to 0.008 ipr from the original 0.003 ipr and depth of cut has been increased from 0.010 inch to 0.040-0.080 inch. (These are the conditions on which the CBN insert edge life of 1000 pieces is based.)

Boring the ID required two passes with carbide inserts; the same operation is performed in one pass with CBN inserts.

All in all, switching from carbide tools to CBN tools has tripled hourly machine output. It has resulted in several quality improvements as well. Carbide inserts tended to bore oversize as their edges became dull. Dull cutting edges increase machining forces, and there is just enough spring in the toolholder/workholder system to result in tool deflection. Further, because of the ability of CBN edges to stay sharp, surface finishes have improved. Typical finishes produced by the carbide inserts were around 125 microinches (AA). Typical finishes produced by the CBN inserts are 32 microinches (AA).

Turning Superalloys. The gas turbine engine industry, the largest user of superalloy parts, has recognized the cost-saving potential of large CBN inserts. Typical operations at General Electric's Evandale, Ohio plant include boring and contour-turning of forged René 95 seal disks, contouring of René 95 compressor disks, and turning of several Inconel 718 parts. Material-removal rates attained on these materials with CBN inserts are five times higher than those attainable with carbide tools.

High material-removal-rate turning of superalloys is essentially a hot machining operation. At the temperatures attained, which are in the 1800-2000 F range, superalloys lose some of their strength and are easier to machine. CBN retains its room-temperature hardness and strength at these temperatures whereas carbide tools tend to fail in seconds.

In the finish turning of René 95 and Inconel 718, the surface finishes produced by CBN are usually 20-40 microinches (AA) as compared to 60-80 microinches for finishes produced by carbide inserts.

An added benefit of machining superalloys with these inserts is an improvement in the metallurgical integrity of the surface and subsurface of the machined part.

All machining operations tend to cause residual stresses in the first few thousandths of an inch under the surface. Tension stresses tend to reduce the fatigue life of the part; compression stresses are beneficial. There's evidence that the compressive stresses produced by turning with CBN inserts extend fatigue endurance limits well beyond those of parts machined with carbide inserts. The end result is an improvement in the reliability and service life of the part — important considerations in the manufacture of gas turbine engines.

Guidelines for Turning. Large CBN turning tools have demonstrated their ability to outperform conventional tools in numerous production turning operations. The following general guidelines reflect what has been learned from production experience in many plants.

► Use a rigid machine that's capable of fully utilizing the productive potential of CBN inserts. Best results are obtained when material-removal rates are an order of magnitude higher than those possible when turning with conventional tools.

► Don't attempt to turn easy-to-cut materials with CBN inserts. They're intended for cutting extremely hard or tough metals only.

► Make the turning setup as rigid as possible. Machining force for high-performance turning with these inserts may be more than double those for turning the same materials with carbide tools at conventional feeds and speeds. Keep tool overhang to a minimum — 1 to 1 1/2 inches at the most. Withdraw the tailstock spindle as much as possible, especially when machining large rolls. Centers and bearings should be in good condition.

► Make sure that the job is within the capacity of the machine. For example, don't put 80-inch diameter rolls on a machine designed to turn 60-inch rolls.

► To minimize edge wear, use negative-rake tools. Standard negative rake angles of 5 deg are usually satisfactory. In some cases, negative rake angles up to 15 deg give better tool life.

► Make sure that inserts have a 5 to 9-deg clearance with the workpiece.

► Choose the largest possible side-cutting edge angle. Avoid SCEA's under 15 deg.

► Avoid intermittent cuts.

► Use plenty of cutting fluid, and make sure that guarding is adequate to contain both coolant and chips. Usually, 20:1 water-soluble oil mixtures give satisfactory results.

► Stop the cut immediately if chatter is audible. Chatter indicates that the tool edge is dull or that the setup is not rigid enough.

The feed and speed recommendations given in the table are for roughing cuts. Finishes produced under these conditions are excellent, provided that the cutting edges are sharp. Which leads to another recommendation — don't attempt to machine with dull CBN inserts. When a uniform flank wear of 0.030 inch has been reached, it's time to index or change the tool.

When taking finishing cuts, higher speeds and lower feeds than used for roughing cuts are normally required. As with rough turning, an order-of-magnitude greater productivity can be obtained when conventional tools are replaced with CBN tools. And the surface finishes produced by sharp CBN tools are sufficiently good that the need for subsequent grinding operations may be eliminated.

Insert Selection. Standard CBN inserts are available as 1/2 and 5/8-inch rounds and squares, and as 1/2-inch triangles. Squares are made with 1/32, 3/64 and 1/16-inch nose radii. Triangles are offered with 3/64 and 1/16-inch nose radii only.

Because of their shape, rounds are the strongest inserts, and they are recommended for severe operations such as the turning of hardened steel and chilled cast iron rolls at heavy material-removal rates. Where part geometry dictates the use of square or triangular inserts, inserts with the larger nose radii are best for roughing, because of their high mechanical strength. Somewhat finer finishes can be obtained when the smaller nose radii inserts can be used.

Because of their high cutting efficiency, the inserts can be effectively put to work on lathes having normal horsepower. And they can be mounted in standard toolholders. Of importance, CBN inserts cost substantially more than conventional inserts. But they can pay for themselves in demanding applications.■

Reprinted courtesy of the General Electric Company

BZN Case Histories And Application Briefs

TURNING HARD FLAME-SPRAYED ALLOYS WITH BZN® INSERTS TRIPLES LATHE OUTPUT AT PRECISION PRODUCTS.

Precision Products Company, Grove City, Ohio, has pioneered in the application of tools with BZN inserts for the turning of small parts from difficult-to-machine materials.

A case in point is an insertable valveseat ring with a layer of flame-sprayed material – Colmonoy No. 43 – with a Rockwell C hardness of 55.

Excess flame-sprayed material was removed in two operations – plunge cutting and finish cutting the face, which has a 45-degree chamfer, and boring the ID – when triangular cemented carbide inserts were used. For these operations, tool life averaged three pieces per edge.

Machining Conditions: face, chamfer, bore

	Carbide insert	BZN insert
Material hardness (R_C)	55	55
Tool shape	TNG-433	BTNG-433
Turning speed	142 sfpm	295 sfpm (90 Mpm)
Feed	0.003 ipr	0.008 ipr (0.2 mmpr)
Depth of cut	0.020-0.040"	0.040-0.080 (1.0-2.0 mm)
Clearance angle	5	5
Back rake angle	5 (negative)	5 (negative)
Tool change time (minutes)	5	5
Tool life (pieces per edge)	3-4	1000
Surface finish	125 rms	32 rms

With General Electric BZN inserts, tool life averages 1000 pieces per edge, and each operation requires only one pass.

Geometry of both tools is the same. The machine is a 7.5-horsepower Clausing lathe. The cutting fluid is a 15:1, water: soluble-oil mixture.

Besides decreasing the frequency of tool changes, BZN inserts have made it possible to greatly increase feeds, speeds and depths of cut. Boring the ID, which took two passes with carbide inserts to achieve straightness, now takes only one pass. Result: Machine productivity is tripled. Added benefit; smoother finishes with BZN inserts.

®Trademark of General Electric Company, USA.

OPERATION

Location: Japan
Machining Operation: O.D. Turning
Part Description: Long Sleeve
Machined Area Dimensions: ⌀ 300 x 2000 mm (11.8″ diam. x 78.7″)
Material Specifications: Ni: 67% Cr: 16% C: 0.5% Others: Si, B, Fe, Cu, Mo
Hardness: HRc 58-62
Prior Machining Method: Cutting
Tool Type: Tungsten Carbide (K 05)

MACHINING CONDITIONS	BZN Compact	Tungsten Carbide Tool
Cutting Speed	35 m/min. (115 sfpm)	< 10 m/min. (< 33 sfpm)
Feed Rate—per Revolution	0.2 mm (0.008″)	0.2 mm (0.008″)
Depth of Cut	0.2 mm (0.008″)	0.5 mm (0.020″)
TOOL GEOMETRY		
Tool Type — Description	BZN Blank Tool	
Nose Radius	1.0 mm (0.040″)	3.0 mm (0.118″)
Back Rake Angle	+ 3°	0°
ECONOMICS		
Cutting Edges per Tool	1	1
Tool Change Time	3 min.	3 min.

REMARKS

- Tungsten Carbide tools were changed an average of three times to finish a part, due to edge breakdown. The BZN Compact Tool cut well with little wear on the edges—even after cutting the entire 2,000 mm length, and was good for keeping tolerance of taper.

OPERATION

Location: France
Machining Operation: Boring (I.D. Turning Operation)
Part Description: Roller Bearing
Machined Area Dimensions: ⌀ 130 mm x 73 mm (5.12″ diam. x 2.87″)
Material Specifications: 100 Cr6 (similar to AISI 52100)
Hardness: 62 HRc
Machine Type: TUAM Size: 25 CV
Prior Machining Methods: Grinding
Wheel Specifications: Aluminum Oxide

MACHINING CONDITIONS	BZN Compact
Cutting Speed	80 m/min. (262 sfpm)
Feed Rate — per Revolution	0.2 mm (0.008″)
Depth of Cut	2.5 mm (0.098″)
Coolant	Soluble Oil
TOOL GEOMETRY	
Tool Type — Description	BSNG-433
Back Rake Angle	-5°
Side Rake	-5°
End Clearance	5°
ECONOMICS	
Cutting Edges per Tool	4
Pieces Machined per Edge	8
Tool Change Time	3 min.
Workpiece Change Time	5 min.

REMARKS

- Turning with the BZN Insert Tool has eliminated the previous average scrap rate of 28 pieces per day.

OPERATION

Location: France
Machining Operation: Facing
Part Description: Ratchet-Reel
Machined Area Dimensions: ⌀ 102 - 80 mm (4″ - 3.15″ Diam.)
Material Specifications: Stellite* III
Hardness: 53 HRc
Machine Type: Iternault Somma **Size:** Cholet 550
Prior Machining Method: Cutting, Grinding
Tool Type: Tungsten Carbide/Aluminum Oxide
Finish Specifications Required: 0.9 μm (35μ″ AA)

MACHINING CONDITIONS	BZN Compact	Tungsten Carbide Tool
Cutting Speed	135 m/min. (443 sfpm)	48 m/min. (157 sfpm)
Feed Rate—per Revolution	0.1 mm (0.004″)	0.1 mm (0.004″)
Depth of Cut	1.5–2 mm (0.060″ x 0.080″)	1.5–2 mm (0.060″ x 0.080″)
Coolant	Soluble Oil	Soluble Oil
Finish Obtained	0.6 μm (23μ″AA)	
TOOL GEOMETRY		
Tool Type — Description	BZN Blank Tool (#6325)	
Nose Radius	1.5 mm (0.060″)	
Back Rake Angle	-5°	
Side Rake	-5°	
ECONOMICS		
Cutting Edges per Tool	1	8
Pieces Machined per Edge	40	0.25
Tool Change Time	15 min.	15 min.
Workpiece Change Time	5 min.	5 min.

REMARKS

- Tungsten Carbide Tool could machine 3 mm maximum. Grinding operation was necessary to obtain finish. With BZN Compact Tool, size and finish were obtained with only one pass.

OPERATION

Location: Japan
Machining Operation: O.D. Turning
Part Description: Drum
Machined Area Dimensions: ⌀ 1575 x 273 mm (62″ diam. x 10.75″)
Material Specifications: Martensitic Wear Resistant Steel
Hardness: Hs 72-74
Machine Type: Vertical Lathe —30 HP
Prior Machining Method: Cutting, Grinding
Tool Type: Tungsten Carbide (K 05)
Wheel Specifications: (Tool Post Grinder) Aluminum Oxide Wheel

MACHINING CONDITIONS	BZN Compact	Tungsten Carbide Tool
Cutting Speed	150 m/min. (492 sfpm)	20 m/min. (65.6 sfpm)
Feed Rate—per Revolution	0.1 mm (0.004″)	0.7 mm (0.027″)
Depth of Cut	0.15 mm (0.006″)	0.5 mm (0.020″)
Finish Obtained	S: 3-7 μm (15-30μ″AA)	S: 50 μm (200μ″AA)
TOOL GEOMETRY		
Tool Type — Description	BZN Blank Tool: 60° (#6320)	
Nose Radius	0.8 mm (0.031″)	0.8 mm (0.031″)
Back Rake Angle	0°	-5°

REMARKS

- Replacement of the tungsten carbide tool with BZN Compact Tool eliminated the 6 hour grinding operation and improved surface finish.
  Material Composition: – Fe, C: 0.4, S: 0.85, Mn: 2.0, Cr: 6.4, Mo: 0.95, W: 1.7.

OPERATION

Location: France
Machining Operation: Facing (Face Turning)
Part Description: Gas Turbine
Machined Area Dimensions: ⌀ 64 mm (2.52" diam.)
Material Specifications: Rene*-77
Machine Type: Cazeneuve
Prior Machining Method: Cutting/Grinding
Tool Type: Tungsten Carbide/Aluminum Oxide
Finish Specifications Required: 0.8 μm RA (32μ"AA)

MACHINING CONDITIONS	BZN Compact	Tungsten Carbide Tool
Cutting Speed	125 m/min. (410 sfpm)	12 m/min. (40 sfpm)
Feed Rate — per Revolution	0.1 mm (0.004")	0.2 mm (0.008")
Depth of Cut	1.4 mm (0.055")	1.4 mm (0.055")
Coolant	Soluble Oil	Soluble Oil
Finish Obtained	0.8μm RA (32μ"AA)	>2μm RA (>77μ"AA)
TOOL GEOMETRY		
Tool Type — Description	BTNG-433	
Back Rake Angle	-5°	
Side Rake	-5°	
ECONOMICS		
Cutting Edges per Tool	3	1
Pieces Machined per Edge	6	0.5-1
Tool Change Time	3 min.	10 min.
Workpiece Change Time	5 min.	5 min.

REMARKS

- BZN Compact Tool reduced cutting time by 90%.
- Subsequent grinding time required after turning with the BZN Compact Tool has also been reduced 50%.

OPERATION

Location: Canada
Machining Operation: Facing
Machined Dimensions: 1.524M (5' diam.)
Material: Monel* 418, Inconel* 600
Hardness: Highly abrasive. Approx. 40 Rc.
Machine Type: Rotary Table Vertical Borer
Type Tool Replaced: Tungsten Carbide
Insert Used: BRNG-43

MACHINING CONDITIONS	Tungsten Carbide	BZN Insert
Rotational Speed (RPM)	3	40
Cutting Speed (Max. Diameter)	14.3 m/min (47 sfpm)	192 m/min (628 sfpm)
Feed/Revolution (inch/rev.)	.254 mm (.010")	.254 mm (.010")
Depth of cut	.508 mm (.020")	.508 mm (.020")
Coolant	Dry	Dry
TOOL GEOMETRY		
Nose Radius	Full Round	Full Round
Cutting Rake	5° Neg.	5° Neg.
ECONOMICS		
Tool Cost per Part	x	5x
Time to machine (include tool change)	15 hrs.	1½ hrs.

REMARKS

- BZN Inserts machined to center regardless of diminishing cutting speed — three indexes necessary to maintain .002 metric flatness tolerance.
- Tungsten Carbide tools indexed approximately fifteen times to maintain same tolerance, but still inferior to BZN Insert on ease of pick up on new cut.
- Major savings: Machining time and quality of surface.

OPERATION

Location: France
Machining Operation: O.D. Turning
Part Description: Punch Axes
Machined Area Dimensions: ⌀ 40 x 25 mm (1.57" diam. x 1")
Material Specifications: T-15
Hardness: 60 HRc
Machine Type: Cazeneuve
Prior Machining Method: Grinding (Aluminum Oxide Wheel)

MACHINING CONDITIONS	BZN Compact
Cutting Speed	90 m/min. (295 sfpm)
Feed Rate—per Revolution	0.6 mm (0.023")
Depth of Cut	0.8 mm (0.032")
Coolant	Soluble Oil
Finish Obtained	1.0 μm Ra (39μ"AA)
TOOL GEOMETRY	
Tool Type — Description	BTNG-433
Back Rake Angle	-5°
Side Rake	-5°
End Clearance	5°
ECONOMICS	
Cutting Edges per Tool	3
Pieces Machined per Edge	35
Tool Change Time	3 min.
Workpiece Change Time	5 min.

REMARKS

- Turning with BZN Compact Tool has allowed for continued production without the purchase of a new grinding machine.

OPERATION

Location: France
Machining Operation: O.D. Turning
Part Description: Machine Tool Control System
Machined Area Dimensions: ⌀ 25 x 50 mm (1" diam. x 2")
Material Specifications: Inconel* 600
Machine Type: Hardinge
Prior Machining Method: Grinding (Aluminum Oxide Wheel)

MACHINING CONDITIONS	BZN Compact
Cutting Speed	125 m/min. (410 sfpm)
Feed Rate—per Revolution	0.04 mm (0.0016")
Depth of Cut	0.005 mm (0.0002")
Coolant	Soluble Oil
Finish Obtained	0.38 μm Ra
Size, Taper Tolerances	< 1μm (< 0.00004")
TOOL GEOMETRY	
Tool Type — Description	BZN Blank Tool (#6320)
Nose Radius	0.6 mm (0.024")
Back Rake Angle	-5°
Side Rake	-5°
ECONOMICS	
Cutting Edges per Tool	1
Pieces Machined per Edge	25
Tool Change Time	15 min.
Workpiece Change Time	10 min.

REMARKS

- Machining at low feed rates, the BZN Compact Tool was able to maintain close tolerances much longer than the aluminum oxide wheel.

OPERATION

Location: France
Machining Operation: O.D. Turning
Part Description: Broach
Machined Area Dimensions: ⌀ 63 mm x 600 mm (2.5" diam. x 23.6")
Material Specifications: Z 80 WD 06
Hardness: 64 HRc
Machine Type: Cazeneuve
Prior Machining Method: Grinding (Aluminum Oxide Wheel)

MACHINING CONDITIONS	BZN Compact
Cutting Speed	79 m/min. (260 sfpm)
Feed Rate—per Revolution	0.8 mm/rev. (0.032"/rev.)
Depth of Cut	1.8 mm (0.071")
Coolant	Soluble Oil

TOOL GEOMETRY	
Tool Type — Description	BZN Blank Tool (#6325)
Nose Radius	1.0 mm (0.040")
Back Rake Angle	-5°
Side Rake	-5°

ECONOMICS	
Cutting Edges per Tool	1
Pieces Machined per Edge	2 (8 passes each)
Tool Change Time	12 min.
Workpiece Change Time	30 min.

REMARKS

- Turning with BZN Compact Tool reduced by 50% the time required to remove "great overthickness" of material, when compared to the length of time it took when grinding with the aluminum oxide wheel.

OPERATION

Location: France
Machining Operation: O.D. Turning
Part Description: Piston Body
Machined Area Dimensions: ⌀ 230 mm x 280 mm (9" diam. x 11")
Material Specifications: Stellite*
Hardness: 52 HRc
Machine Type: Cazeneuve
Prior Machining Method: Cutting, Grinding
Tool Type: Tungsten Carbide/Aluminum Oxide Wheel

MACHINING CONDITIONS	BZN Compact	Tungsten Carbide Tool
Cutting Speed	108 m/min. (354 sfpm)	43 m/min. (141 sfpm)
Feed Rate	0.2 mm (0.008")	0.2 mm (0.008")
Depth of Cut	1.5 mm (0.060")	1.5 mm (0.060")
Coolant	Soluble Oil	Dry
Finish Obtained	0.8 μm Ra (32μ"AA)	

TOOL GEOMETRY	
Tool Type — Description	BZN Blank Tool (#6325)
Nose Radius	1.0 mm (0.040")
Back Rake Angle	-5°
Side Rake	-5°

ECONOMICS	
Cutting Edges per Tool	1
Pieces Machined per Edge	12
Tool Change Time	15 min.
Workpiece Change Time	10 min.

REMARKS

- The grinding operation has been eliminated, and the time required for the turning operation has been reduced by 70%, since the user began turning with the BZN Compact Tool.

OPERATION

Location: France
Machining Operation: O.D. Turning
Part Description: Tooling
Machined Area Dimensions: ⌀ 40 mm x 100 mm (1.57" diam. x 3.94")
Material Specifications: 100 Cr6 (similar to AISI 52100)
Hardness: 59 HRc
Machine Type: Cazeneuve
Prior Machining Method: Grinding (Aluminum Oxide Wheel)

MACHINING CONDITIONS	BZN Compact
Cutting Speed	90 m/min. (295 sfpm)
Feed Rate—per Revolution	0.1 mm (0.004")
Depth of Cut	3 mm (0.118")
Coolant	Soluble Oil

TOOL GEOMETRY	
Tool Type — Description	BSNG-433
Back Rake Angle	-5°
Side Rake	-5°
End Clearance	5°

ECONOMICS	
Cutting Edges per Tool	4
Pieces Machined per Edge	35
Tool Change Time	3 min.
Workpiece Change Time	10 min.

REMARKS

- Use of the BZN Compact Tool eliminated rough grinding operation on parts which were both out of round and also 3 to 5 mm (0.118"-0.196") oversize.

OPERATION

Location: France
Machining Operation: I.D. Boring
Part Description: Punch Rings
Machined Area Dimensions: ⌀ 41.3 mm x 60 mm (1.62" diam. x 2.36")
Material Specifications: 285 WD V060602
Hardness: 62 HRc
Machine Type: Les Inovation Mecaniques
Prior Machining Method: Grinding (Aluminum Oxide Wheel)

MACHINING CONDITIONS	BZN Compact
Cutting Speed	104 m/min. (341 sfpm)
Feed Rate—per Revolution	0.08 mm (0.003")
Depth of Cut	0.7 mm (0.028")
Coolant	Soluble Oil

TOOL GEOMETRY	
Tool Type — Description	BZN Blank Tool (#6320)
Nose Radius	0.8 mm (0.031")
Back Rake Angle	0°
Side Rake	0°

ECONOMICS	
Cutting Edges per Tool	1
Pieces Machined per Edge	25
Tool Change Time	15 min.
Workpiece Change Time	10 min.

REMARKS

- Use of the BZN Compact Tool eliminated the grinding operation.

BZN® COMPACTS EDGE OUT CARBIDES FOR MACHINING CHILLED CAST IRON ROLLS AT UNITED STATES STEEL'S CANTON PLANT

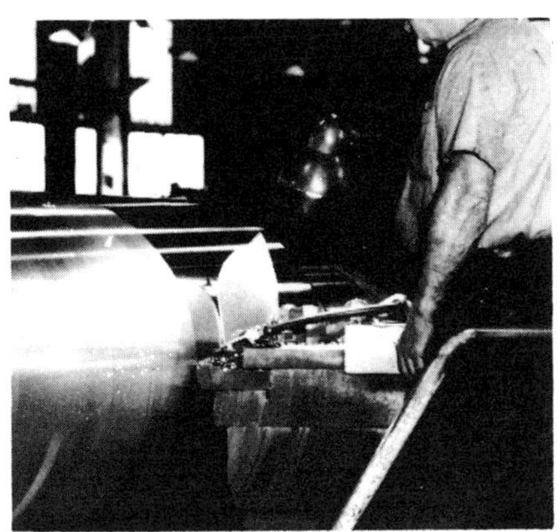

Until recently, machining of chilled gray cast iron steel-mill rolls was a slow, costly process at United States Steel's Canton, Ohio, plant.

Chilled cast iron is as hard as some hardened tool and die steels (Rockwell C 61-62). And it contains cementite particles that are as hard as the carbides used in cemented carbide tools and dies.

When Grade C-3 cemented carbide inserts were used to turn the roll bodies, cutting edges dulled so fast that it was impossible to attain satisfactory material-removal rates.

The machine—a 100-horsepower lathe—had plenty of power. But cutting speeds had to be held down to 45-50 sfpm to protect the tool. As a result, the material-removal rate was only 2.0 cubic inches per minute—far below the potential material-removal capability of the lathe.

Because of tool wear, the operator had to stop cutting metal frequently in order to change tools and make machine adjustments to compensate for tool edge wear.

Even with frequent tool adjustments, it was impossible to hold size with carbide tools. Consequently, rolls had to be transferred to a grinder for final sizing—an 8-hour operation.

In order to increase productivity and reduce costs, United States Steel tried out BZN insert tools. These tools, designed for efficient cutting of hardened tool and die steels, difficult-to-machine cast irons and tough superalloys, have solved the problem.

General Electric BRNG-53 BZN inserts are used. These are negative-rake, ⅝-inch-diameter rounds, $\frac{3}{16}$ inch thick.

With BZN inserts, United States Steel is cutting at much higher speeds—275-300 sfpm. The material-removal rate has been more than doubled to 4.8 cubic inches per minute.

Despite this high material-removal rate, life of the BZN insert tools is four times the life of the carbide tools.

Because of the slow wear of BZN tools, size control is easy. Even on the longest (130-inch long) roll body, size is consistent within 0.005 inch from one end to the other.

With good size control, United States Steel has been able to eliminate the grinding operation.

So besides more than doubling lathe output, the BZN inserts are saving 8 hours of grinding time per roll.

Machining Conditions and Economics

	Grade C-3 Carbide Tool	BRNG-53 BZN Insert Tool
Depth of cut, inch (mm)	0.125-0.180 (3.2-4.6)	0.090-0.100 (2.3-2.5)
Speed, sfpm (mpm)	45-50 (14-15)	275-300 (84-91)
Feed, ipm (mmpr)	0.020-0.030 (0.5-0.8)	0.012-0.020 (0.3-0.5)
Material-removal rate, cubic inches per minute (cc per min)	2.0 (32.8)	4.8 (78.7)
Turning time, hours	44.1	37.5*
Grinding time, hours	8	not required
Total machining time, hours	52.1	37.5
Machining time savings, hours	—	14.5
Relative tool life	X	4X

*35 hours = removing scale with tungsten carbide
2½ hours = finish cut with BZN inserts

Ⓣ Trademark of General Electric Company, U.S.A.

TURNING HARDENED STEEL ROLLS WITH BZN® INSERTS SIGNIFICANTLY REDUCES MACHINING TIME AT TOOL STEEL GEAR AND PINION

Manufacturing engineers at Tool Steel Gear & Pinion Company, Sharonville, Ohio, used to consider grinding the only possible way to machine hardened steel rolls.

Grinding produced satisfactory finishes and adequate dimensional accuracy, but it was a slow operation. Rough grinding a typical roll with a ground length of 80 inches and a rough-ground diameter of 25 inches, for example, took approximately 6 hours.

Conventional turning tools wear so fast, that machine operators would have spent more time changing tools than cutting metal.

The material was a modified 52100 steel, hardened to Rockwell C 68. About 0.090 inch of stock was removed from the diameter in rough grinding.

When the company tried turning the same roll with a standard round General Electric BZN insert, grinding suddenly became an obsolete process for the rough machining operation. However, rolls are still required to be finish ground.

Rough Turning Conditions using BZN insert

Material	Modified 52100 steel
Material hardness (R_C)	68 HR_C
Operation	Turn OD
Tool	BRNG-43
Turning speed	275 sfpm (84 Mpm)
Depth of cut	0.045" avg (1.1 mm)
Feed	0.022 ipr (0.6 mmpr)
Rake angle	5 (negative)
Clearance angle	5
Coolant	20:1 soluble oil

Turning instead of grinding the O.D. reduced machining time by approximately four hours. Overall time reduction for all machining operations was reduced by 10-15%, depending on the part geometry.

A 25-horsepower American Tool Works lathe is used for the turning operation. The cutting fluid is a 20:1, water:soluble-oil mixture.

Besides turning hardened rolls with BZN inserts, Tool Steel Gear & Pinion plans to use them for turning other products. In fact, all hardened parts that are currently being ground are considered to be potential applications for tools with BZN inserts.

BZN COMPACTS APPLICATION DATA

NI HARD CAST IRON MACHINING

COMPANY A-S-H PUMP
HAMBURG, PA USA

MATERIAL TYPE NI HARD CAST IRON (53-56 Rc)

MACHINING APPLICATIONS METAL REMOVAL

MACHINE TYPES VARIOUS VERICAL AND HORIZONTAL LATHES

PARTS MACHINED LARGE INDUSTRIAL PUMP IMPELLERS, CASINGS, LINERS

PREVIOUS METHOD AL_2O_3 GRINDING WHEELS

BZN COMPACT TOOL TYPE BRNG-43 INSERT

TOOL GEOMETRY INSERTS USE FOLLOWING EDGE PREPARATION:
10 CHAMFER (K LAND)
0.6 - 0.8 MM WIDE
0.13 RADIUS HONE

STANDARD SHARP EDGE TENDS TO ALLOW EDGE TO FRACTURE AND TOOL PERFORMANCE IS INCONSISTENT

MACHINING CONDITIONS:

CONTINUOUS CUT

SPEED	100 M/MIN
DOC	0.7 - 2.0 MM
FEED	0.25 - 0.5 MM
COOLANT	NONE

INTERRUPTED CUT

SPEED	30 - 60 M/MM
DOC	0.7 - 2.0 MM
FEED	0.25 - 0.5 MM
COOLANT	NONE

BZN COMPACT TOOL LIFE 6-8 HOURS, THEN BLANK IS GROUND TO SMALLER IC

COMMENTS: GRINDING NI-HARD IS SLOW AND LABOR INTENSIVE. A-S-H HAS DISCONTINUED ALL GRINDING OPERATIONS AT HAMBURG OPERATION. BZN INSERTS ARE USED FOR STRAIGHT TURNING, BORING AND SURFACING OF PUMP COMPONENTS. A-S-H PERSONNEL ESTIMATE AN ANNUAL SAVINGS OF $400 - 500,000 WITH BZN INSERTS.

WORLD MARKET FOR HARDENED FERROUS AND
HIGH STRENGTH/HIGH TEMPERATURE ALLOYS

Market Segment	Key Parts	Typical Material
Automotive	Gears Shafts Valves & Seats	5620H 8620 Carborized High Cobalt Steels Cast Iron Nickle Chrome Steels High Cobalt
Aerospace	Blades Discs Shafts	Inconel 718 In-100 Hipped Rene 95 Hipped 4140, 4340 Steels
Steel Mill	Hardened Rolls	Cast Iron
Bearings	Races Rolls Balls	Bearing Steels 5400 4120 4300
Pumps & Valves	Wear Parts Surfacing	Ni-Hard Cast Iron Stellite Waspalloy Chromalloy

Presented at the Industrial Diamond Association of America's Diamond—Partners in Productivity Conference, 1974.

DIAMONDS IN GLASS INDUSTRY

WILLIAM M. DeANGELIS, GENERAL MANAGER
PENSAR, INC.
NORWALK, CONNECTICUT

Over the last few years the Corning Glass Company has introduced a glass ceramic material to the industry which can be worked with ordinary machine tools; that is, machined on a lathe with an ordinary tool bit, cut with a hack saw, milled with an end mill and drilled with a twist drill, much the same as one would work cast iron or with aluminum, and as a matter of fact, the machining characteristics are very similar to that of aluminum. However, most of the glasses which are used in the scientific and industrial community are not so submissive as the one just discussed. Yet, if glass is to be used in more applications, techniques must be developed to allow the glass worker to meet the more stringent requirements, which industry and research demand.

The principal area to be covered by this paper is the forming of small holes in glassy materials with mention of the many related techniques which have been used over the years to form openings in tubes, rod and plate, with particular discussion of our experience utilizing the various types of diamond drills, including the rotary ultrasonic milling machine.

Most glass workers are familiar with the technique of forming holes in the walls of tubing and plate by heating the glass to a temperature where the glass is soft enough to move so that the insertion of a piece of carbon or tungsten into the heated area forms a hole. Unfortunately, since a fire is used, the area adjacent to the hole is also heated, and the immediate area being deformed. This does not land itself to close tolerance hole location or size. Prior to the advent of the diamond drill and still in many places in industry today, a sharpened carbide securely mounted into the chuck of a drill press well flushed with coolant has long been the mainstay of the man trying to drill a hole. This process works fairly well for holes larger than 2mm, and smaller than 6mm. However, due to the point pressure required and limited contact area, there is usually considerable chipping at the entrance of the hole and extensive chipping on the exit. If there are secondary operations such as for the grinding of the surface or fire polishing, perhaps this is no problem. However, the last operation where tight tolerances are required both to hole location and size, the carbide hole drilling technique will not prove satisfactory.

With the advent of the diamond drill, hole drilling has become easier. Diamond impregnated tools and diamond plated tools can be finished to close tolerances and the size of the hole more closely approximates the size of the tool. Diamond drills provide the user with the choice of types which would have economic advantages depending on the number of holes which have to be drilled and the finish which is desired on the internal diameter of the hole. If the user wants a few drilled holes, it is possible to select a plated drill. For production an impregnated drill, more expensive perhaps, but capable of hundreds or even thousands of holes of a range of sizes from .010" to as large as you make like. The principal difficulty experienced has been keeping the face of the drill lubricated, cooled if you like. Solid drills in particular have this problem since the working portion of the drill is the least likely to be in contact with the coolant. If the drill is of a plated variety the heat may become so intense as to cause the plating to strip from the matrix leaving the operator with only the non cutting metal, much like drilling through hard material with a section of uncharged brass pipe. Removal of the diamond abrasive can also occur in the impregnated drills by the same forces of face pressure and resultant heat generated by the rotational force of the drill. It is very important to keep the diamonds in the matrix cool if you expect a reasonable tool life. The major advantage of impregnated drills is that they can be redressed by grinding across the cutting face of the drill so that when the dressing is done, providing fresh diamonds are removing a portion of the matrix which supports them.

One answer to the problem of coolant has been supplied by the core drill. In the core drill the coolant is forced through a coupling device held on the spindle, then through the tube containing the bonded abrasive past the cutting face along the outside of the core and out of the work piece. Diamond drills, for all they can provide in increased drilling speed, hold size and location uniformity, still leave the problem of breakout, especially if it is difficult to back up the material being drilled. There is a newly added problem, that of the broken core lodging in the drill causing the tool to be clogged, damaged or made otherwise useless. Furthermore, core drills are usually not available in sizes less than .030". In conventional ultrasonic drilling, a tool is mounted to a transducer so that ultrasonic vibrations of the tool itself impart to an abrasive slurry which wears into the hard material. This is a kind of an extension of a method which used a brass tool with an abrasive back to drill holes, a basic technique which has been used throughout the course of history. The difference is that the ultrasonic circuit is tuned to the mass and length of the tube so that maximum energy is focused at the work point. The tool, in intimate contact with abrasive slurry abrases the surface which is in contact with the tool. Unfortunately, not only the work piece is abraided, whether it is glass, quartz, ceramic or some other crystalline material, but as the work piece wears in the shape of the tool, so also does the tool wear. If the tool is shaped like a rod for the purpose of drilling round holes, it will soon acquire a tapered configuration so that the hole size will change and all sharp corners soon become rounded.

On the other side of the coin, ultrasonic drilling allows a number of holes to be drilled at the same time with relative center locations as described as the tool. It also allows one to grind any shape where a reverse image tool can be provided. Metal tools can be duplicated easily by electro forming. Simple hole drilling tools can be manufactured of cold rolled steel, tolerances held to .005" or less. The surface of the work area, will have the characteristics of the grain size of the abrasive used. And of course, diamond slurry can be used when required. There are two other considerations when using conventional ultrasonic drilling; abrasive slurries are difficult to handle and the abrasive must be distributed evenly over the face of the tool to maintain adequate cutting speed. There are limitations on tool size, since the ultrasonic energy of the transducer is limited. Manufacturers will state the maximum size of the tool which their system can utilize. Larger tools will either not function efficiently or can break the transducer and that is an expensive repair bill. Another consideration is that the abrasive packing reduces the cutting speed. The slurry should be kept wet and if water is used sometimes the wetting agent with a slightly alkaline PH which increases the surface hydrolysis to stop the speed of cutting action on glassy materials. Since it is the abrasive that does the cutting the abrasive should be fresh, since used abrasives tend to cut more slowly. If it is possible, light metal such as titanium, is used for tools and tool holders.

If conventional ultrasonic tools are to be used for hold drilling, the problem of the tapered hole must be considered and frequent tool dressing will be required. The smallest size hold which can be drilled with conventional ultrasonics is limited by the diameter of the tooling and the technique used to apply slurry to the tool and to the work without deforming the tool. Grinding speed can be expected to be slow with small diameters since flexibility causes loss of energy.

Use of rotary motion in conjunction with ultrasonincs provides the technician with a new tool to hold tight tolerances. The use of the rotary tool with ultrasonics means that holes can be drilled longer with less face pressure resulting in faster drilling, reduced tool wear and straighter holes. This means that it becomes possible to build long straight holes since less force is required which would bend the delicate shanks of drills.

Small holes can be drilled quite effectively with the use of high speed spindles and sensitive drill presses without the use of ultrasonics or other complicated devices. To be assured, the success of the user of diamond tools is dependent upon the technique. The technique which is found to be effective by a number of users is an intermittent drilling with extremely light face pressure for a very small time cycle, a time cycle where perhaps two incursions per second are made with an operator observing a dial indicator to advise the depth of each incursion as well as the desired final depth to be achieved. This sort of drilling is usually done submerged so that both the material and the drill are well cooled at all times.

Drilling with small drills from .020" to .040" requires a high speed sensitive drill press and a sturdy, well-placed base. The recommended speeds for drills of the plated variety in this size range is from 8 to 12 thousand rpm. When impregnated drills are used, higher speeds can also be used. While higher speeds have the advantage of drilling at faster rates, they also require more careful usage, since the twisting movements and shock which occurs as diamond and matrix strike material to be drilled can tear the diamonds away. As the drill gets smaller, the number of cutting points are reduced, if the grit size is constant, and the loss of coolant or excessive face pressure can quickly damage the drill.

Lubrication, whether lubricating oil, kerosene or water with a wetting agent is a necessity. Research indicates thatthe longest diamond life and most effective coolant is oil. The higher the concentration, the better. Since oils may not be desirable since they require extensive cleaning, our own firm has used ethylene glycol with success. This is water soluable and inhibits rust.

In summary, small holes can be drilled with success at a very low cost per unit when the proper equipment and tooling are used with the proper technique which consists of a careful, patient, confident, experienced operator using appropriate drilling speeds and coolants.

WILLIAM M. DeANGELIS

A native of Pennsylvania, William DeAngelis received a B.S. from the University of Pittsburgh in 1955. After graduation, he joined Fischer & Porter as a salesman in precision glass parts. In 1972 he founded Pensar, a firm which manufactures state of the art precision glassware for instrumentation and electronics and diamond cutoff and grinding machinery.

CHAPTER 3

ABRASIVE MACHINING

Reprinted from Philips Technical Review, Volume 38, 1978/79

Grinding brittle materials

A. Broese van Groenou and J. D. B. Veldkamp

When a metal is ground the abrasive particles 'plough' through the material to form well-defined, 'clean' grooves. If the material is a brittle one, however, there is chipping and cracking in the neighbourhood of the abrasive particles. This is an advantage if the material is to be removed as efficiently as possible, but a disadvantage if a good surface finish is required. In setting up a grinding programme a better understanding of the formation of grooves, cracks and fractures is therefore highly desirable. The investigation described below contributes to this understanding. Since grinding is a 'multi-point scratching process' it will only give a simplified picture of the effects to be studied; 'single-point' experiments were found to be of greater value in the investigation.

Introduction

Grinding is necessary as the final process operation for many ceramic and glass products. It is often difficult to produce shapes accurately by the ceramic process of pressing and firing, and grinding is then the only way of bringing the product to the required shape and size. Let us look at two examples. In d.c. motors for windscreen wipers the stator magnet consists of segments of Ferroxdure (*A* in *fig. 1*), which are ground to the correct radii of curvature. In television receivers the magnetic deflection system includes a Ferroxcube yoke (*B*) fitted around the picture tube; this yoke is ground to the correct shape. Sometimes layers more than 1 mm thick have to be removed in this 'coarse grinding'.

'Finish grinding' is generally used where the optical or magnetic quality or the strength of the surface is important. For example, the magnetic quality of the pot cores used in the *LC* circuits for carrier telephony (*C*) is strongly dependent on the nature of the surface at the air gap. In magnetic heads for a video-tape recorder (*D*) a good surface is also a first requirement, both for the magnetic quality and to keep the spacing between head and tape as small as possible. Television picture-tube screens are ground to give a good optical finish, but the strength aspects are also important here. The strength of a brittle object is often dependent on the depth of surface cracks; finish grinding removes the cracks and increases the strength.

Grinding is a costly operation; the machines are expensive and the grinding takes a long time. The cost of grinding the magnet segments in fig. 1 is 20% of the production costs, and for the picture-tube yokes the corresponding figure is no less than 40%.

Detailed analysis shows that grinding is a complicated process. As the abrasive particles scratch the surface of the material several different microprocesses come into play (flow, cracking, fracture), and the result depends on a number of factors, including the shape, the force and the velocity of the particles, the humidity of the environment, the hardness and the crack resistance of the material and the distribution of microcracks, cavities and other flaws in the material. In view of the importance and the expense of grinding it seemed useful to make a closer analysis of this complicated process. This article gives a general picture of our investigations of the process. They mainly relate to the grinding of ferrites and glass; because these materials are very hard diamond was always used for the abrasive particles. In general terms the objective of the

Dr A. Broese van Groenou is with Philips Research Laboratories, Eindhoven; Dr Ir J. D. B. Veldkamp, formerly with these Laboratories, is now with the Philips Glass Division, Eindhoven.

investigation was to find out how the *specific energy*, i.e. the energy required for the removal of unit volume of the material, and the *damage to the surface* of the workpiece depend on the setting of the grinding machine and the properties of the material.

grinding conditions formed an important part of our investigations.

An essential feature of the grinding process is the force F that the wheel exerts on the workpiece (see fig. 2). This force largely determines the amount of

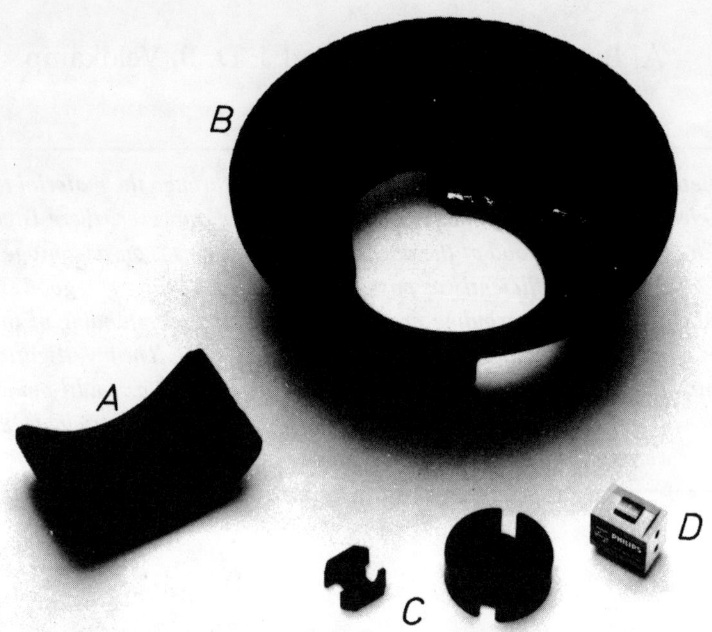

Fig. 1. Some examples of ground ceramic products. *A* Segment of the stator magnet in a windscreen-wiper motor; Ferroxdure. *B* Yoke for the magnetic deflection system of a television receiver; Ferroxcube. These products are brought to the required shape by coarse grinding. *C* Half pot cores for tuning circuits in carrier telephony; Ferroxcube. The surface of the central cylinder and the corresponding surface of the other half form an air gap whose purpose is to reduce the temperature coefficient to a low value; these two surfaces are finish ground to give good magnetic quality. *D* Magnetic head for video tape recorder; Ferroxcube. The gap width is ½ μm. Finish grinding is required here to give good magnetic quality and a well-defined spacing between head and tape.

The characteristic quantities of the grinding system

Let us consider a surface grinder as a representative example of a grinding system (*fig. 2*). The rate of volume removal Z — the volume ground away per second — is the product of the depth of cut a, the width b_c of the contact zone between the grinding wheel and the workpiece, and the speed v_w of the workpiece (the 'rate of feed'):

$$Z = ab_cv_w. \tag{1}$$

The specific energy e is the ratio of the power P required for grinding to the rate of volume removed:

$$e = P/Z. \tag{2}$$

The specific energy appears on initial consideration to be a constant of the material (roughly speaking e increases with the hardness of the material), but it can after all be affected by the grinding conditions, sometimes very strongly. Indeed, a study of the effect of the damage to the surface and the wear of the wheel. The force ratio

$$f = F_t/F_n, \tag{3}$$

where F_t is the tangential component of F and F_n the normal component, is often introduced.

The grinding power P is given by

$$P = F_tv_s, \tag{4}$$

where v_s is the circumferential velocity of the wheel (the 'wheel speed'). For constant e it can be seen from (4) that a high value of v_s is desirable: for the same power (the same rate of volume removal) smaller forces are necessary, so that less damage and wear may be expected. In practice, there is in fact a tendency to operate grinding wheels at high speeds.

From (2) and (4) we find that the 'reduced' force F_t' (the tangential force per unit width, F_t/b_c) can be

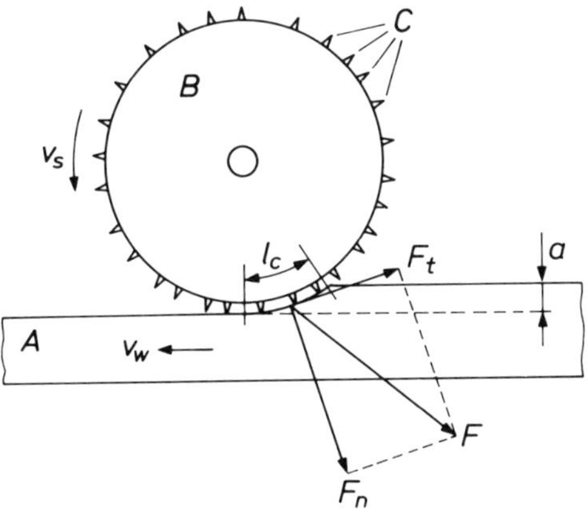

Fig. 2. Diagram representing a surface grinder. The workpiece A is passed under the grinding wheel B, with circumferential velocity (wheel speed) v_s, at a rate of feed v_w. C diamonds, a depth of cut, l_c length of the contact zone; the width of this zone (the dimension at right angles to the paper) is b_c. The wheel exerts a force F on the workpiece; F_n is the normal component of F, and F_t the tangential component.

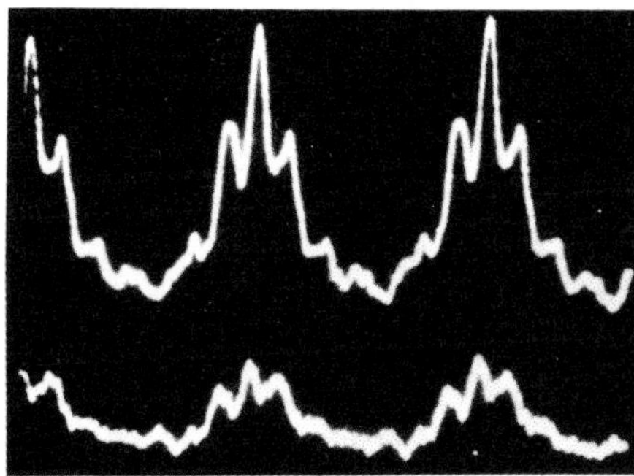

Fig. 3. The normal force F_n (*upper trace*) and the tangential force F_t (*lower trace*) during a little more than two revolutions of the grinding wheel, in an experiment with $v_s = 58$ m/s, $v_w = 60$ mm/s, $a = 0.7$ mm, $h_{eq} = 0.75$ µm. The variations arise because the wheel is not truly circular.

expressed as

$$F_t' = eh_{eq}, \quad (5)$$

where

$$h_{eq} = Z/v_s b_c. \quad (6)$$

The specific energy e can therefore be found by measuring the tangential force.

The quantity h_{eq} is useful for comparing different grinding systems: it is the machine-related quantity that determines the reduced force for given e. From (6) it follows that for every revolution of the wheel a layer of grindings of thickness h_{eq} would be applied to the circumference of the wheel if the grindings were uniformly distributed and remained densely packed. This can be seen by multiplying the numerator and denominator of the right-hand side of (6) by the time T for one revolution; the numerator then gives the volume produced in time T and the denominator gives the product of the length v_sT and the width b_c of the layer. The quantity h_{eq} is therefore called the *continuum chip thickness*. The value of h_{eq} is of the order of 0.01 µm for finish grinding and 1 µm for coarse grinding.

Grinding experiments at high wheel speeds

Typical wheel speeds for ordinary grinding are from 20 to 30 m/s. Since high speeds are of interest, we carried out experiments at speeds of about 60 to 100 m/s[1]. The grinding forces were measured with a dynamometer. Strong periodic variations in the forces are unavoidable in such experiments because grinding wheels are not truly circular, see *fig. 3*.

Fig. 4 gives the results of a large number of experiments with Sr Ferroxdure (Sr hexaferrite). The peak values of the reduced forces $\hat{F}_t'$ and $\hat{F}_n'$ are shown in a log-log plot as a function of h_{eq}. Even though there is a large spread in the measured points these results lead to the conclusion that at a given value of h_{eq}, the quantities $\hat{F}_t'$ and $\hat{F}_n'$, and hence the specific energy, do not depend systematically on v_s. We have found this result for other ceramic materials and for most glasses. Some low-melting-point glasses do have a speed dependence at high wheel speeds. This can be attributed to heating effects; we shall not pursue the matter further.

The slope of the line $\hat{F}_t'(h_{eq})$ in fig. 4 does not differ much from 1, so that e is also almost independent of h_{eq} (or F_t'). This is *not* a rule, however: e usually does depend on h_{eq}. The effect is known as the 'size effect'. A striking example of this is given by MnZn ferrite, in which e varies by a factor of 5 over the same range of h_{eq} considered in fig. 4 (see *fig. 5*).

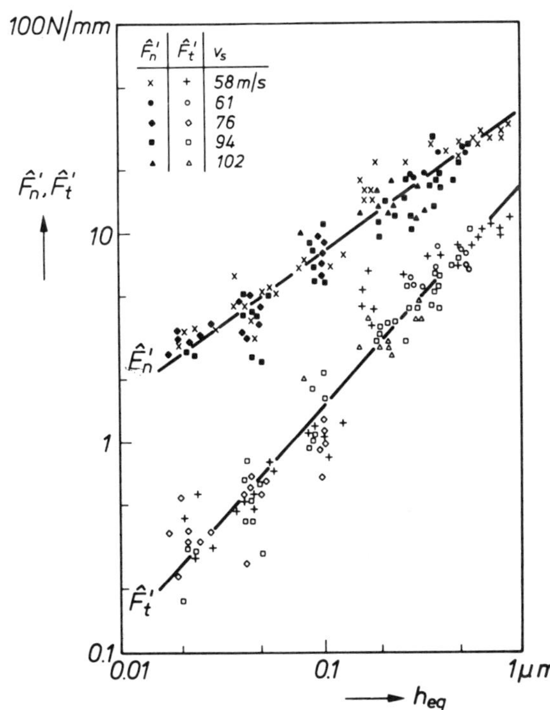

Fig. 4. Peak values of the 'reduced' forces, $\hat{F}_n'$ and $\hat{F}_t'$, as a function of the continuum chip thickness h_{eq}, for speeds of 58 to 102 m/s in grinding Sr Ferroxdure.

[1] A. Broese van Groenou, J. D. B. Veldkamp and D. Snip, J. Physique **38**, C1/285, 1977.

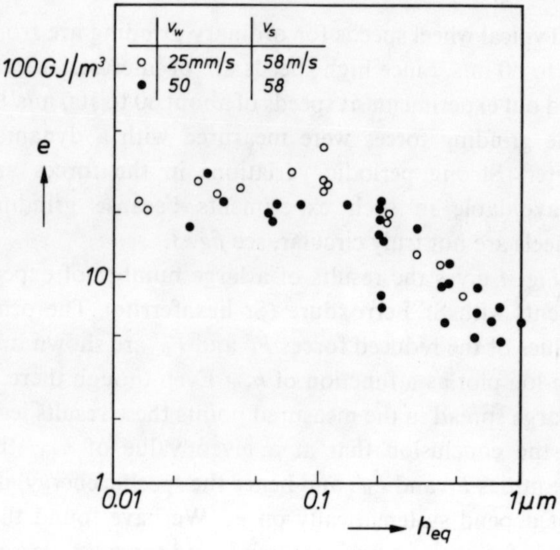

Fig. 5. Specific energy e as a function of the continuum chip thickness h_{eq} for MnZn ferrite. The measured points were obtained at a wheel speed of 58 m/s and at two rates of feed, 25 and 50 mm/s.

The microphotographs in *fig. 6* give a clue to the explanation of the size effect in MnZn ferrite. They show parts of the zone of first contact on the initially polished surface of a slightly tilted sample that was brought into contact with the wheel. This zone shows concisely what happens on a microscale as the depth of cut gradually increases (increasing h_{eq}). The scratches from the individual diamonds start as well-defined, 'clean' grooves which gradually get deeper. Then material starts to chip away from the edges of the groove and eventually the chipped areas from neighbouring scratches join together. In this final stage the area 'cracked' and loosened by one diamond can be removed by another neighbouring diamond with the expenditure of very little energy, so that the specific energy is relatively low.

These photographs also show clearly that grinding is a multi-point process, and that scratching with a single diamond should be the most suitable experiment for obtaining a better understanding of the microprocesses. The experimental conditions can then be clearly defined, e.g. for the geometry of the diamond and the force on the diamond. The basic experiment is *slow scratching*, since heating effects are then avoided.

The scratching experiment; ploughing

Fig. 7 is a photograph and *fig. 8* a diagram of the apparatus we used for our scratching experiments [2]. The sample S is displaced horizontally underneath the diamond D by the micromanipulator Mi, which is driven by the motor Mo. The diamond is attached to the arm A of a balance. This arm is connected to the frame by a leaf spring LS, so that it can be swung up and down but is stiff horizontally. The balance is brought to equilibrium and displaced vertically until the diamond just touches the sample; the desired load L is then applied. The tangential force on the sample is measured by the strain gauges SG on the bending element BE (fig. 8*b*).

In the experiments to be described here the diamonds were ground to the shape of a square pyramid; the scratches were made with one plane leading as in *fig. 9*. The scratching speed was 1 μm/s. Measurements were made on the materials in *Table I*. The range of hardness values was about 9 times. *Fig. 10* shows two scratches in MnZn ferrite, and *fig. 11* shows the associated tangential forces. When the load is small a clean groove

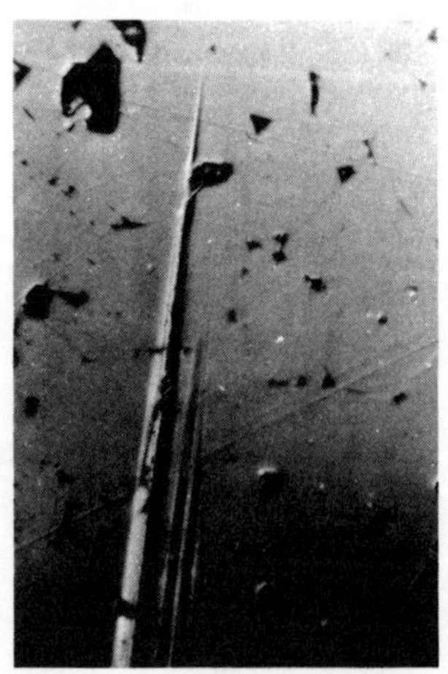

a

b

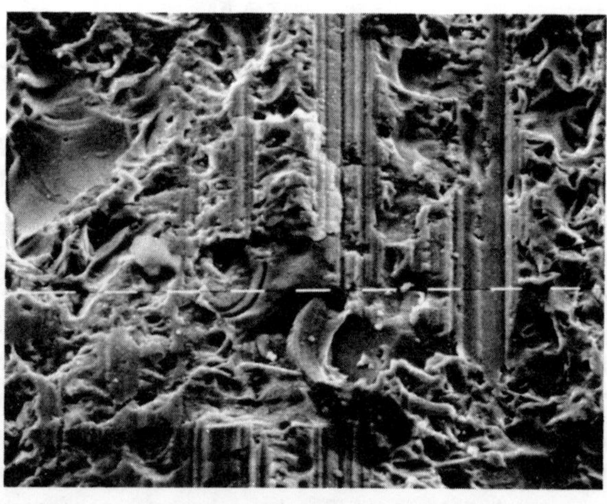

c

Fig. 6. Scratches on the etched surface of a MnZn ferrite, made by the diamonds on the grinding wheel. The start of the scratches is made visible by setting up the sample so that it is slightly tilted during its brief contact with the wheel. *a*) The start of a scratch: a 'clean' groove. *b*) Scratches with chipped edges after several passes of the same diamonds. *c*) The effects of individual diamonds run together.

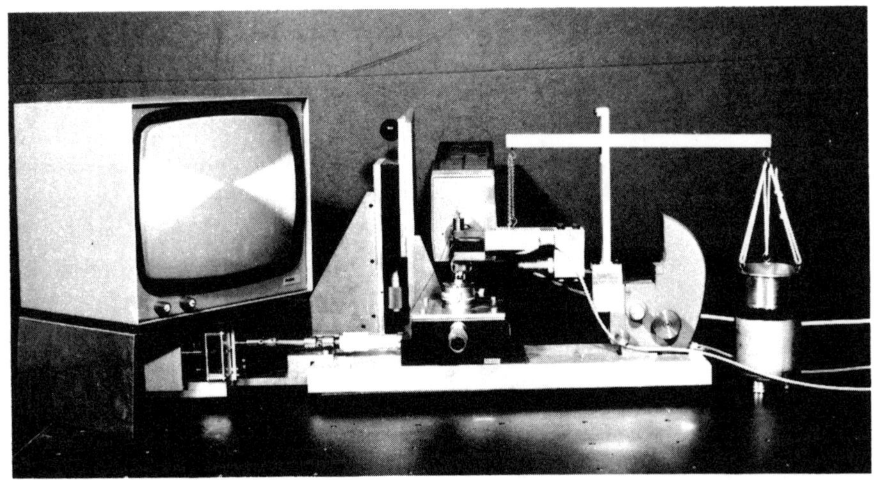

Fig. 7. The scratching apparatus, explained in more detail in fig. 8. The scratching point can be made visible on the monitor on the left.

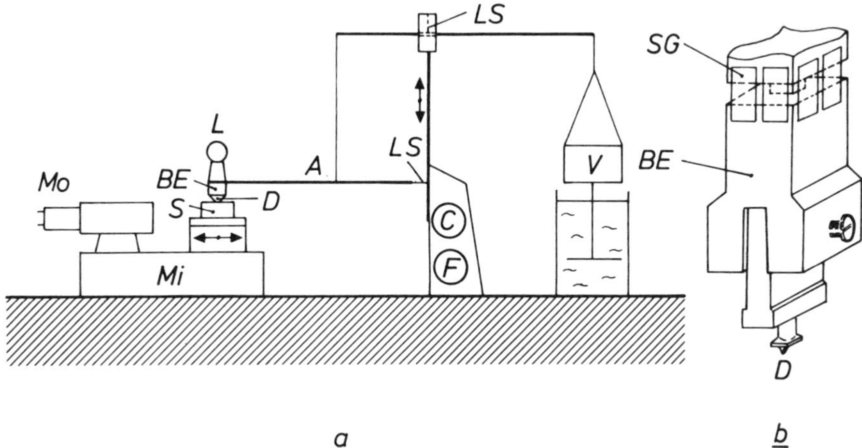

Fig. 8. *a*) Diagram of the scratching apparatus. The sample *S* is displaced underneath the diamond *D* by the motor *Mo* and the micromanipulator *Mi*. The diamond is attached by the bending element *BE* to the balance arm *A*, which can be swung up and down but is stiff horizontally. After the balance has been brought to equilibrium and adjusted vertically the load *L* is applied. *C*, *F* coarse and fine vertical adjustment. *V* counterweight and damper. *LS* leaf springs. *b*) The bending element *BE* with the diamond *D* and strain gauges *SG* for measuring the tangential force. This figure is at true size.

is produced (fig. 10, *lower picture*); with a higher load the groove is wider and there is cracking and chipping (fig. 10, *upper picture*). The large force fluctuations that appear at the higher loads can be connected with the onset of cracking and chipping, but can also be the result of the diamond meeting pores or grain boundaries. In this section we shall only consider the groove.

The formation of grooves in metals has been studied extensively [3]. It is a process of plastic deformation: the diamond ploughs up the material in front of it and pushes it aside. *Fig. 12* shows the plastic nature of the deformation. It may perhaps at first sight seem surprising that such a process can also take place in brittle materials like Ferroxdure and glass. It becomes easier to understand, however, when the pressure

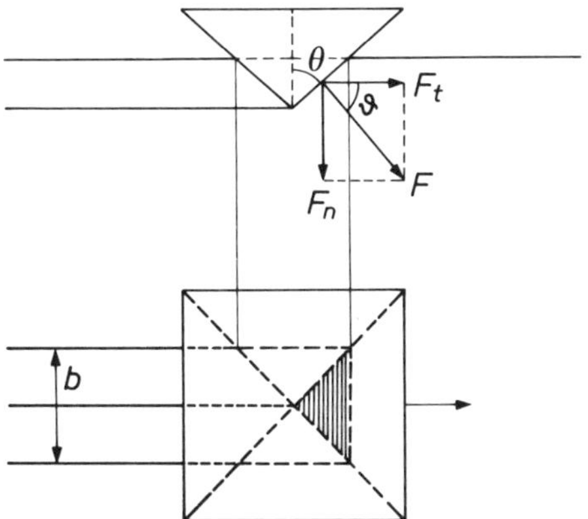

Fig. 9. Geometry of diamond and groove; vertical longitudinal section and plan view. The diamond is a square pyramid (half apex angle θ) that moves with one plane leading. F is the force the diamond exerts on the sample, F_t is the tangential component and F_n the normal component, b the groove width. ('Tangential' and 'normal' relate to the surface of the sample, not to the leading plane of the diamond.) In the ploughing model in its simplest form F is perpendicular to the leading plane.

[2] A. Broese van Groenou, N. Maan and J. D. B. Veldkamp, Philips Res. Repts **30**, 320, 1975.

[3] See for example F. P. Bowden and D. Tabor, Brit. J. appl. Phys. **17**, 1521, 1966.

Table I. The materials investigated. Their hardness H_V is shown on the right in GNm^{-2}.

ZnO	2.6
Lead glass	5.0
Sr hexaferrite (a Ferroxdure)	6.5
MnZn ferrite (a Ferroxcube)	8.0
NiZn ferrite (a Ferroxcube)	9.5
Al_2O_3	23

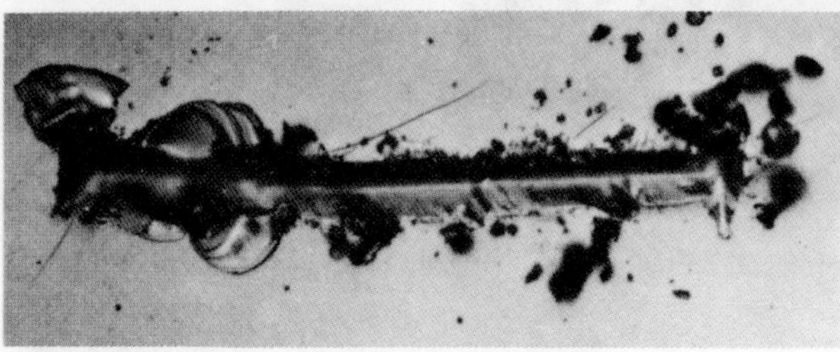

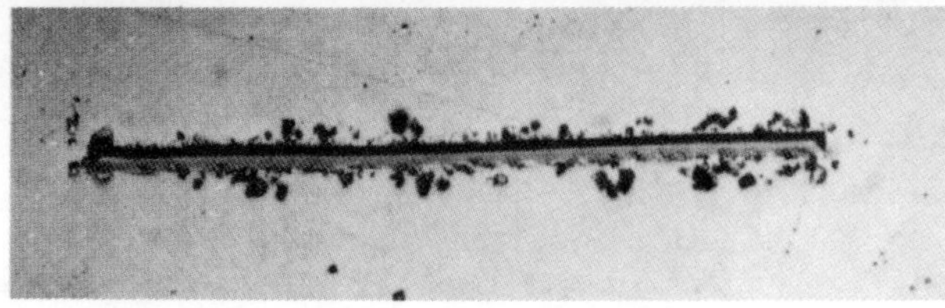

Fig. 10. Slow scratches in a sample of MnZn ferrite made at a load of 0.1 N (*below*) and 0.5 N (*above*). Half apex angle of the diamond: 60°.

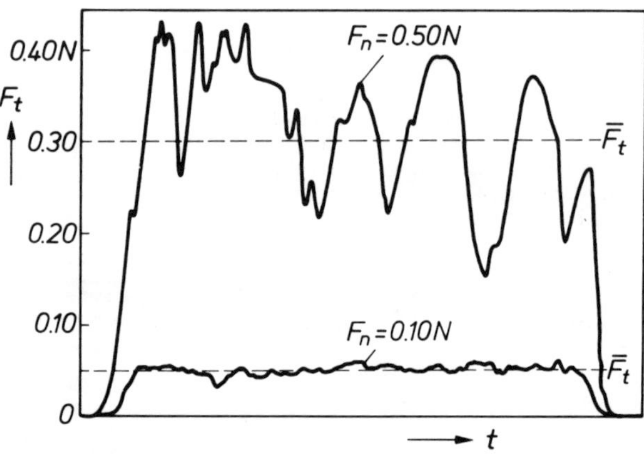

Fig. 11. Tangential forces F_t in making the scratches of fig. 10.

Fig. 12. Cross-section and longitudinal section of a groove made in two-coloured plasticine by a square glass pyramid ($2\theta = 120°$). The plastic deformation is confined to a thin layer below the surface.

beneath the contact area is calculated: it is of the order of 100 000 atmospheres. The plastic nature of the deformation can also be seen from photographs like those in *fig. 13*.

We shall now show that the ploughing model, which was formulated for metals [3], is broadly applicable to groove formation in the materials we have studied. We shall only consider the simplest form of this model here, in which, applied to the situation of fig. 9, *the leading plane takes up all of the pressure and friction plays no part.* This model leads to a few simple rules concerning the force ratio $f (= F_t/F_n)$, the groove width b and the '*specific groove energy*' e_g, which can easily be tested.

We shall consider the force ratio first. Under the conditions assumed the total force must be perpendicular to the leading plane of the diamond, which means that

$$f = \cot \theta. \tag{7}$$

The force ratio would thus be completely determined by half the apex angle θ of the diamond, and not dependent on the load or on properties of the material such as hardness or microstructure. This is generally confirmed by the $\bar{F}_t$-F_n diagram of *fig. 14* for two diamonds with different apex angles and for all the materials of Table I; for both apex angles a straight line through the origin gives a fairly good representation of all the measured points. However, f is somewhat smaller than $\cot \theta$. This difference, which will not be taken into account from now on, can be attributed to the Coulomb friction, which we have neglected [2].

Now let us consider the groove width. This depends on the forces and the specific groove energy e_g of the material, i.e. the energy required for making the groove divided by the volume of the groove, or expressing it another way, the horizontal force F_t divided by the cross-section A_t of the groove:

$$e_g = F_t/A_t. \tag{8}$$

(e_g can be different from the specific *grinding* energy e because any material that may disappear in the form of chips is not included. We should note here that even when there is considerable chipping the groove width can generally still be determined easily; see for example the upper picture of fig. 10.)

From the geometry of fig. 9:

$$F/A = F_n/A_n = F_t/A_t, \tag{9}$$

where A is the contact area and A_n and A_t are the projections of A on the surface of the sample and on the plane perpendicular to the groove direction. In the

[4] N. Maan and A. Broese van Groenou, Wear **42**, 365, 1977.

ploughing model A_t is identical to the groove cross-section. The central quantity F_n/A_n in (9) is by definition the *scratch hardness* H_s, which is thus equal to e_g in the simple ploughing model, but is often not the same in practice and in less simple models. Since A_n is equal to $\frac{1}{4}b^2$ (see fig. 9), we can use (8) and (9) to find a simple relation linking the specific energy, the groove width and the load:

$$e_g\,(=H_s) = 4F_n/b^2. \qquad (10)$$

The groove width b must therefore be proportional to $F_n^{\frac{1}{2}}$, and independent of the apex angle. The measured results in *fig. 15* illustrate this relation for several materials.

Finally, hard materials would be expected to have a higher specific energy than soft ones. This trend is confirmed in *fig. 16*. In the simplified ploughing model e_g would have to be equal to the pressure F/A at which the material just starts to flow (see fig. 9), and would therefore be proportional to the Vickers hardness H_V. In fig. 16 it can be seen that $e_g \approx 2.5\,H_V$ at a load of 0.1 N — a load per diamond not unusual in grinding.

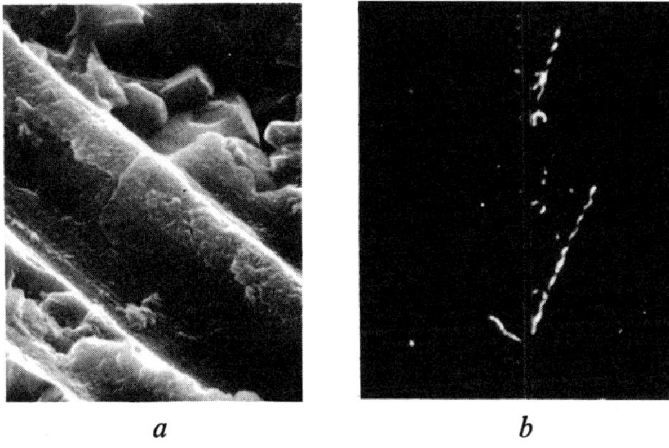

Fig. 13. *a*) Plastic deformation in ground Ferroxdure (magnification 2000 ×). *b*) Plastic deformation, for a scratch in glass ('shavings' are formed; magnification 900 ×).

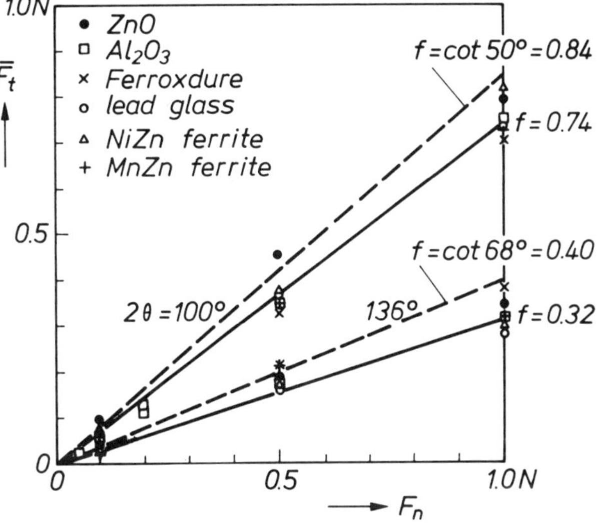

Fig. 14. The average tangential force $\bar{F}_t$ as a function of the load F_n for two diamonds ($2\theta = 100°$ and 136°) and six different materials. In the simple ploughing model (no friction) F_t/F_n would be equal to cot θ (the corresponding values are 0.84 and 0.40, *dashed lines*).

Size effect

As in grinding, the specific energy in scratching is not a true constant of the material: e_g decreases with increasing load, as can be seen from fig. 16. Again, as in grinding e_g is usually considered as a function of the groove depth d (a monotonically increasing function of the load) here. A size effect is then generally found; our results demonstrate this in *fig. 17*. We shall not discuss the explanation of this effect here.

Ploughing and grinding

Let us now consider how realistic it is to interpret 'grinding' as 'ploughing'. To allow the grinding results to be compared with the scratching results we must first convert the continuum chip thickness h_{eq} into an average groove depth d_{av} for the diamonds on the grinding wheel, starting with the assumption that grinding really is ploughing. If C is the number of effective diamonds per unit area of the grinding wheel, b_c the width and l_c the length of the contact zone, then

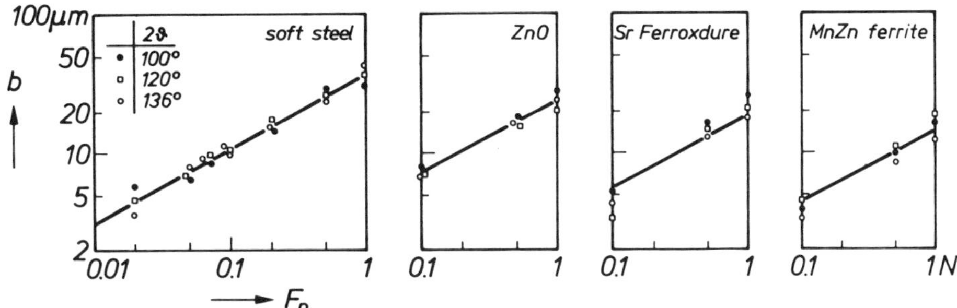

Fig. 15. Groove width b as a function of load F_n for three diamonds in mild steel [4], ZnO, Ferroxdure and MnZn ferrite. The solid lines have a slope of 1 in 2 ($b \propto F_n^{\frac{1}{2}}$). The 'best' straight line for Ferroxdure and MnZn ferrite would be a little steeper (size effect).

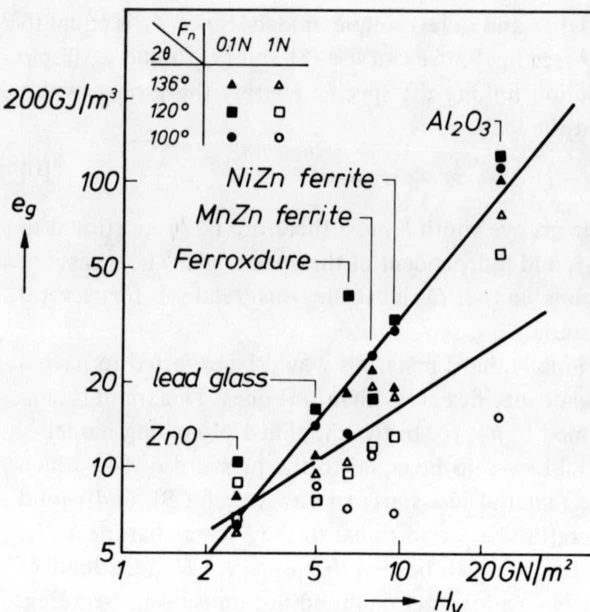

Fig. 16. The specific groove energy e_g for various materials, arranged in order of hardness H_V. In general e_g is higher for harder materials.

is difficult to estimate. Much more closely defined results are obtained from experiments at 5 m/s with a flat grinding wheel (*fig. 18*) [6]. At a given load F_n and wheel speed v_s we measured the tangential force F_t and the rate of volume removal Z (this was obtained from the reduction in height of the sample). Using equations (5) and (6) we calculated e and h_{eq} from the measured quantities. Counting diamonds under the microscope gave C, and $\cot\theta$ was equated to the measured force ratio, so that h_{eq} could be converted into an average groove depth.

The specific grinding energy and the specific groove energy, both as a function of groove depth, can now be compared. The results for Sr Ferroxdure and MnZn ferrite can be seen in *fig. 19*. In view of the great difference in speed (a ratio of $1:10^6$), the agreement between groove and grinding energies for Ferroxdure is remarkable. There is a considerable discrepancy for MnZn ferrite, however: the specific grinding energy is only about a tenth of the specific groove energy.

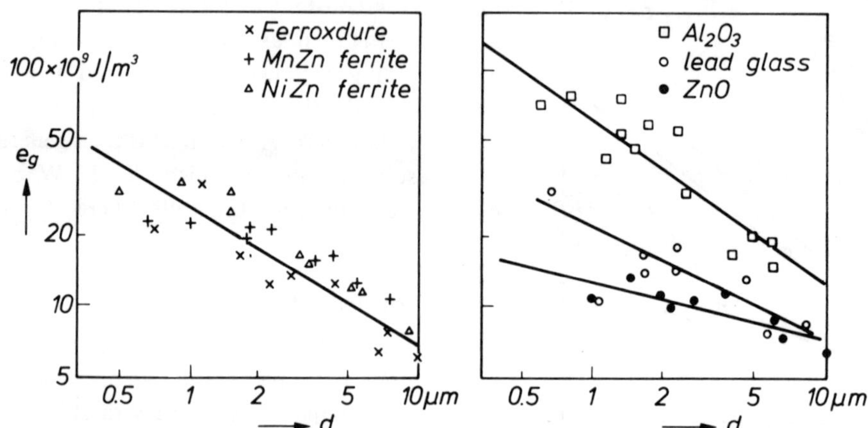

Fig. 17. Size effect for six materials; the specific groove energy e_g depends on the groove depth d.

the number of diamonds in contact with the workpiece at a given moment is equal to Cb_cl_c. The rate of volume removal per diamond is A_tv_s (A_t is the groove cross-section) and the total rate of volume removal Z is therefore $Cb_cl_cA_tv_s$. If we also assume that all the diamonds are pyramids of apex angle 2θ, then A_t is equal to $d_{av}^2 \tan\theta$. Combining these relations with eq. (6) we have [5]:

$$d_{av} = \left(h_{eq}\,\frac{\cot\theta}{Cl_c}\right)^{\frac{1}{2}}. \qquad (11)$$

The high-speed grinding experiments described earlier are not suitable for comparison with scratching experiments: because the wheel is not truly circular (see fig. 3) the number of 'effective' diamonds is rather small and uncertain, so that the average groove depth

MnZn ferrite is much more brittle than Ferroxdure. The results therefore give a more detailed confirmation of the picture of grinding that was outlined before: for

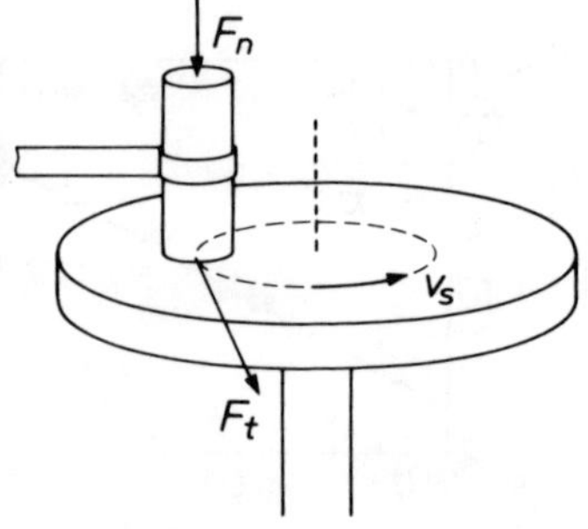

Fig. 18. Grinding with a flat grinding wheel. The sample is fixed in the horizontal direction and pressed vertically against the wheel by a force F_n. F_t tangential force; v_s speed of the wheel at the sample.

'tough' materials grinding is indeed a process of plastic deformation (ploughing), with the specific energy determined by the hardness; in brittle materials, on the other hand, material is removed by chipping as well as by ploughing.

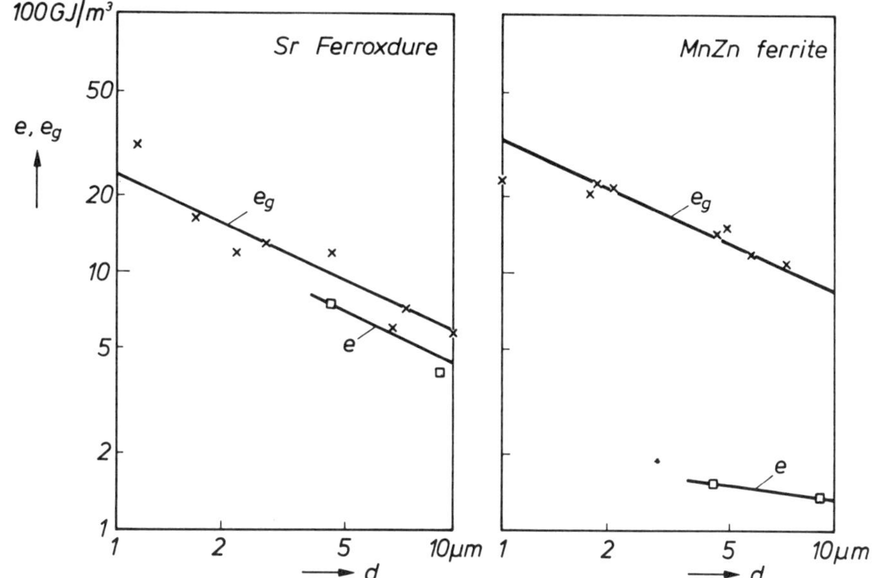

Fig. 19. The specific grinding energy e and the specific groove energy e_g as a function of the groove depth for Ferroxdure and MnZn ferrite.

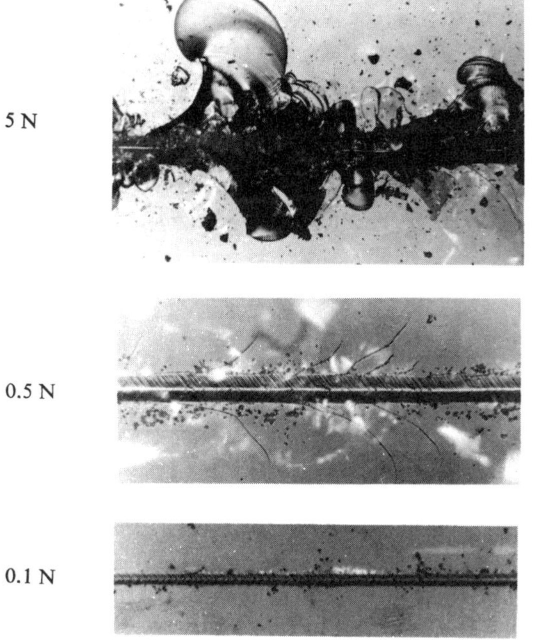

Fig. 20. Effect of the load on the damage in SF58 glass. *From bottom to top:* clean groove (F_n = 0.1 N), lateral cracks (0.5 N), chipping (5 N); v_s = 400 µm/s.

Cracking and chipping

When the load is increased during scratching a clean groove is in general obtained at first, then cracks start to form and finally there is cracking and chipping (*fig. 20*). Three different sorts of cracks can clearly be distinguished: median cracks, subsurface cracks and lateral cracks [7] (*fig. 21*).

The nature of the damage does not in general only vary with the load, but also with the scratching speed.

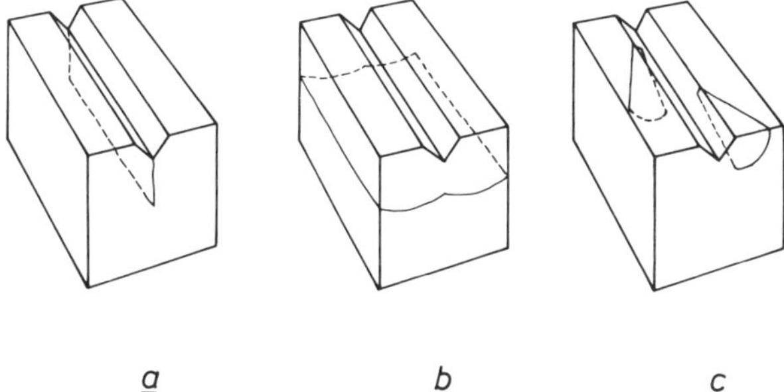

Fig. 21. Types of cracks produced in scratching. *a*) Median crack. *b*) Subsurface crack. *c*) Lateral crack.

This is demonstrated in *fig. 22* for MnZn ferrite and in *fig. 23** for BK7 glass. On increasing the load in glass, median and subsurface cracks are obtained at first, accompanied later by lateral cracks and finally by chipping as well.

We shall give most of our attention to the lateral cracks. These always precede the chipping as the load is increased. They can be directly observed at the

[5] M. C. Shaw, in: New developments in grinding, ed. M. C. Shaw, Carnegie Press, Pittsburgh 1972, p. 220.
[6] J. D. B. Veldkamp and R. J. Klein Wassink, Philips Res. Repts **31**, 153, 1976.
[7] J. D. B. Veldkamp, N. Hattu and V. A. C. Snijders, in: Fracture mechanics of ceramics, vol. 3, ed. R. C. Bradt, D. P. H. Hasselman and F. F. Lange, Plenum Press, New York 1978, p. 273.

Fig. 22. *Photographs on left:* Effect of the scratching speed on the nature of the damage in MnZn ferrite. At $v_s = 4$ μm/s there is cracking and chipping, at 40 μm/s only cracking; $F_n = 0.5$ N. *Photographs on right:* Effect of the load. From bottom to top: a clean groove ($F_n = 0.1$ N), groove with cracks (0.7 N), groove with chipping (3 N); scratching speed 400 μm/s. *Centre:* Nature of the damage in an F_n-v_s diagram. The diagram can be roughly divided (by boundaries *I* and *II*) into regions of formation of clean grooves (▽), grooves with cracks (⩓) and grooves with cracking and chipping (◡).

surface and can easily be counted and measured. The subsurface cracks are best known in glass, where they are readily visible because of light reflections at the surface of the crack. Median cracks have been investigated in detail some time ago [8].

The boundary *I* in figs 22 and 23 is of particular importance for finish grinding: if the operating point is kept to the right of and below this boundary there will be no damage to the surface. For coarse grinding it is possible to take advantage of chipping by working above boundary *II*. In glass chipping starts in the region between *I* and *II*: the subsurface cracks allow chips to form easily on repeated passes of the scratching particles.

The speed effect expressed in figs 22 and 23 is usually only seen clearly at the lower speeds; at speeds above about 1 mm/s the surface damage is generally only dependent on the load. We shall now discuss the crack formation as a function of the load at these higher speeds, with special attention to the way in which the *minimum force for crack formation* (boundary *I*) depends on the properties of the material.

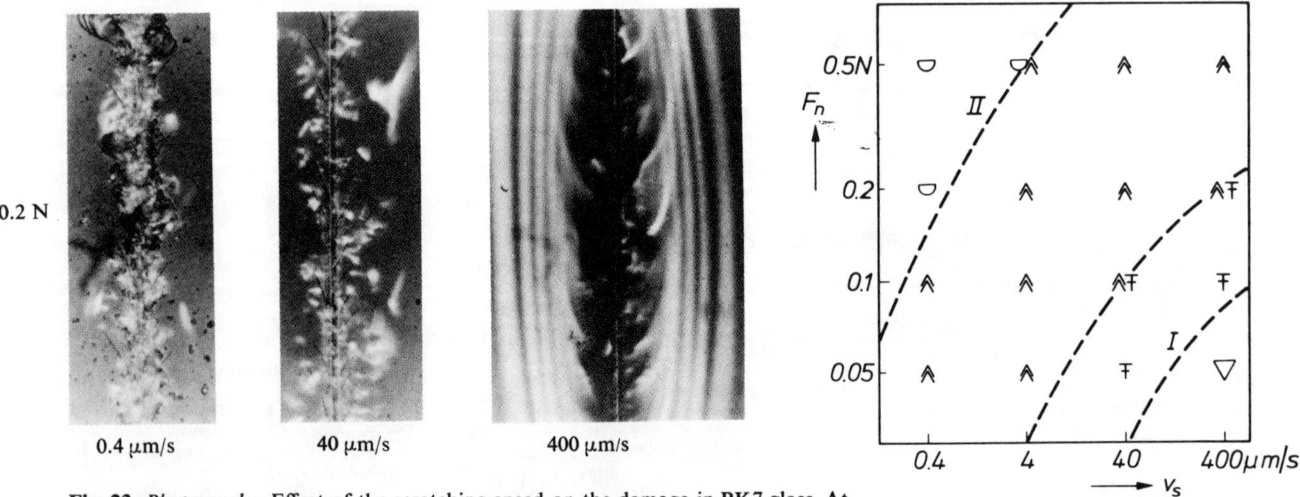

Fig. 23. *Photographs:* Effect of the scratching speed on the damage in BK7 glass. At $v_s = 0.4$ μm/s there is chipping, at $v_s = 40$ μm/s there is lateral cracking and at 400 μm/s only subsurface cracks are formed; $F_n = 0.2$ N. In the F_n-v_s diagram we find from lower right to upper left: clean grooves (▽); grooves with median and subsurface cracks (∓); next there are also lateral cracks (⩓); in the last region there is chipping as well, but there are no more subsurface cracks (◡).

Crack formation as a function of load

A diagram of the model situation for crack formation [9] is shown in *fig. 24*. An infinite sheet of material is loaded by a uniform tensile stress σ in a direction in the plane of the sheet. A planar crack forms across the sheet at right angles to the direction of the stress; the width of the crack is $2c$. The stress in the region of the crack concentrates round the tips of the crack (R). The intensity of the (very inhomogeneous) stress field around either tip of the crack is proportional to σ and to $\sqrt{c}$; it is specified completely by the 'stress-intensity factor' $K = \sigma\sqrt{\pi c}$.

If the stress σ in this situation is increased, K eventually reaches a value K_c, characteristic of the material, at which the material splits at the tips of the crack. We then have an 'explosive' situation; the splitting continues and the tip of the crack propagates in the material at approximately the velocity of sound: the material *fractures*. For simplicity we shall assume that the crack does not propagate at all for $K < K_c$; we can then speak of 'the' cracking condition

$$\sigma \sqrt{\pi c} = K_c. \qquad (12)$$

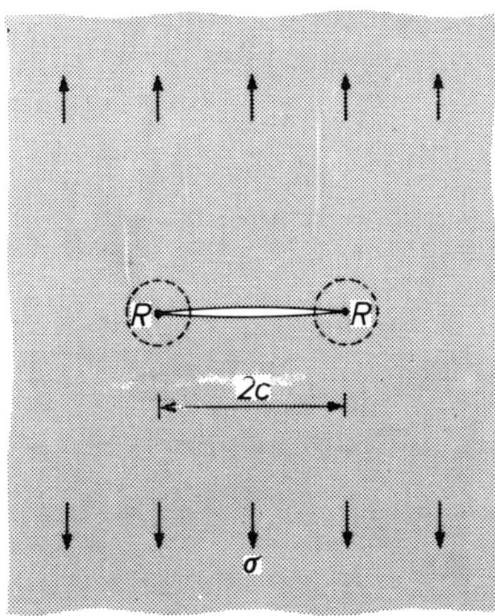

Fig. 24. Model situation for crack formation: an infinite sheet in the plane of the paper is loaded by a uniform tensile stress σ in a direction in this plane. At right angles to this direction there is a planar crack of width $2c$ across the sheet. The stress in the region of the crack concentrates round the tips of the crack (R). The intensity of the stress field at the tip is given by the stress-intensity factor $K = \sigma\sqrt{\pi c}$.

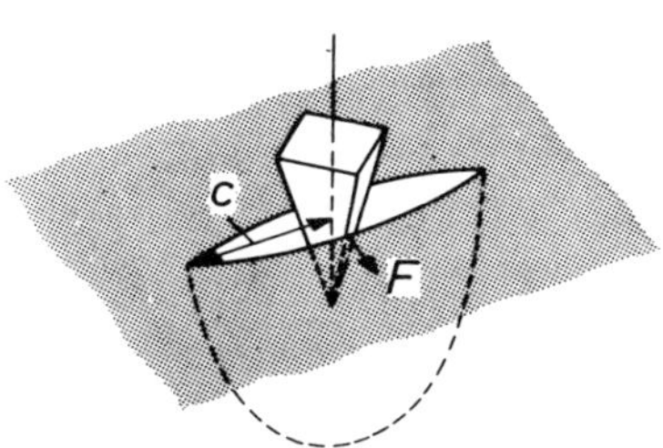

Fig. 25. Diagram of a 'half-penny' crack of radius c caused by a Vickers indenter. The wedge action pushes the surfaces of the crack apart with a force F.

To predict whether the tip of a crack will propagate in a more complicated situation, and if so how far, it is necessary to calculate the stress field and find the values σ and c in fig. 24 should have to give the same stress field around the tips of the crack, since given certain conditions the field around the tip of a crack always takes the same form. The cracking condition again follows from (12) with the values found for σ and c.

The lateral cracks in scratching are related to the 'half-penny' cracks that sometimes form around an indenter in Vickers hardness measurements[10] (*fig. 25*); we shall now consider these briefly by way of introduction. The situation here, for a well-developed crack, is characterized by two quantities: the radius of the crack c and the force F with which the diamond pushes the surfaces of the crack apart. Dimensional considerations now lead us to expect that the stress σ required for (12) is proportional to the characteristic force (F) divided by the characteristic area (c^2). The cracking condition thus takes the form $(F/c^2)\sqrt{c} \propto K_c$, or:

$$F = \alpha K_c c^{\frac{3}{2}}, \qquad (13)$$

where α is a constant. This result is in agreement with a more rigorous analysis of the problem [10]. The situation is not explosive: at a given F the crack grows until the condition (13) is satisfied. The linear relationship between F and $c^{\frac{3}{2}}$ has been experimentally confirmed for this type of cracking [10]. At the minimum force for crack formation (F_{min}) cracking just starts; c is then equal to the half-diagonal $\frac{1}{2}b\sqrt{2}$ of the indentation, and this half-diagonal is again related to F_{min} through the *hardness H* of the material:

$$H = F_{min}/b^2. \qquad (14)$$

Combining (13) and (14):

$$F_{min} \propto K_c^4/H^3. \qquad (15)$$

Let us now return to the lateral cracks associated with scratching (*fig. 26*). The moving diamond is accompanied by a stress field in the material; if the cracking condition is satisfied at a flaw in the material a crack forms. The forward-tilted direction of the lateral cracks (see figs 10, 20-23) corresponds to the

[8] K. Peter, Glastech. Ber. **37**, 333, 1964.
[9] See for example B. R. Lawn and T. R. Wilshaw, Fracture of brittle solids, Cambridge University Press, 1975.
[10] B. R. Lawn and E. R. Fuller, J. Mat. Sci. **10**, 2016, 1975.

calculated stress field below an obliquely loaded point [11]. The horizontal force F_t opens the crack, and is therefore related to F in fig. 25. The widths of the groove and of the cracking and chipping regions for a coarse-grained NiZn ferrite are shown in *fig. 27* as a function of the load. In agreement with eq. (13) the width of the cracking region, $2c$, is approximately proportional to $F_n^{\frac{2}{3}}$.

For scratching as shown in fig. 26 we have set up a crack-formation model, and from this we have obtained a relation between c and F_t [7]:

$$F_t = \alpha K_c c\sqrt{c + \tfrac{1}{2}\beta b}, \qquad (16)$$

where b is the groove width (which is related to F_t by (10) and (3)). If the cracks are large ($c \gg b$), eq. (16) becomes equivalent to eq. (13). The constants α and β are determined by the shape and orientation of the diamond.

We have measured b and c for a square pyramidal diamond with an apex angle of 136° and a leading plane, for a series of loads. Eq. (16) gives a good representation of the measured results for various materials, with $\beta = 12.4$. The constant α was determined by measuring K_c by one of the standard methods for one of the materials (glass) [9]; the result was $\alpha = 2.24$.

It is now possible to derive the minimum force for crack formation from (16), in the same way as above for 'half-penny' cracks from (13). The result is almost the same:

$$F_{t\,\min} = C_s K_c^4/H_s^3, \qquad (17)$$

where H_s is the scratch hardness (see eq. (10)); C_s is determined by α, β and the force ratio f:

$$C_s = \alpha^4(1 + \beta)^2/f^3.$$

Inserting the values of α and β given above and $f = \cot 68° = 0.40$, we find $C_s = 0.7 \times 10^5$.

Equations (15) and (17) show that a material that is very crack-resistant in tensile tests (high K_c) can nevertheless crack easily in indentation and scratching if it is also very hard.

The high sensitivity of $F_{t\,\min}$ to K_c and H_s explains the very different scratch and cracking behaviour of different materials. *Table II* gives K_c and H_s and the minimum load ($F_{n\,\min} = F_{t\,\min}/0.4$) calculated from them by means of (17) for a number of materials. *Fig. 28* presents this information in a K_c-H_s diagram. In our scratching experiments the load varied from 0.05 to 5 N; the load per diamond in very heavy grinding is estimated at no more than 1 N (solid line in fig. 28). Table II predicts, in agreement with our results, that under these circumstances crack formation cannot be avoided in MgO, MnZn ferrite and television glass, that Ferroxdure and Al_2O_3 must be heavily loaded to

Table II. The minimum load for crack formation $F_{n\,\min}$, calculated from the critical stress-intensity factor K_c and the scratch hardness H_s by using eq. (17), with $C_s = 0.7 \times 10^5$ and eq. (3), $F_n = F_t/f$. The value used for C_s should apply to diamonds with an apex angle of 136° and a leading plane. This value is very sensitive to the force ratio f; this was taken as cot 68° = 0.4. If this should fall to say 0.3, a value often found in practice, C_s and $F_{t\,\min}$ become more than twice as large.

	K_c ($MNm^{-\frac{3}{2}}$)	H_s [*] (GNm^{-2})	$F_{n\,\min}$ (N)
MgO, single crystal {100}	0.8 [9]	17 [13]	0.015
MnZn ferrite	1.2 [6]	22 [7]	0.035
Television glass	1.0 [6]	16 [7]	0.042
Al_2O_3	4.8 [6]	70 [13]	0.27
Ferroxdure	2.5 [6]	24 [13]	0.50
Steel (high C content, hardened)	20 [12]	10 [4]	28000

[*] At a scratching speed of 1 mm/s.

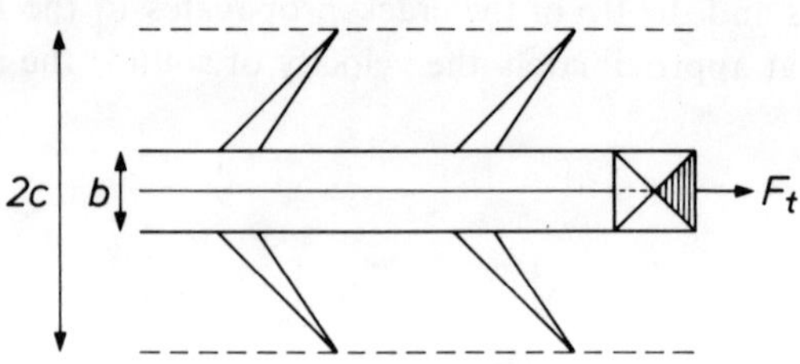

Fig. 26. Diagram of lateral cracks produced in scratching. F_t tangential force, b width of the groove, $2c$ width of the cracking region.

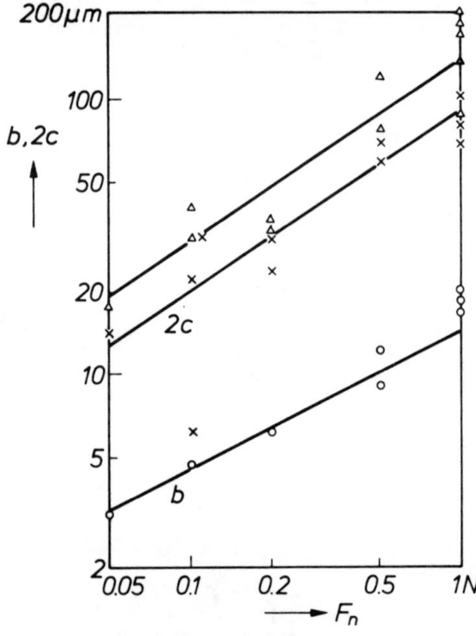

Fig. 27. The width b of the groove (○), the width $2c$ of the cracking region (×) and the width of the chipping region (△) as a function of the load F_n for a coarse-grained NiZn ferrite. The groove width b is approximately proportional to $F_n^{\frac{1}{2}}$, and the width of the cracking region $2c$ to $F_n^{\frac{2}{3}}$.

[11] G. M. Hamilton and L. E. Goodman, Trans. ASME E (J. appl. Mech.) **33**, 371, 1966.
[12] P. Kenny and J. D. Campbell, Progr. Mat. Sci. **13**, 135, 1967.
[13] Unpublished results of measurements made at these laboratories.

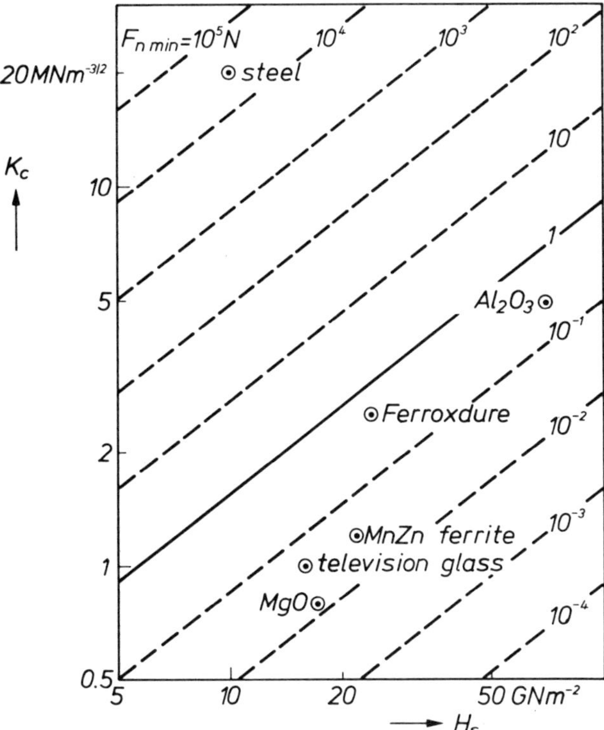

Fig. 28. The K_c-H_s diagram with lines of constant $F_{n\,min}$ ('minimum load for crack formation'), as given by eq. (17) with $C_s = 0.7 \times 10^5$ and $f = 0.4$. The solid line gives the estimated load per diamond in very heavy grinding ($F_n = 1$ N). The points for different materials represent the values of Table II.

produce cracks and that cracking in steel is impossible. The cracking boundary in Table II for MnZn ferrite and glass agrees in order of magnitude with figs 22 and 23.

As the load increases, not only does the width of the cracking region increase, but the number of cracks per unit length often increases as well. This can largely be explained by the increasing number of flaws in the material ('crack nuclei') at which the cracking condition is satisfied.

The growth in number and magnitude of the lateral cracks eventually leads to chipping; the boundary II in figs 22 and 23 has then been reached. The interaction with other cracks probably comes into play here. Chipping is therefore difficult to analyse quantitatively, and we shall not discuss it further in this article.

Although we assumed for simplicity that a crack does not propagate if K is smaller than K_c, it does in fact usually do so in the situation of fig. 24, but at a velocity v_c that decreases very strongly as K decreases (*fig. 29*). If this relation between K and v_c is taken into account, the equations (13) to (17) remain valid, provided K_c is replaced by K. This is then the value of the stress-intensity factor at the tip of the crack at the associated crack velocity. One of the things explained by this relation is the speed dependence of boundary *I* in figs 22 and 23, if we assume that the crack velocity can be identified with the scratching speed. This assumption appears to be justified for the conditions in our experiments [7].

The knowledge of K as a function of the crack velocity (fig. 29) is essential for calculating the strength and life of objects that are mechanically loaded. The well-known methods [8] for determining $K(v_c)$ are very labour-intensive, however. If the highest accuracies are not required, scratching experiments offer a simple alternative: K can be determined via (16) from a measurement of the width of the cracking region ($2c$) and the groove width (b) at a given load ($F_n = F_t/f$) and the desired velocity. We have been able to determine the K-v_c curve for a number of materials in this way [7].

Conclusions

The objective of the investigation, in brief, was to establish how specific energy and surface damage are related to machine setting and properties of the material. The conclusions are as follows.

The specific energy e is primarily a function of the material: the harder the material, the greater e. The effect of the machine setting on the specific energy can be considerable, however, and here a second material property that shows up clearly in scratching experiments often has an important part to play. This is the minimum force per diamond for crack formation ($F_{t\,min}$). For simplicity we shall also refer to *the* force per diamond in grinding as well as in scratching in order to compare it with $F_{t\,min}$, but in fact this concept is not closely defined, because the diamonds on a wheel do not protrude by the same amount.

If we examine this effect of the machine setting on the specific energy and on the surface damage, we find that in practice the wheel speed v_s has no significant effect. It is therefore always advisable to try to use higher speeds, since the power required for a given rate of volume removal (the product of the wheel speed and the horizontal grinding force) can then be supplied at a smaller force, thus limiting surface damage and the wear of the wheel. This is true if heating effects do not

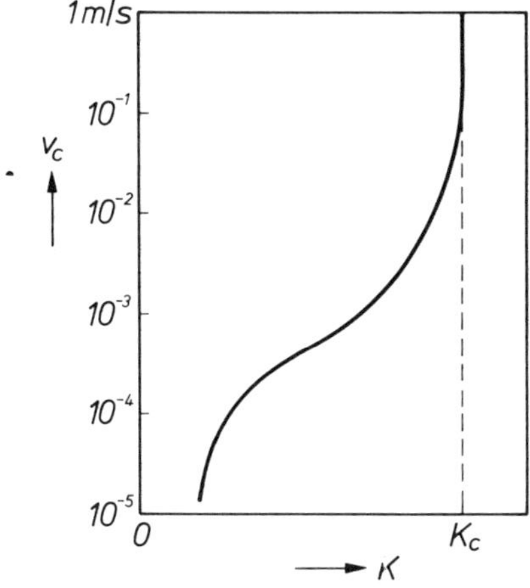

Fig. 29. Diagram showing the relation between the crack velocity v_c and the stress-intensity factor K. The v_c-scale is logarithmic, the K-scale linear. At $K = K_c$ the tip of the crack propagates at approximately the velocity of sound.

come into play. Pronounced speed effects have been observed in scratching, but these are at speeds of microns to millimetres per second, which are of no significance in grinding.

The specific energy and the surface damage are strongly dependent on the grinding forces, however. If the tangential force per diamond, F_t, is smaller than $F_{t\,min}$, grinding is a form of 'ploughing'. The effect of the force on the specific energy is then still relatively small; the surface irregularities remain limited to grooves that are impressions of the diamond points. If F_t is larger than $F_{t\,min}$, however, effects of a different order occur because of cracking and chipping. Much more material is then removed per joule than in ploughing, and the surface damage is much greater.

A high speed and a large force are therefore both desirable for rapid and efficient removal of material; a large force is particularly useful if $F_{t\,min}$ is exceeded. On the other hand, surface damage in the form of cracks can be prevented by making F_t smaller than $F_{t\,min}$; and even then the roughness decreases as F_t becomes smaller. In many cases it will therefore be preferable to grind in two stages: coarse grinding (fast) first, until the correct shape has been obtained; this leaves a damaged layer of thickness equal to the crack depth; then this layer is ground away with a force per diamond small enough (i.e. the removal is slow enough) to give the required quality of finish. If the grinding is done in several stages the operation can be made even more efficient.

The minimum force for crack formation, $F_{t\,min}$, depends very strongly on both the crack resistance and the hardness (see eq. 17), and consequently differs greatly from material to material (see Table II and fig. 28). For example, $F_{t\,min}$ in steel is so large that cracks will never form; Ferroxdure and Al_2O_3 must be very heavily loaded to take advantage of chipping. In materials like MnZn ferrite, glass and MgO, on the other hand, $F_{t\,min}$ is so small that other processes (e.g. lapping, polishing) have to be used to produce a good finish.

Summary. In setting up a grinding machine the 'rate of volume removal' Z and the 'continuum chip thickness' h_{eq} are important quantities. At a given 'specific energy' e for the material, h_{eq} determines the force exerted on the workpiece by the grinding wheel, and hence the surface damage. A high circumferential velocity v_s for the wheel has the advantage that the grinding power P ($= eZ$) required is obtained with a small force. The specific energy (generally high for hard materials) is usually dependent on the machine setting. The influence of v_s, at a given h_{eq}, is very small. The influence of h_{eq} (or the force) on e is also very small in 'tough' materials, but is considerable in brittle materials, because of chipping. The single-point scratching experiment has been used for further investigations of groove formation, cracking and chipping. The groove formation corresponds to the 'ploughing' model; grinding a tough material is 'ploughing'. The minimum force required for crack formation in scratching depends strongly on the properties of the material. This explains the very different scratching and grinding behaviour of the materials investigated.

Presented at SME's International Tool And Manufacturing Engineering Conference, May 1977

Grinding Damage In Ceramics

By B. G. Koepke
And R. J. Stokes
Honeywell Corporate Research Center

In this paper we discuss surface and subsurface damage introduced in ceramic workpieces by abrasive machining (ie. grinding) operations. The effects of machining parameters and workpiece properties on the nature and extent of the damage are first reviewed and then discussed in terms of the mechanics of the material removal process. Finally some examples are given of how the grinding operation can influence the mechanical and physical properties of a finished component.

I. INTRODUCTION

Recently interest in understanding what abrasive machining does to the surface of a ceramic, apart from simply shaping and smoothing it has arisen for many reasons [1-4]. First, ceramics offer the user perhaps the widest range of physical and mechanical properties of any class of materials and as a result are continually being put into service in new and demanding roles. Second, more and more ceramic components have shapes, surface conditions and tolerances not obtainable by primary fabrication techniques; they must be machined. Finally, since many physical and mechanical properties of ceramics are highly sensitive to the surface condition of a component one must be concerned whether or not the machining operation itself affects the ultimate usefulness of a ceramic component. The answer to this question lies in an understanding of the effects of the machining operation on the regions at and beneath the surface of the component. We address the topic of surface and subsurface machining damage in this paper.

Discussion will be limited to surface grinding since this machining operation is the one used most widely in industry. We will furthermore concentrate on semifinish grinding since the damage introduced then is persistent and usually causes the majority of flaws in a finished part. In what follows we discuss the nature and extent of damage that can be introduced into a ceramic surface by grinding; the origin of the damage in terms of the mechanics of the grinding process; some practical implications of these observations.

II. NATURE AND EXTENT OF GRINDING DAMAGE

Before discussing how a grinding wheel affects a ceramic workpiece, it is first instructive to consider how different ceramics will react to the wheel. A convenient means of classifying ceramics is according to their deformation characteristics under ordinary stresses at room temperature [5,6]. In this way ceramics are classified as ductile, semibrittle and brittle. Since most important technical ceramics fall into the latter two classes we will only consider them here. A semibrittle ceramic is one which exhibits a limited amount of deformability (ie. dislocation mobility) at room temperature. Examples of semibrittle ceramics are MgO, LiF and NaCl.

Brittle ceramics, as expected, are those that show no ductility under ordinary stresses at room temperature. Examples of brittle ceramics are Al_2O_3, BeO, and SiN.

A. Grinding damage in semibrittle ceramics

The alteration of both the surface and subsurface regions of a semibrittle ceramic by grinding can be extreme. The effects grinding has on semibrittle ceramics can be effectively illustrated if MgO single crystals are used as the workpiece. MgO crystals are particularly useful in revealing grinding damage because they can be cleaved on {100} cube planes and etched to reveal dislocation etch pits[7]. The presence of etch pits indicates that plastic deformation has taken place. In a typical experiment MgO crystals are machined at different rates of material removal and then examined for damage. The schematic of a grinding wheel in Figure 1 shows that the rate of material removal can be systematically changed by varying the depth of cut, d, and/or the workpiece feedrate V. Also shown on Figure 1 are some geometrical aspects of MgO crystals. If {100} cleavage planes always form the sides of a workpiece one can machine a number of different surfaces and still cleave a surface perpendicular to the ground surface for etching. The {110} slip planes of MgO are also indicated on Figure 1.

If a 46 grit vitrified bond alumina wheel is used to grind MgO, the open structure of the wheel loads and the surface burnishes. The top photo in Figure 2 shows how smooth the burnished surface can be. The parallel lines indicated by the arrows on the photo are grinding cracks that formed when the hot, as-ground surface quenched following passage of the wheel. The cracks are perpendicular to the grinding direction. The subsurface damage introduced by the alumina wheel is shown in the bottom three photos. Each photo shows a surface that has been cleaved transverse to a {100} ground surface and etched. We see that grinding introduces a dense layer of dislocation etch pits just under the ground surface of the MgO crystal. The high density of etch pits indicates the region just under the ground surface was intensely plastically deformed by the wheel. The rate of material removal used to grind the samples shown in the bottom photos increased as one moves left to right in a clockwise fashion. It is clear the depth of damage is a strong function of the severity of the grinding operation. The depths of deformation can be substantial and can exceed 150μm in extreme cases.

When MgO crystals are ground with a 100 grit resinoid bond diamond wheel the surface is quite rough as shown in the top photo in Figure 3. In this case material is removed almost entirely by brittle cleavage fracture accounting for the surface appearance. Subsurface regions are still deformed, however, as shown by the bottom photos in the figure. The bottom photos show subsurface damage in crystals ground in the different orientations shown in Figure 1. We note that in orientations where the {100} cleavage planes are not parallel and perpendicular to the ground surface, extensive subsurface cracking can result.

The above experiments[8] on ground semibrittle MgO crystals reveal a number of things concerning grinding damage in ceramics. First considerable plastic flow accompanies grinding. Secondly, the mechanisms of material removal can be plastic flow and/or brittle fracture, depending on the conditions at the wheel-workpiece interface and on the workpiece orientation. Finally, extensive subsurface deformation and fracture introduced during semifinish grinding must be removed in subsequent finishing operations if surface integrity is required.

When polycrystals are ground, the resulting damage is a combination of the effects observed with single crystals. Figure 4 shows scanning electron micrographs of polycrystalline MgO ground with the same 46 grit alumina and 100 grit diamond wheels used for the single crystals[9]. The flat areas on the ground surfaces show where material has been removed by plastic flow. The faceted areas are where material has been removed by brittle fracture which has occurred mostly along the grain boundaries. (ie. intergranular fracture.) The depth of grinding damage in large grained MgO polycrystals is about the same as that found in single crystals[10]. In fine grained material the damage is generally confined to a depth of about two grain diameters[11,12].

B. Grinding damage in brittle ceramics

Brittle ceramics are, as expected, almost exclusively machined with diamond wheels. Types of grinding damage in brittle ceramics are essentially the same as those observed in semibrittle materials. As expected, brittle transgranular and intergranular fracture are routinely observed on ground surfaces of brittle ceramics such as alumina (Al_2O_3). Figure 5 shows electron micrographs of the ground surface of an AVCO fine grain alumina workpiece[9]. The faceted regions show where material was removed mainly by intergranular fracture. A surprising aspect of these finished surfaces are the smooth regions where material was removed by plastic flow, just like the softer ceramics. This was an unexpected result but as discussed in the next section, can be rationalized.

Studies of subsurface damage in ground single crystals of alumina (ie. sapphire) have shown that grinding can introduce both plastic deformation and twinning[4,13,14]. Isolated regions of deformation in crystals ground with a 180 grit diamond wheel can extend 30μm beneath the surface. Based on the results shown for MgO and the existence of similar types of damage in sapphire crystals we expect the depth of grinding damage in a polycrystalline brittle ceramic to again be on the order of one to two grain diameters.

Let us summarize our observations of grinding damage at this point.

- In semibrittle ceramics grinding introduces a substantial deformed layer beneath the ground surface. The deformed region can be up to 150μm in depth.

- Ground surfaces of semibrittle ceramics contain cleavage cracks, intergranular fracture and plastic flow. The relative amounts of each depend on the wheel type and rate of material removal.

- In brittle ceramics (eg. alumina) grinding introduces cleavage and transgranular fracture into the machined surface as expected. An unexpected result was that grinding also can introduce large amounts of plastic flow into a brittle ceramic.

- In brittle ceramics the heavily deformed region is shallow but isolated slip bands can extend as far as 30μm beneath the surface[4,13,14].

These results all depend on the conditions at the wheel-workpiece interface. To better understand the observations we now discuss the mechanisms of chip formation, particularly the stresses, strain rates and temperatures existing when a chip is formed.

III. Chip formation during ceramic machining

Some indications of the conditions existing during chip formation can be obtained from grinding force measurements. These measurements are typically made with a grinding dynamometer[15]. The dynamometer measures, independently, the horizontal, F_H, and vertical, F_V, forces on the workpiece during a grinding pass. This information can be used to obtain the actual stress on an individual chip.

The forces generated when a ceramic is ground differ somewhat from those experienced during grinding of metals. First the forces are lower. Secondly, in contrast to metals, the vertical forces exceed the horizontal forces. Both observations are a direct result of the large amount of brittle fracture that accompanies chip formation in ceramics. In ceramics, as in metals, the grinding forces increase with the rate of material removal.

To calculate the stress on a chip we need to know the geometry of chip formation; ie., the size and shape of an individual chip and the relation of chip size to grinding parameters such as the distribution of cutting points on the wheel surface, the workpiece feed rate and the wheel depth of cut.

The shape of an idealized grinding chip is represented by the dashed area in Fig. 6. The workpiece moves forward a distance, PQ, in the time an abrasive grain on the wheel surface cuts the length of the chip. PQSR is approximately triangular in shape as shown. The length, ℓ, and maximum thickness, t, of the chip are given[16] in terms of determinable quantities, by

$$\ell = \sqrt{Dd} \qquad (1)$$

and

$$t = \left[\frac{4v}{\pi cnrd}\sqrt{\frac{d}{D}}\right]^{\frac{1}{2}} \qquad (2)$$

where

D = wheel diameter

d = wheel depth of cut

v = workpiece feed rate

n = wheel speed in rpm

r = ratio of mean scratch width to mean scratch depth

c = areal density of cutting points on the wheel surface.

Typical values of the chip length and chip thickness are 2mm and 1.0μm respectively for a 25μm cut taken at a feed rate of 25 cm/min with a 100 grit diamond wheel. Note that grinding chips are quite thin with respect to their length.

Once the chip geometry is defined, it is a simple matter to estimate the stress on the chip[17,18]. A useful quantity in this respect is the specific grinding energy, U,[15]. U is the energy expended in removing a unit volume of material from a workpiece.

$$U = \frac{F_H \text{ x distance abrasive grain moves in unit time}}{\text{rate of material removal}} = \frac{\pi D n F_H}{Vdb}$$

where b is the width of the cut. If $\dot{W}$ is the work expended in unit time to remove material then

$$\dot{W} = UVdb.$$

R is the rate of chip formation thus

$$R = VbC$$

The work expended per chip is $\frac{\dot{W}}{R}$ so the mean force on the chip is

$$\overline{F}_{chip} = \frac{\dot{W}}{Rl}$$

The mean volume of a chip is the rate of material removal divided by the rate of chip formation.

$$\overline{V}_{chip} = \frac{Vdb}{Vbc}$$

To get the mean cross sectional area of a chip the chip volume is divided by the chip length, l.

$$\overline{A}_{chip} = \frac{Vd}{Vcl}.$$

Thus we see the mean stress on a chip is equal to the specific grinding energy.

$$\overline{\sigma}_{chip} = \frac{\overline{F}_{chip}}{\overline{A}_{chip}} = U$$

Specific grinding energies for ceramics are on the order of

10^{11}-10^{12} erg/cm^3 [3]. These values are quite large and, in fact, are on the order of the theoretical shear strength of a ceramic. It is not surprising, therefore, to find that plastic deformation accompanies chip formation even when brittle ceramics are machined.

The extremely high stresses are not the only cause of the deformed surface regions, however. The temperatures in the near surface regions are expected to get quite high during chip formation. Temperature measurements are difficult but estimates have been made which exceed, 1000°C [19]. This temperature is on the order of the softening temperatures of many ceramics. The situation is further complicated by the fact that the strain rates during chip formation are exceedingly high. The strain rate can be estimated by

$$\dot{\gamma} = \frac{V}{t} \simeq 10^{8} \text{ sec}^{-1}$$

under ordinary conditions. The extremely high strain rate is most certainly accompanied by adiabatic conditions during chip formation. A further complicating factor is that chips are formed at high strain rates under what may be considered impact conditions. There are, no doubt, shock waves introduced during chip formation that affect material removal. Some elegant experiments have, in fact, shown that a measurable amount of material is removed after an abrasive grain cuts a chip [20]. This effect is attributed to spalling as a result of reflected shock waves. The exact conditions of chip formation during ceramic grinding are hard to characterize, of course, and will be the subject of much future research before they are completely known.

In view of the extreme conditions existing during chip formation, the existence of plastically deformed surface layers on ground brittle ceramics is not surprising. The deformed layers are quite shallow because the extremely high shear stresses are expected to exist only to a depth on the order of the chip thickness. Large amounts of brittle fracture are expected. The impact of the abrasive grains on the workpiece propagates fracture to depths greater than the chip thickness. Subsurface intersection of the cracks accounts for much of the observed surface roughness. We emphasize again that most indications are that subsurface damage appears to be limited to a depth of one or two grain diameters in polycrystalline ceramics.

IV. PRACTICAL IMPLICATIONS

An understanding of the nature and extent of the damage introduced into ceramics by semifinish grinding serves two purposes. First it guides the subsequent operations which must be performed to finish the part. As an example, mechanical failure as a result of machining induced flaws in a ceramic component is minimized only if the finishing operations remove material beyond the depth of the largest flaws introduced during grinding. It has been suggested that the size of the flaws introduced by abrasive machining are a direct function of the abrasive grit size [4]. It is essential that subsequent operations use successively finer grit sizes to minimize this damage. It is usually not wise to eliminate intermediate steps since the rate of material removal during polishing, for instance, is so low that unacceptable times would be needed to polish through the damage. Furthermore, the plastic deformation occurring during polishing may conceal surface flaws. From the standpoint of mechanical strength the machining direction on the component can be important. Recent work has shown that components machined in a direction lying perpendicular to a tensile service stress were considerably weaker than those machined in a direction parallel to the stress[21].

The above statements pertain mainly to the elimination of subsurface fractures introduced by grinding. The consequences of plastically deformed layers introduced by grinding and lapping are also notable. Figure 7 shows the residual stress in a ground magnesium oxide crystal as revealed by polarized light[22]. The residual stress (in this case compressive) is a direct result of the plastic deformation occurring under the grinding wheel and can be shown to be on the order of 10^5 psi ($1.1 MNm^{-2}$).

Stresses of this magnitude can cause warping of precision components. They can also result in measurable changes in the physical properties of many ceramics[3,4,23]. An example is shown in Figure 8[3]. The figure shows B-H loops of a torroidal nickel-zinc ferrite sample before and after annealing. A decrease in permeability induced by machining can be restored by annealing and vice versa. These effects are also more severe after lapping and polishing than after only grinding. This is due to the cracks introduced by grinding which relieve the residual stresses that cause the observed degradation in magnetic properties.

V. SUMMARY

In this paper we have reviewed the damage that can be introduced into a ceramic workpiece by a semifinish grinding operation.

The damage consists of

- Brittle transgranular and intergranular fractures extending to a depth of about two grain diameters below the surface in polycrystalline ceramics.

- Heavy surface and subsurface deformation. In semibrittle ceramics the deformed layer can be over 100μm deep. In brittle ceramics the heavily deformed layer is quite shallow, ie. on the order of a few μm.

The damage, particularly the unexpected large amounts of plastic flow, can be rationalized in terms of the extreme conditions (stress, temperature, strain rate) existing at the wheel-workpiece interface during chip formation.

Finally some examples were discussed in which a realization of the existence of the grinding damage led to procedures that could improve the performance of machined ceramic components.

VI. ACKNOWLEDGMENTS

A large amount of the work contained herein was supported by the Office of Naval Research under Contract No. N00014-69-C-0123. The authors are indebted to Dr. A. N. Diness, Office of Naval Research, for his continued interest and to Dr. R. B. Maciolek, Honeywell Corporate Research Center, for his critical review of the manuscript.

VIII. REFERENCES

1. Ceramic Processing, National Academy of Sciences Publication 1576 (1968).

2. The Science of Ceramic Machining and Surface Finishing, National Bureau of Standards (U.S.) Spec. Publ. 348 (1972).

3. B. G. Koepke and R. J. Stokes, Naval Research Reviews, 27, 1 (1974).

4. P. F. Becher, in Ceramic Fabrication Processes, Treatise on Mater. Sci. and Tech., Vol. 9 p. 217, Academic Press, N.Y. (1976).

5. J. B. Wachtman, Jr., Bull. Am. Ceram. Soc., 46, 756 (1967).

6. R. J. Stokes, Ceramic Microstructures, R. M. Fulrath and J. A. Pask, ed., p. 379, Wiley and Sons, N.Y. (1968).

7. R. J. Stokes, T. L. Johnston and C. H. Li, Phil. Mag., 3 718 (1958).

8. B. G. Koepke and R. J. Stokes, J. Mater. Sci., 5, 240 (1970).

9. B. G. Koepke, p. 317 in reference 2.

10. R. W. Rice, J. Am. Ceram. Soc., 56, 536 (1973).

11. W. B. Harrison, J. Am. Ceram. Soc., 47, 574 (1964).

12. A. G. Evans and R. W. Davidge, Phil. Mag., 20, 373 (1969).

13. P. F. Becher, J. Am. Ceram. Soc., 57, 107 (1974).

14. P. F. Becher, ibid, 59, 143 (1976).

15. E. R. Marshal and M. C. Shaw, Trans. Am. Soc. Mech. Eng., 74, 51 (1952).

16. W. R. Backer, E. R. Marshal and M. C. Shaw, Trans. Am. Soc. Mech. Eng. 74, 61 (1952).

17. G. S. Reichenbach, J. E. Mayer, S. Kalpakcioglu and M. C. Shaw, Trans. Am. Soc. Mech. Eng., 78, 847 (1956).

18. W. R. Backer, discussion to reference 17. ibid 857.

19. P. J. Gielisse, et. al., Final Tech. Rep., NASC Contract N00019-72-C-0202, Univ. of R.I., (1972).

20. O. Imanaka, S. Fujino and S. Mineta, P. 37 in reference 2.

21. R. W. Rice, p. 365 in reference 2.

22. E. Bernal G. and B. G. Koepke, J. Am. Ceram. Soc., 56, 634 (1974).

23. R. J. Stokes, p. 343 in reference 2.

VIII. FIGURE CAPTIONS

1. Grinding and crystal geometry used when grinding MgO single crystals.

2. Surface and subsurface damage in MgO crystals ground in a {100}<100> orientation with a 46-grit alumina wheel.

3. Surface and subsurface damage in MgO crystals ground with a 100 grit diamond wheel.

4. Scanning electron micrographs showing polycrystalline MgO surfaces ground with two different wheels.

5. Electron micrographs of the surfaces of AVCO alumina samples ground with a 100-grit diamond wheel showing extensive plastic flow.

6. Schematic showing shape of idealized grinding chip.

7. Ground MgO crystals photographed in transmitted polarized light showing highly stressed machined surface.

8. Effect of surface condition on the B-H loop of a Ni-Zn ferrite torroid showing degradation of magnetic properties resulting from ceramic machining (Courtesy J. R. Kench, Honeywell Corporate Research Center.)

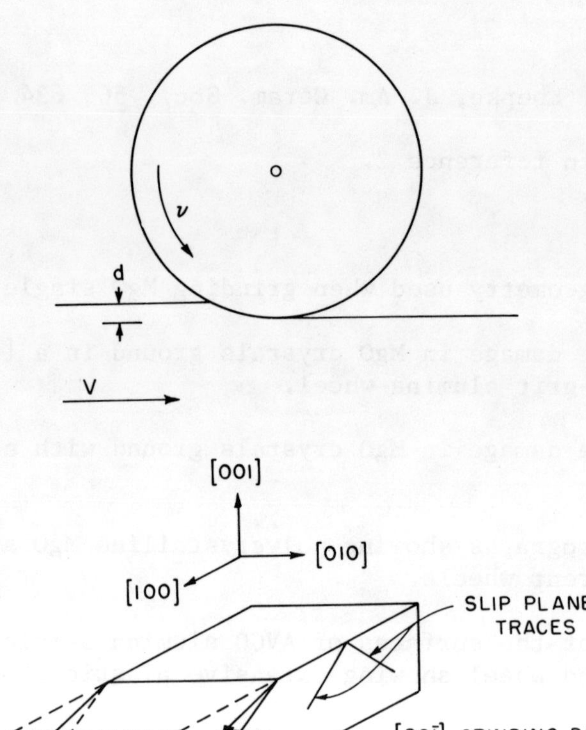

GRINDING AND CRYSTAL GEOMETRY USED WHEN GRINDING MgO SINGLE CRYSTALS

Figure 1

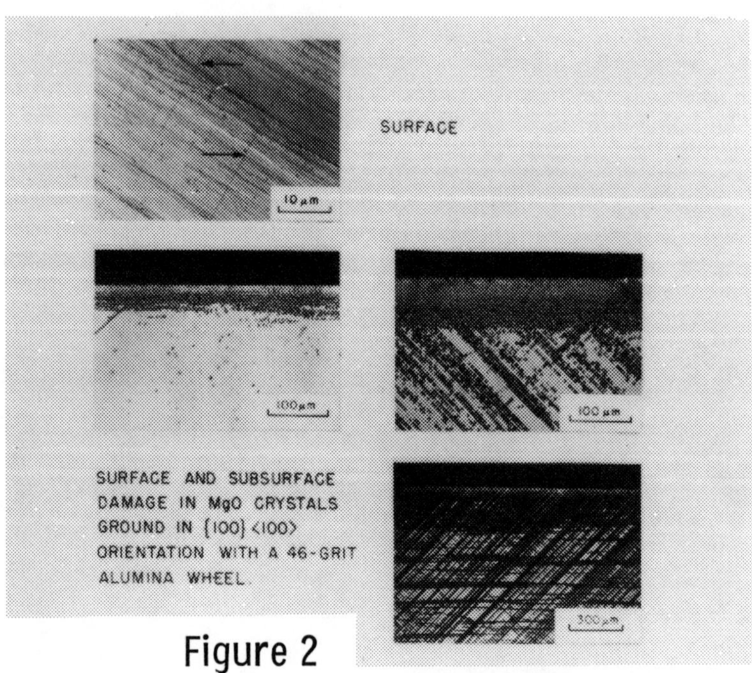

SURFACE AND SUBSURFACE DAMAGE IN MgO CRYSTALS GROUND IN {100} <100> ORIENTATION WITH A 46-GRIT ALUMINA WHEEL.

Figure 2

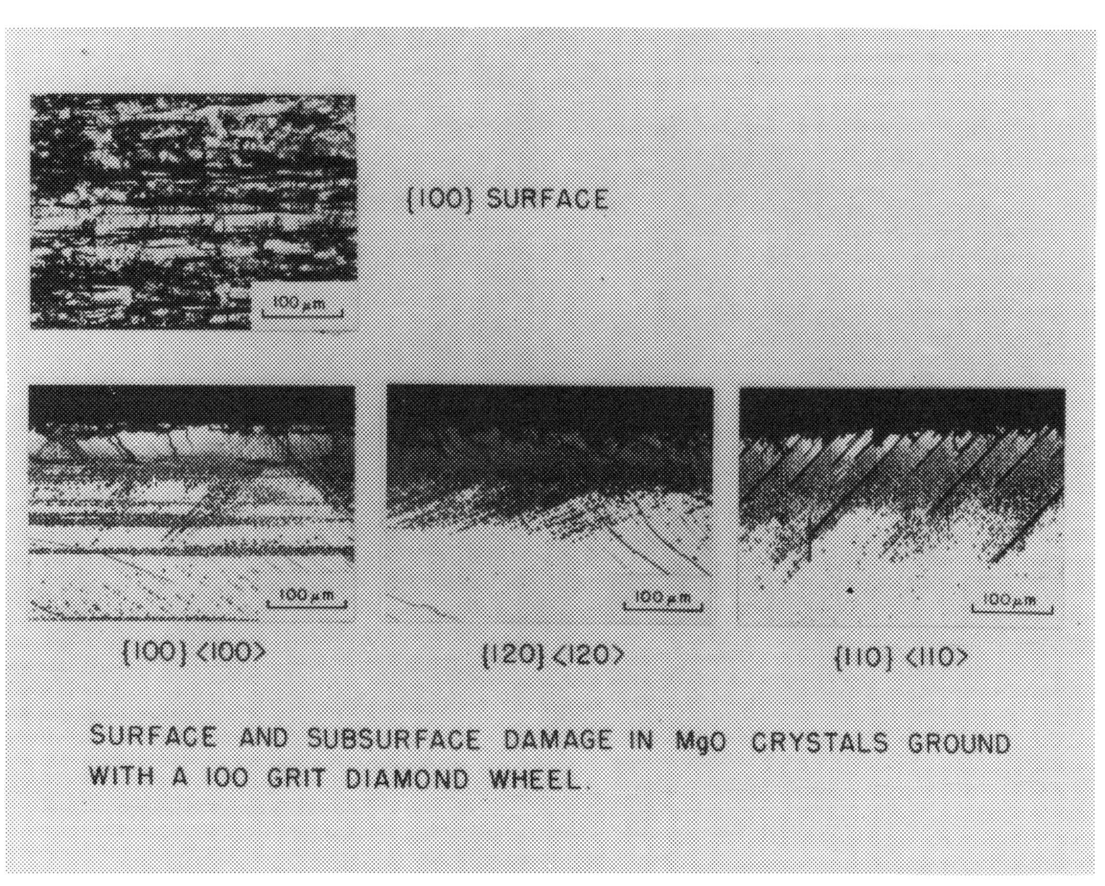

Figure 3

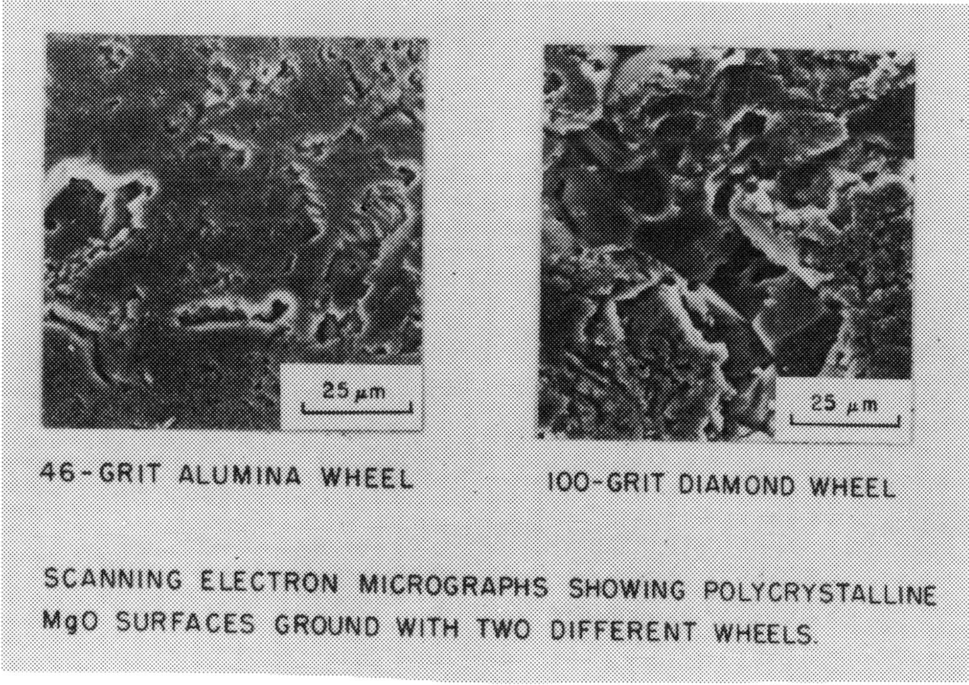

Figure 4

ELECTRON MICROGRAPHS OF THE SURFACES OF AVCO ALUMINA SAMPLES GROUND WITH A 100-GRIT DIAMOND WHEEL SHOWING EXTENSIVE PLASTIC FLOW.

Figure 5

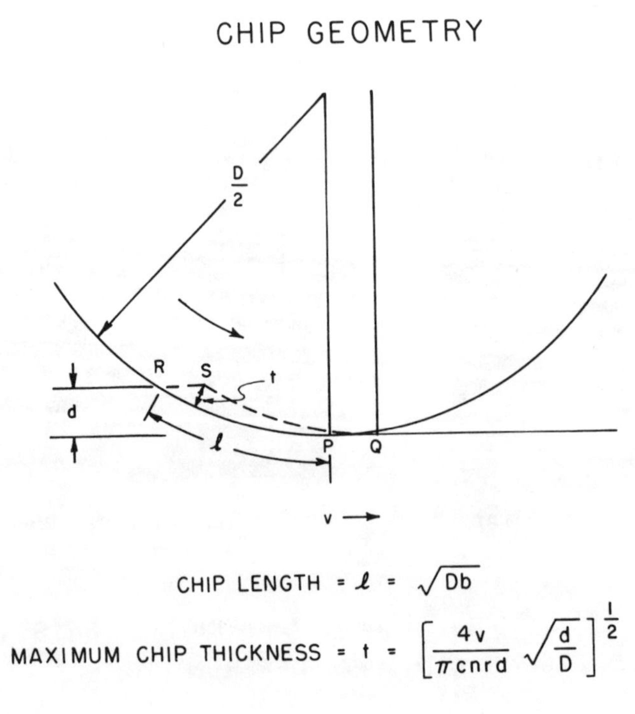

Figure 6

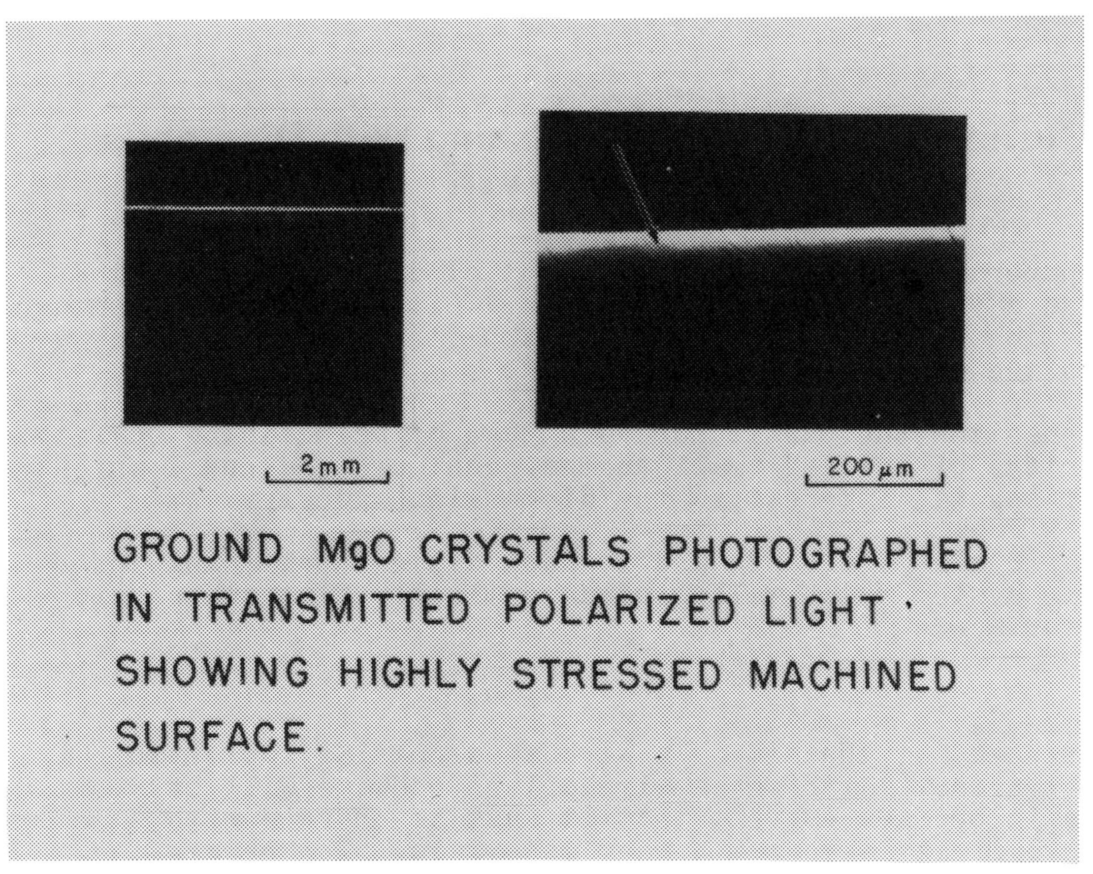

Figure 7

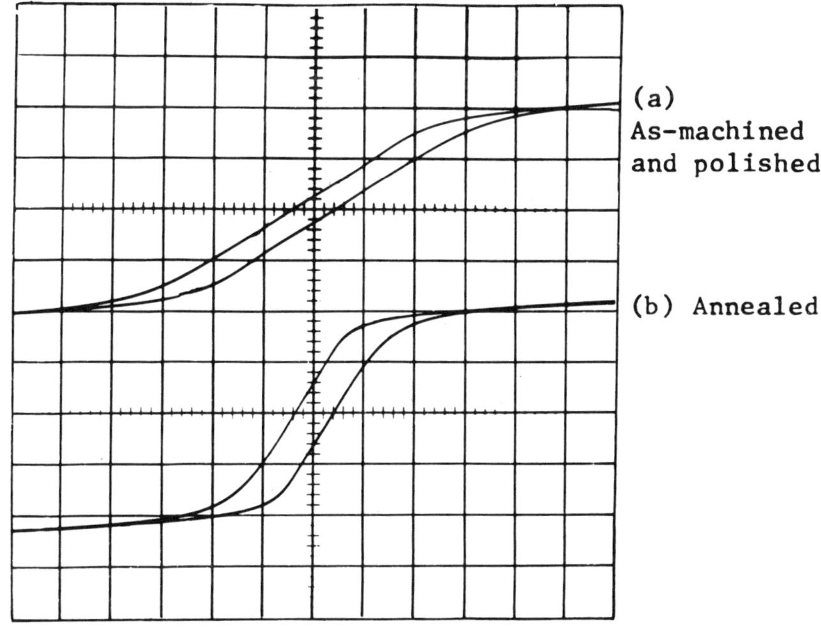

Figure 8

Presented at SME's Westec '79, March 1979

Cutting With Wires And Polishing With Diamonds

By Howard B. McLaughlin
Laser Technology Inc.

This paper discusses the development of diamond impregnated wire as a versatile tool for cutting. It offers the advantage of cutting all types of materials from the most fragile crystal to the hardest carbide.

It emphasizes the particular advantage of maintaining surface integrity, of cutting small angles, contours and radii which are of special interest to the die making industry.

The paper also discusses a six-wheel diamond polishing and lapping system which will produce optically flat and scratch-free surfaces.

Introduction

If you have not thought of a wire as a cutting tool, you are not alone. Although a wire driven endwise will cut soft things it becomes evident that it must be charged with some abrasive if it is to cut harder materials. The harder the material to be cut the further up the hardness scale the abrasive must be. The upper end of the hardness scale is diamond so this paper is going to describe "Cutting with Diamond Wire" as it applies to the tool industry.

Definition

A wire saw is a device which pulls a very taut wire endwise over a specimen to be cut. When an abrasive is impregnated into the wire any material can be cut which has a hardness below that of the abrasive. All solid materials can be cut using diamond wire.

History

Cutting materials with wires and ropes dates back to antiquity. For example, the ancient Egyptians employed stone hammers and chisels to remove and trim huge blocks used in the construction of the pyramids. It is believed that they used some abrasive in cutting and drilling the pink and gray granite. Figure 1 shows an artists conception of how a large rock might be cut by applying an abrasive slurry to the cut. There is no written proof of just how these cuts were made to prepare the blocks for the pyramids. But reflect for a moment on the magnitude of this job. Some of these stones were 45 feet wide x 45 feet high x 90 long and have an estimated weight of 22,000,000 pounds. There is nothing available today which can lift one of these stones. Yet the ancients moved them over 200 miles and piled them up over 500 feet high. Who can argue with the artist who thinks they were cut with a rope or wire?

Figure 1 ANCIENTS CUTTING STONE

A Mechanical Stone Saw. As the wire saw envolved it was equipped with a mechanical means for driving the wire. Although the old stone saws were used to cut relatively inexpensive materials, this changed as the demand for more sophisticated sectioning increased. The saws were refined to greatly improve their accuracy.
No longer could a large wire be used because it would cut away too much material. When cutting expensive materials such as platinum, the difference between the amount wasted in cutting a few slices is greater than the price of the saw.

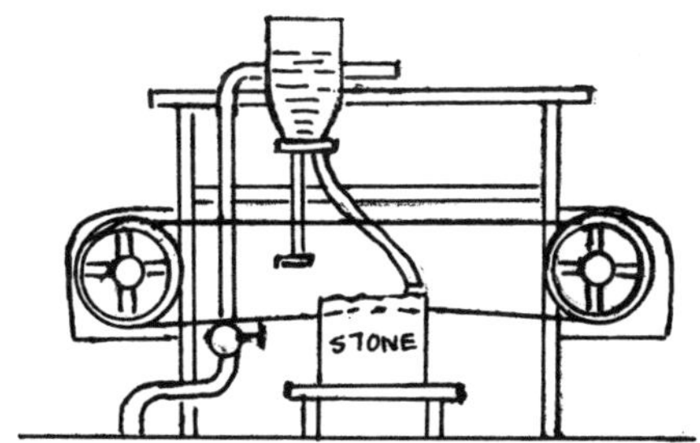

Figure 2 TYPICAL STONE CUTTING SAW

With the development of small diamond impregnated wires a tool was envolved which would cut every known material in this world and also things from outer space. One company specializes in cutting meteorites and 23 wire saws were used in the laboratories where the moon rocks were analized.

Figure 3

DIAMOND WIRE
CUTS EVERYTHING

<u>How diamond wire is made</u>. Any material which can be drawn into a wire can be made into diamond wire. If the wire is made of a hard material, it must be plated. Tungsten for instance would make a great wire but for one problem. It becomes brittle and breaks.

Some materials which are soft, stretch excessively.

The requirements for a good cutting wire are (1) it must have high tensile strength, (2) it must hold the diamonds well, (3) it must last until it is worn out.

The wire shown in Figure 4 has an alloy core with a tensile strength of over 460,000 P.S.I. It is plated with copper. The copper holds the diamond particles and the core will withstand the required tension for straight accurate cuts.

When this wire is made to .008" diameter, it can be operated at 3500 grams of tension. An ordinary wire under this tension will either stretch excessively or pull a part.

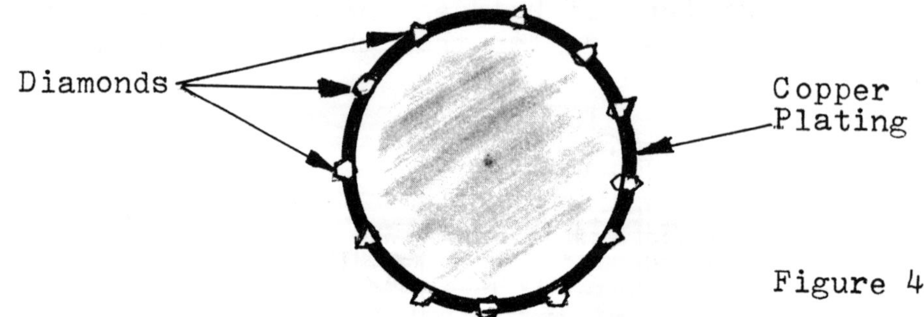

Figure 4

To utilize the phenomenal cutting capabilities of diamond wire special motor-driven wire saws were designed. These saws accomodate 100 feet of cutting wire in one continuous length. The wire is fastened at both ends of a large spool which drives it forward and backward. This reciprocating motion enables the wire to cut in both directions.

This method has several advantages:

1) When the wire reverses, the material which was removed by the cutting operation in one direction is taken from the kerf by the cutting action in the opposite direction. Perhaps it is this feature which prevents the removed material from adhering to the wire.

2) Since there are no joints or splices in the wire, it can be operated at very high tension. An .008" wire when spliced together like a band saw blade can be run around 500 grams of tension. The wire especially designed for wire sawing can operate at 3500 grams or 7 times the tension. The higher the tension the straighter the wire becomes. The straighter the wire, the more accurate the cut.

3) If an internal cut is to be made, the wire is removed from one end of the capstan (spool) and threaded thru a small hole drilled in the material. The loose end of the wire is reattached to the capstan and and the cutting operation is resumed.

Wire Sizes. A diamond wire can be made in a large variety of sizes. Figure 5 shows 6 standard sizes of diamond wire. The table lists the standard sizes, the diamond particle sizes in both inches and millimeters. It also shows the kerf sizes left by the cutting of each size wire.

Other sizes can be made to order. If a cutting problem requires a wire of some specific size, it can be made accurately to that size. In fact it can be made to cut at the upper end of the tolerance. As the wire wears the kerf (the cut left after the wire passes thru) becomes smaller. If for example the allowable tolerance is .0005", the wire will be spent by the low limit of the tolerance.

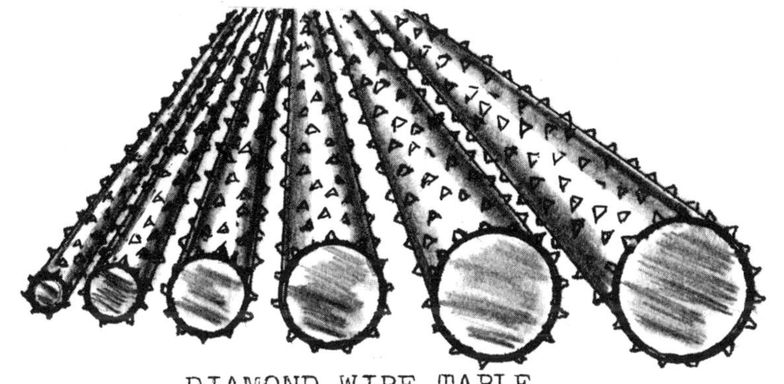

Figure 5

DIAMOND WIRE TABLE

WIRE SIZE		DIAMOND SIZE	KERF SIZE	
Inch	MM		Inch	MM
.003	.076	8 micron	.00325	.082
.005	.127	20 micron	.0055	.124
.008	.203	45 micron	.009	.228
.010	.254	45 micron	.011	.279
.012	.305	60 micron	.0135	.343
.015	.381.	60 micron	.0165	.419

Features of Diamond Wire

1) It cuts all semiconductor materials which are now on the market and many more which have been cut experimentally.

2) Cuts with practically no breakout. Breakout in cutting is one of the perplexing problems in the fabrication of glass, quartz, crystals, gemstones and many more friable materials.

Diamond wire saws cut all of these materials with almost no breakout.

3) Cutting causes little or no hair line cracks or craze.
This is of particular value to the semi conductor industry. Hair cracks and craze caused by conventional cutting must be polished off if the semi conductor is to perform to specification. The wire saw causes the least damage to crystals of any cutting method.

4) Generates little or no heat.
Most parts in the gem and semiconductor industries are secured for cutting with wax. The sawing operation requires the wire to pass through the part, through the wax and into a ceramic substrate. There is never enough heat generated by this operation to melt the wax. Nothing will adhere to the diamond wire.

5) A diamond wire cuts all metals.
It cuts hard metals better than soft ones. Although the diamond wire will cut lead, tin, copper and soft steel, it is not recommended for these materials.

6) The wire cuts angles leaving only a 5 mil radius.
This feature is of great importance in the die making industry. The wire will cut into a corner leaving a radius of ½ the diameter of the wire. For most extrusion die designs, a .004" or .005" radius is acceptable in the finished product. This enables the diamond wire to cut graphite electrodes with an accuracy and surface finish, which can be transferred directly to an EDM machine without subsequent hand work.

7. Cuts all metals including tungsten carbide.
Tungsten carbide cuts quite well with diamond wire. The saw is used to cut form tools. The finish is ready for production. The front of the tool is elevated to provide with clearance.

8) Cuts all gem materials.
Diamond wire will cut gem materials of all hardness up to and including diamond.

9) Cuts thin wall tubing.
Cutting thin wall tubing in production is a difficult problem. Most cutting methods have a tendency to mash or distort the hole or leave an objectional burr. The cost of deburring can often be considerably higher than the tubing. A diamond wire saw cuts this material almost burr free.

10) Cuts the moon rocks.
The moon rocks were subdivided to be sent to laboratories all over the world. Diamond wire saws were used to cut these rocks.

11) Cuts extremely fragile crystals.
When a wire saw is strung with a small wire and operated at low speed, it will cut slices so fragile they cannot be handled without danger of breakage.

12) Can cut very thin slices.
Opal for example is fragile but diamond wire will cut slices .003" thick.

13) Cuts high priced materials with very little waste.
If slices of .020 thickness are cut with a wheel, .020" thick the yield is 50%. If the same .020" slice is cut with a .005"

diamond wire, the yield is 75%.
14) Leaves a kerf slightly larger than the wire.
See Figure 5.
15) Cuts water soluble crystals.
In this case plain wire is used which is dampened by a wick or sponge.
16) Makes mechanical and electrical cut away sections.
17) Cuts potted assemblies for inspection.
When an electrical assembly fails after being potted, there is usually no suitable way to open the unit to determine the cause of failure. With a diamond wire saw, the assembly can be cut into slices thru all of the components including the potting material.

All kinds of conglomerates can be cut. Several oil companies use wire saws to cut oil well cores.

Diamond wire cuts honeycomb almost entirely without burrs. Nylon, paper, titanium, inconel, stainless steel and aluminum honeycomb can be successfully cut without distorting the cells.

In 1967 diamond impregnated wire was selected by Industrial Research as one of the most significant new technical products.

Eighteen different models of wire saws have been developed to use diamond impregnated wires. These range from table models for crystals and small specimens to guillatine saws having a cutting capacity of 24" x 24" or larger. They are offered for both manual and automatic operation.

The most recent developments are contouring wire saws used primarily to cut electrodes for EDM die making.

One model is designed for use with a Linemaster. A photo electric sensing head follows the layout lines of an enlarged drawing (up to 50X). This motion is transmitted to the saw in a selected reduced scale, by means of servo motors driving precision lead screws attached to X and Y tables. The saw then cuts a continuous path profile of the electrode.

Another model is operated by a keyboard computer. The X and Y tables are equipped with electronic scales which can detect a movement of .0001". These minute table movements are transmitted to the computer which sends signals to stepper motors to satisfy the program. The computer has memory storage provided on tape cassettes. Each cassette will accomodate the program for as many as 100 or more parts depending upon their complication.

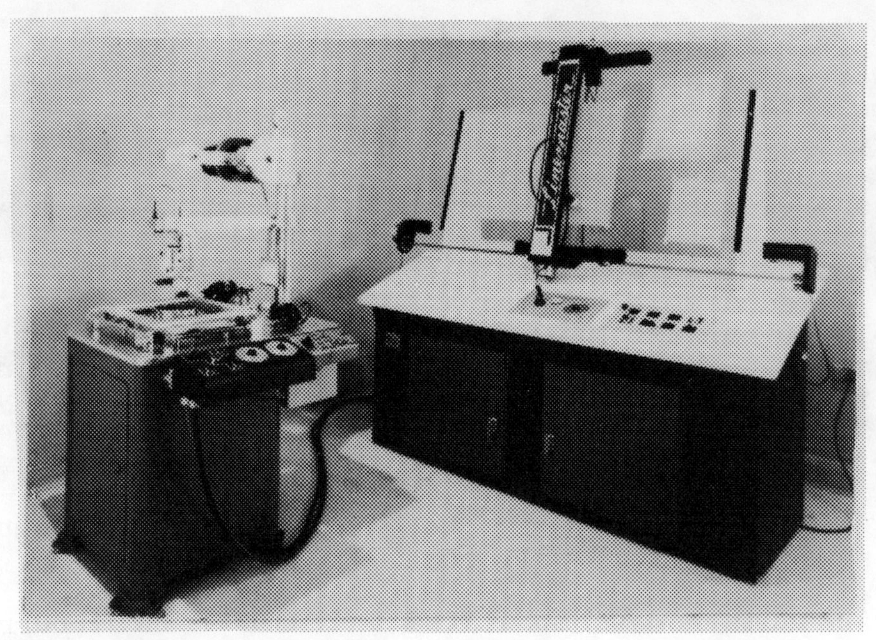

Figure 6

CONTOURING WIRE SAW
WITH LINEMASTER
TRACING SYSTEM

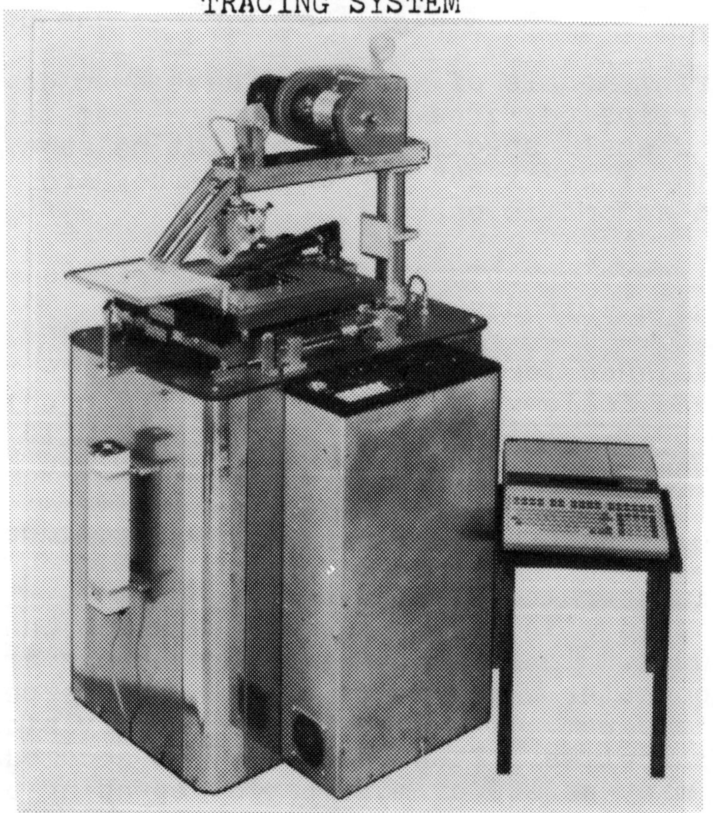

Figure 7

CONTOURING WIRE SAW
WITH COMPUTER
PROGRAM CONTROL

FEATURES OF LASTEC CONTOURING WIRE SAW

1. It turns internal corners leaving only a .005" radius.
2. It can cut insulators as well as conductors.
3. It can cut material up to 2" thick - thicker if necessary.
4. It makes 8 to 10 graphite electrodes per day.
5. The operator need not be an experienced tool maker.
6. Maximum number of electrodes per sheet of graphite.
7. Operation is entirely automatic.
8. Electrode ready for use off the saw.
9. A wire lasts an average of 15 hours for cutting graphite.
10. The saw cuts 4 sq. in. per hour.
11. The cost of the wire is less than $1.00 er sq. in.
12. Multiple electrodes can be made by stacking the graphite.
13. A self-contained vacuum system traps the process dust.
14. Wire is aligned with a Shadowgraph.
15. Cabinet is noise insulated.
16. Wire can be installed in less than 5 minutes.
17. Only tool required is a wire cutter.
18. Storage spool holds 10 wires.
19. Cuts all grades of graphite.
20. Offset provided to increase or decrease size of electrode.
21. Little or no heat is generated while cutting.
22. Parts are held down by double backed tape.

POLISHING WITH DIAMONDS

The Principle of Polishing.

The difference between polishing and lapping as the term applies to flat surfaces, is a matter of degree of fineness. Just where lapping stops and polishing begins depends upon the particle size of the abrasive used. Many kinds of abrasives can be used to lap and polish. The basic rule is that the abrasive used must be as hard or harder than the part to be polished.
When two discs are run together in an offset position and an abrasive is placed between them they make a polishing machine. If one of the discs is made of softer material than the other, the softer one will become impregnated with the abrasive and will polish the harder one.
To effectively polish to a scratch-free surface, the work should be done in a clean room with air locks on the doors and the air entering the room is filtered.
If a clean room is not available, the next best method is to use an environmental polisher (such as described here) which has a large transparent dome which effectively covers the polishing wheels.
When polishing is done in a dusty atmosphere, the air-bourne dust particles are often harder than the material being polished. Under these conditions, the material can never be polished to a scratch free surface.

Polishing In Steps.

When only two polishing wheels are available, and the material to be polished must be taken from the course to the fine, the fine wheel will take many hours to remove the course scratches. A polisher having 6 wheels is used, the parts to be polished are taken from the course to the final polish in 6 steps. The 6-wheel Environmental Polishing Machine is a complete sequential polishing system which will produce optically flat and scratch-free surfaces.
The six 10" lapping wheels are used in the following order:

No. 1 Wheel, copper, charged with 45 micron diamonds.
No. 2 Wheel, cast iron, charged with 20 micron diamonds.
No. 3 Wheel, cast iron, charged with 9 micron diamonds.
No. 4 Wheel, tin, charged with 3 micron diamonds.
No. 5 Wheel, tin, charged with 1 micron diamonds.
No. 6 Wheel, tin, charged with 1 micron aluminum oxide.

A super finish can be obtained by converting the No. 6 wheel to a chemical polisher.
The Chemical Polisher (used in place of the 1 micron No. 6 wheel) consists of an optically flat tin wheel mounted in a PVC pan which rotates with the wheel. A cavity around the periphery of the wheel holds the chemical polishing solution. The solution is picked up by a plastic pump (without moving parts), to deliver the fluid to the wheel surface. The chemical polisher completely removes all 1 micron scratches.

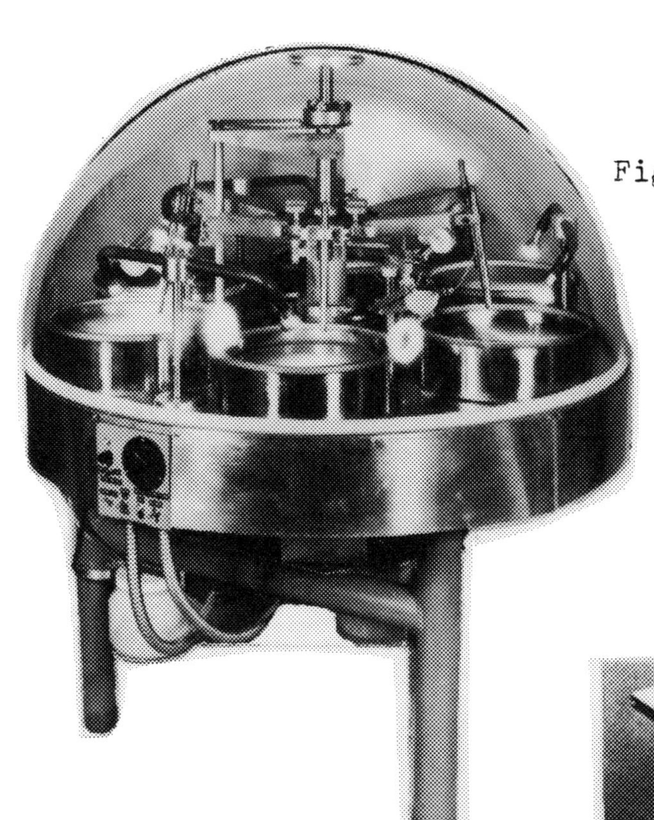

Figure 1

6-WHEEL POLISHING MACHINE
WITH ENVIRONMENTAL DOME

Figure 2

CHEMICAL POLISHER
TO PRODUCE A SUPER FINISH

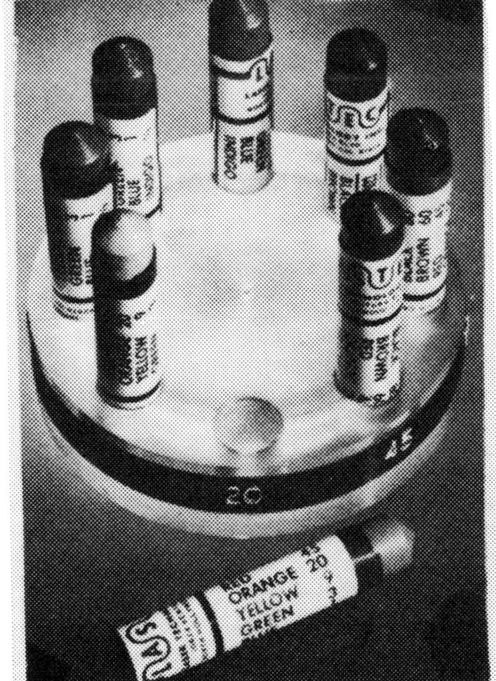

Figure 3

DIAMOND STICKS
TO CHARGE POLISHING LAPS

Available in micron sizes
from ½ to 80 microns.

Reprinted from Industrial Diamond Review, December 1976

CDA*—the new high performance diamond abrasive for grinding carbide

BY **A T NOTTER** AND **M O NICOLLS** DE BEERS TECHNICAL SERVICE CENTRE CHARTERS ASCOT UK

The ideal characteristics of a diamond abrasive for grinding cemented tungsten carbide have been discussed in the previous article from the De Beers Diamond Research Laboratory. The article below gives examples of the performance of De Beers new CDA diamond abrasive in its four versions in wet and dry grinding of various grades of cemented tungsten carbide, based on an extensive series of tests carried out at De Beers Technical Service Centre, Ascot (slightly abridged version of a paper presented at De Beers Seminar 'High Performance Abrasives & Tools for Machining Metals & Non-metallic Materials', 27–28 October 1976, Zurich, Switzerland)*

*Registered Trade Mark

It is difficult enough to establish the most important characteristics of a good grit for grinding cemented tungsten carbide, but even more so to make a grit containing them all. De Beers scientists and engineers have spent two years of development and intensive research culminating in the production of Carbide Diamond Abrasive, a specially synthesized grit containing all the essentials for grinding tungsten carbide:

Blockiness
An irregular blocky shape is essential to avoid the problems of orientation associated with elongated or flat particles.

Friability
Friable particles will break down in a controlled way rather than pull out.

Mosaic structure
To achieve a free-cutting action, the breakdown of each grit particle must be such that new and sharp cutting edges are continuously presented. A mosaic structure ensures this.

Surface roughness
A rough surface with re-entrant angles provides the necessary mechanical keying in resin bonds.

Homogeneity
All grit particles in a given size range should have similar strength characteristics.

Consistency
Whether coarse or fine, the grit particles should exhibit the same physical characteristics.

Reliability
From day to day, year to year, the toolmaker should be assured of unvarying quality.

This report gives an introduction to the characteristics of CDA*, and gives examples of the performance of the grit in its four versions (CDA, CDA55N, CDA30N and CDA50C) in the wet and dry grinding of various grades of cemented tungsten carbide.

Uncoated CDA

Uncoated CDA is mainly used where a soft grinding action is required. CDA is blocky yet friable, and the mosaic structure promotes controlled breakdown of each particle to ensure a constant supply of new, sharp cutting edges. Fig 1 shows the consistent particle shape of CDA; the colour ranges from light to dark grey.

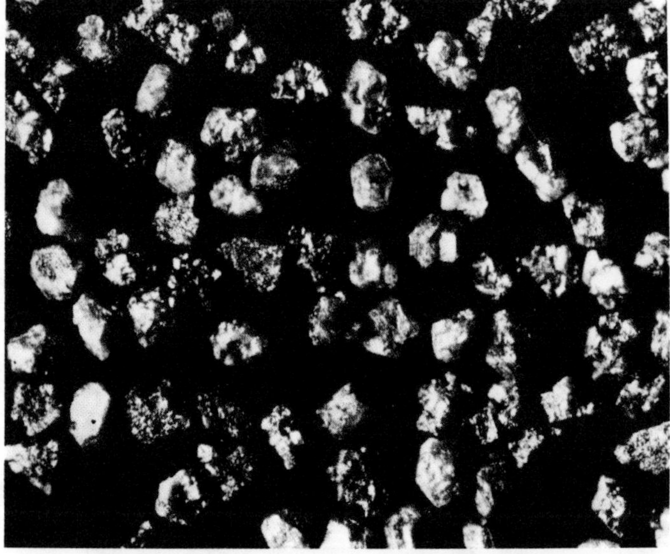

Fig 1 *CDA 60/80 US mesh. CDA is blocky yet friable, and the mosaic structure promotes controlled particle breakdown to give a free-cutting effect*

Fig 2 is a scanning electron micrograph which highlights in detail some of the desirable characteristics, present in CDA, for grinding tungsten carbide: the blocky shape and sharp cutting edges; the rough and irregular surface, together with the re-entrant angles which provide the necessary mechanical keying in resin bonds. As a result, the particles do not pull out easily. Fig 3 shows that the important characteristics such as blocky shape, structure and surface features, are to be found in all CDA particles. This ensures a product with consistent appearance and performance.

CDA in metal cladding (CDA55N, CDA30N and CDA50C)

The metal cladding acts firstly as a mechanical cage, which grips the friable grit particle and helps to control the rate of breakdown and prevent premature diamond loss during the grinding process. The rough surface of a good metal cladding further assists the retention of the diamond particle by keying into the bond. Finally, the cladding acts as a 'heat sink.' Without the cladding, the transient high temperature developed at the grinding point would be conducted by the diamond particle—which is an extremely good heat conductor—to the resin bond, which could be charred and weakened.

CDA55N

CDA55N (Carbide Diamond Abrasive 55% Nickel) is clad to 55% by weight of high nickel alloy, for use in wet and dry grinding of tungsten carbide in resin bonds. Again the shape is very consistent and the cladding does not remove the sharp cutting edges. The tolerance in cladding is guaranteed to ±0·5% in weight.

The scanning electron micrograph (Fig 4) shows the rough surface of the 55% nickel cladding, which ensures good retention of the particles in the bond and reduces premature pull-out. It also ensures that the particles are retained in the wheel after breakdown, thereby helping to control particle loss, and creating new cutting edges.

Another scanning electron micrograph of a few more particles (Fig 5) again shows the roughness of the cladding, which differs very little from particle to particle, indicating the consistent quality of CDA55N.

Fig 2 *CDA 100/120 US mesh, illustrating the rough surface of this grit type and re-entrant angles which ensure excellent bond retention*

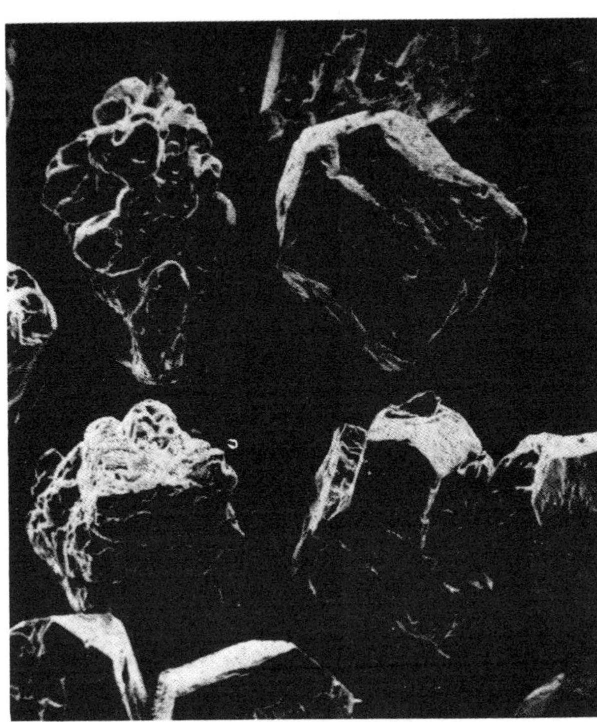

Fig 3 *CDA 100/120 US mesh. This scanning electron micrograph shows the consistent appearance of the product*

Fig 4 *This scanning electron micrograph of 100/120 US mesh CDA55N shows the rough surface of the nickel cladding*

Fig 5 *The consistent rough surface of each particle can be seen in this scanning electron micrograph of 100/120 US mesh CDA55N*

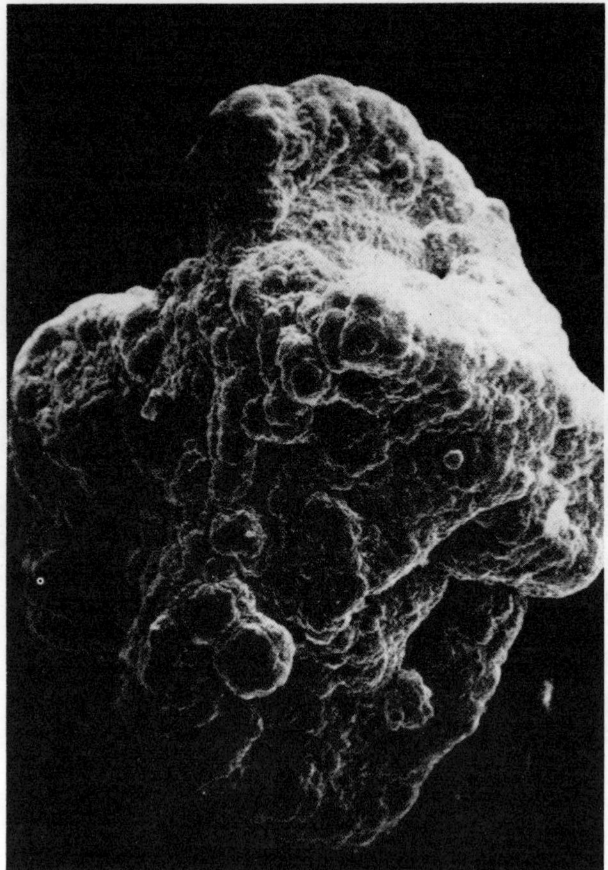

Fig 6 *The rough surface of the copper cladding can be seen in this scanning electron micrograph of 100/120 US mesh CDA 50C*

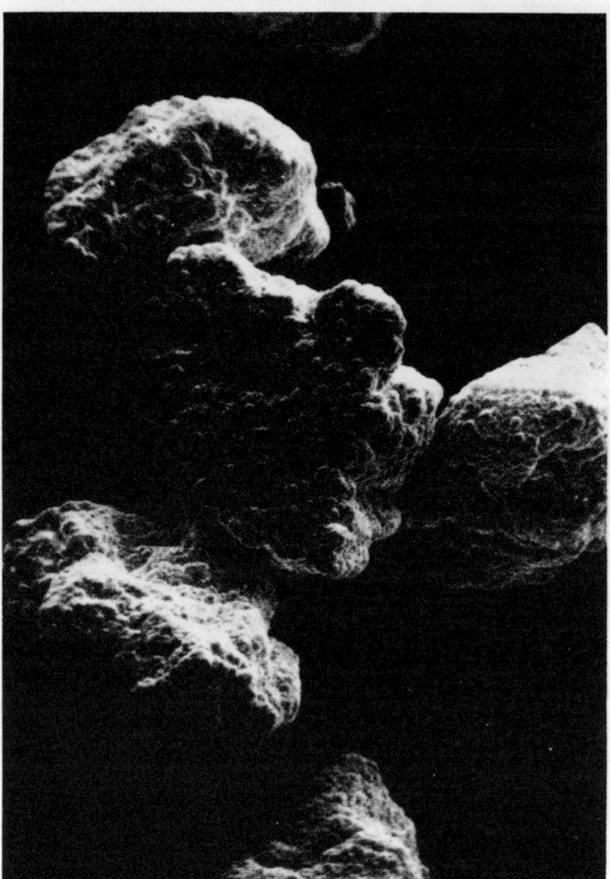

Fig 7 *This S.E.M. shows some more CDA50C particles (100/120 US mesh), demonstrating the consistent roughness of the cladding*

CDA30N

CDA30N has the same appearance as CDA55N. The reduced nickel coat of CDA30N has proved to be effective in certain cases where the 55% nickel cladding could result in smearing of the metal which produces higher grinding forces in dry grinding.

CDA50C

CDA50C (Carbide Diamond Abrasive 50% Copper cladding) is recommended for dry grinding of tungsten carbide in a resin bond. The copper cladding is a more conductive metal layer than nickel alloy for certain specialised bonds.

Fig 6 is a scanning electron micrograph which shows the rough surface of the 50% cladding which is tightly controlled to ±1·0% in weight. The cladding has in no way masked the underlying shape of the CDA particle and the sharp cutting edges and re-entrant angles are therefore left to carry out their important functions.

Fig 7 shows some more CDA50C particles all demonstrating that the roughness of the cladding is consistent throughout the product. This is evidence of the extremely tight controls which are placed on the production of the grit.

Grinding tests

The grinding tests have all been carried out with resin bond wheels and the results are expressed in three ways:

1. Grinding ratio (G-ratio)—the relationship between the volume of carbide removed and the volumetric wear of the wheel matrix.
2. Power drawn by the spindle motor of the grinding machine.
3. Temperature of the wheel matrix face during dry grinding.

The results shown in the graphs, to be discussed later, were obtained by averaging-out the figures from over 1800 individual wheel tests. De Beers had to be certain that they tested CDA wheels against wheels typical of the best of those used throughout industry. To achieve this, major diamond tool manufacturers were asked to make wheels incorporating their best bond for cemented carbide grinding, together with their favoured grit type. These wheels are specified as 'Industry Norm' in the graphs. In addition, De Beers supplied equivalent diamond types, CDA, CDA55N, CDA30N and CDA50C to be incorporated in identical bonds. These wheels were then tested against the comparable 'Industry Norm' wheels under identical test conditions.

Wet surface grinding with CDA and CDA55N

Fig 8–13 show the performance of CDA and CDA55N in wet grinding ISO grade P20 tungsten carbide. Both G-ratio and power drawn on the spindle motor are related to 'Industry Norm'.

The wheels were all made by well-known diamond tool manufacturers from various countries; and the tungsten carbide was obtained from several carbide manufacturers. Different surface grinding machines were used, and all these variations helped to ensure results meaningful to industry.

The wheel matrix surfaces were microscopically examined after each test to study features such as grit particle pull-out. The tungsten carbide workpieces were measured for surface roughness and the surfaces examined for structural damage.

The wheels used in the programmes were of the following specifications:

1. 1A1 resin bond: 125 mm × 6·35 mm × 31·75 mm (nominal 5 in. × $\frac{1}{4}$ in. × 1$\frac{1}{4}$ in.)
   Concentration: 100
   Grit size: Various
   Grit type: CDA and CDA55N 'Industry Norm'

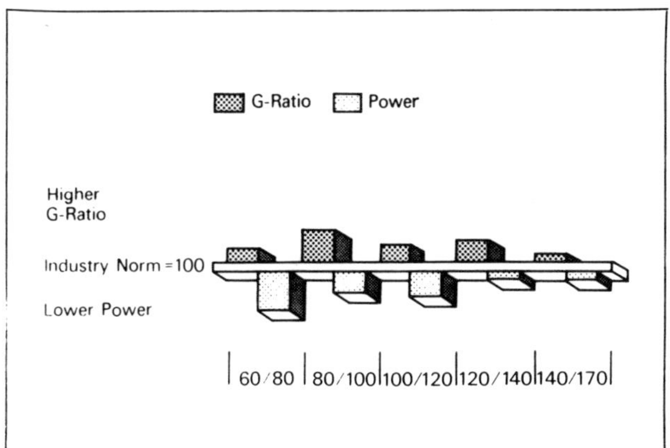

Fig 8 *CDA—relative performance in wet surface grinding. 1A1 wheels, 125 mm diameter 60/80 to 140/170 US mesh*

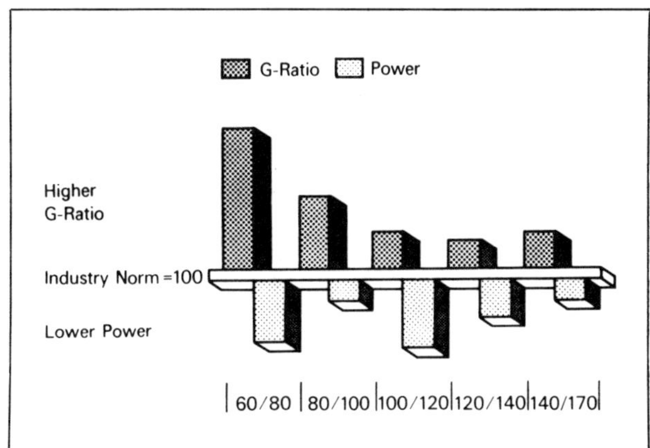

Fig 9 *CDA55N—relative performance in wet surface grinding. 1A1 wheels, 125 mm diameter, 60/80 to 140/170 US mesh*

2. 1A1 resin bond: 200 mm × 10 mm × 31·75 mm (nominal 7¾ in. × ⅜ in. × 1¼ in.)
   Concentration: 100
   Grit size: 100/120 US mesh
   Grit type: CDA55N 'Industry Norm'

As already mentioned CDA would be used when a fairly soft grinding action is required. When grinding a sensitive carbide, this type of action would help to avoid heat checks and cracks. Fig 8 shows the relative G-ratio and relative power drawn on the spindle motor in such a situation. In all grit sizes tested good results have been achieved compared to 'Industry Norm'. As the power is always less than for 'Industry Norm', the CDA grinds more freely, ensuring the least possible damage to the tungsten carbide.

A large programme was carried out using CDA55N in grit sizes 60/80 to 140/170 US mesh, which are the sizes mainly used by industry in wet surface grinding. The vast majority of carbides can be ground extremely successfully with CDA55N in resin bond wheels. Fig 9 shows the relative G-ratio and relative power levels achieved. There are two distinct advantages for CDA55N over the 'Industry Norm': firstly, CDA55N yields a higher G-ratio, and secondly it invariably draws less power on the spindle motor, thus making CDA55N a very economical and free-cutting nickel clad grit.

Fig 10 shows the power charts taken with two 125 mm (nominal 5 in.) diameter wheels with 60/80 US mesh, over four tests. In each case, the first chart is taken from the Conditioning Test and generally tends to be more erratic since the wheel is still generating the correct matrix profile during this time. The figures from the Conditioning Test are, therefore, ignored when working out an average power level.

The average power for CDA55N (Fig 10) was 760 watts as against 860 watts for the 'Industry Norm' wheel. A further point is that there was a considerable variation from minute to minute in the level of power drawn by the 'Industry Norm' wheel, whereas the CDA55N was much more consistent. The grinding sound from the 'Industry Norm' wheel was very harsh and the breakdown of the matrix inconsistent.

Table 1 lists the tightly controlled parameters at which all 125 mm (nominal 5 in.) diameter wheels have been tested.

In evaluating CDA and CDA55N, no new grinding technique has been introduced. The parameters have been carefully established over the many thousands of tests carried out by De Beers in the past, and are confirmed by industry as yielding the most acceptable results. This is important since it lends confidence to the results and they can be meaningfully compared to results achieved in the past, under the same test conditions, with other grit types.

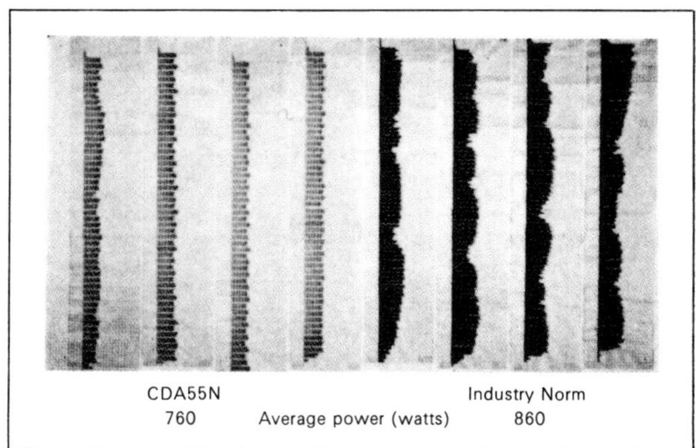

Fig 10 *Power charts—1A1 wheels 125 mm diameter, 60/80 US mesh in wet surface grinding*

TABLE 1

Wet surface grinding—CDA and CDA55N

Test conditions – 1A1 wheels 125 mm × 6·35 mm × 31·75 mm

Machine	Jones & Shipman '540' surface grinder
Spindle speed	3800 r.p.m.
Wheel peripheral speed	25·3 m/s
Downfeed	0·025 mm
Total downfeed	2·0 mm
Crossfeed	1·5 mm
Table speed	16 m/min
Specimen	P20 tungsten carbide
Specimen face size	156 mm × 80 mm
Coolant	Syntilo No. 3
Coolant ratio	1:80 dilution in water
Coolant flow rate	3 litres/min

Work was also carried out using 1A1 wheels 200 mm diameter × 10 mm wide (nominal 7¾ in. × ⅜ in.) in 100 concentration, grit size 100/120 US mesh. This wheel size could not be ignored as it is one of the most widely used in industry. Fig 11 shows the results for wheels made with four bonds from four different commercial wheel manufacturers; each bond was recommended as being suitable for metal clad grits in wet surface grinding of tungsten carbide. This shows that CDA55N does not require any specialised bonds or manufacturing techniques. Each bond containing CDA55N gave higher relative G-ratios whilst drawing relatively less power than the 'Industry Norm'.

Fig 12 shows the power charts taken with two 200 mm (nominal 7¾ in.) diameter wheels, with 100/120 US mesh, over four tests each. The most striking feature, comparing these charts against those for the 125 mm (nominal 5 in.) diameter wheels, is the greater consistency of the 200 mm diameter wheels throughout each test. It is invariably the case that wheels of 125 mm diameter yield lower G-ratios and give more erratic power profiles than wheels of 200 mm diameter.

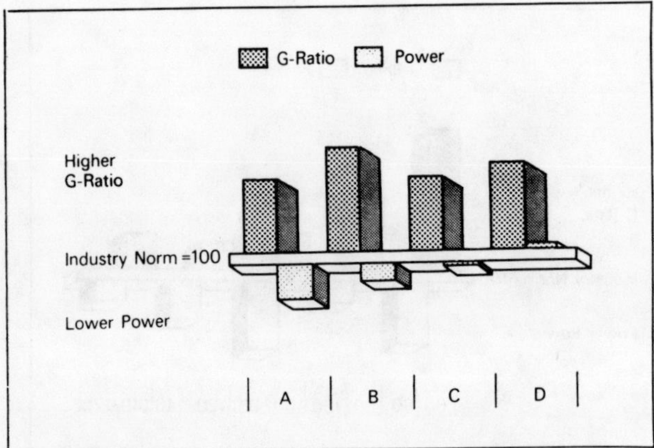

Fig 11 *CDA55N—relative performance in wet surface grinding. 1A1 wheels, 200 mm diameter, 100/120 US mesh*

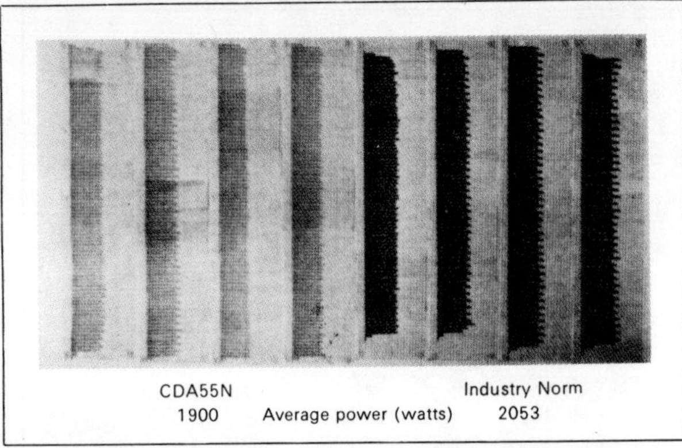

Fig 12 *Power charts—1A1 wheels 200 mm diameter, 100/120 US mesh in wet surface grinding*

The differences between the CDA55N and the 'Industry Norm' 200 mm charts are rather more marginal than is the case with the 125 mm charts. Fig 12 shows that the power level for CDA55N on average is 1900 watts, ignoring the Conditioning Test figures, whereas that of the 'Industry Norm' is 2053 watts. In terms of consistency from minute-to-minute and test-to-test, the wheel containing CDA55N is marginally better than the 'Industry Norm'.

Table 2 lists the parameters at which all 200 mm diameter wheels were tested which, as with the 125 mm diameter wheels, involve no new techniques or changes from established procedures.

From a test point of view, it is important to achieve sufficient wheel wear to yield meaningful results. Therefore, it is generally necessary to test a 200 mm diameter wheel for a much longer period than a wheel of 125 mm diameter; this accounts for the fairly large total downfeed per test shown in these parameters.

As with 125 mm diameter wheels, these parameters yield

TABLE 2

Wet surface grinding—CDA55N

Test conditions – 1A1 wheels 200 mm × 10 mm × 31·75 mm

Machine	Norton S3 surface grinder
Spindle speed	2250 r.p.m.
Wheel peripheral speed	23·9 m/s
Downfeed	0·025 mm
Total downfeed	4·0 mm
Crossfeed	2·0 mm
Table speed	16 m/min
Specimen	P20 tungsten carbide
Specimen face size	156 mm × 80 mm
Coolant	Syntilo No. 3
Coolant ratio	1:80 dilution in water
Coolant flow rate	3 litres/min

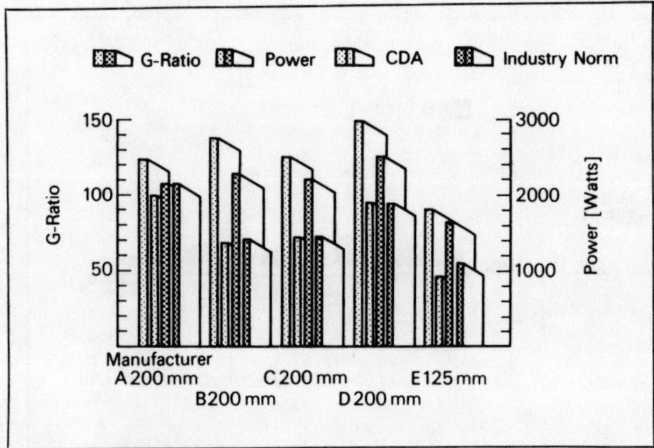

Fig 13 *CDA55N—comparison of 125 and 200 mm diameter 1A1 wheels, 100/120 US mesh*

the highest G-ratio commensurate with the highest possible removal rate.

Fig 13 shows the G-ratio achieved with eight pairs of wheels, 200 mm × 10 mm, and two pairs 125 mm × 6·35 mm; five pairs containing CDA55N and five pairs being 'Industry Norm'. Each manufacturer made one pair of 200 mm wheels in each grit type and one of the manufacturers also made the pairs of 125 mm wheels. Fig 13 shows the actual performance of CDA55N against 'Industry Norm'. It also shows the differing performance between 200 mm diameter wheels and 125 mm diameter wheels, both having the same grit size. In every case CDA55N outperforms the 'Industry Norm' in G-ratio and power drawn at the motor. It shows the advantage for larger sized wheels, i.e. 125 mm diameter gives a G-ratio of 90, whereas a 200 mm diameter wheel gives a G-ratio of at least 125, with the additional advantage of higher removal rates.

A simple conclusion can be drawn from this picture regardless of wheel size: CDA55N always yields a higher level of G-ratio than 'Industry Norm' for the same level of power drawn on the motor.

Dry grinding with CDA, CDA55N, CDA30N and CDA50C

So far our evaluations with CDA and CDA55N have been in wet grinding. A further programme has been carried out, with more to follow, in dry grinding with CDA, CDA55N, CDA30N and CDA50C, both in commercially and DRL-made wheels on different machines.

Dry grinding raises certain problems which do not exist to the same extent in wet grinding. Firstly, there is the temperature problem of the wheel matrix. Should this increase to a level at which the bond begins to suffer, then however good the grit, the wheel will not grind effectively. Secondly, there is the problem of wheel peripheral speed in dry grinding. It is found, in general, that with ISO 'P' grade carbides, a low peripheral speed, 12 m/s to 15 m/s (2400–3000 s.f.p.m.) is desirable; less power is drawn on the motor and less temperature is generated at the matrix face. There is, therefore, less danger of heat damaging the carbide. ISO 'K' grades, being to a large extent straight tungsten carbides, do not generally suffer from excessive heat, and the wheel seems to work better at peripheral speeds between 15 m/s and 20 m/s (3000–4000 s.f.p.m.). These points have been borne in mind when testing wheels under dry conditions, and two wheel peripheral speeds have been used.

As in wet grinding, the wheel matrix surfaces have been microscopically examined after each test, and the tungsten carbide pieces were examined, in particular for heat damage.

All results were achieved using wheels of the following specification:

11V9 resin bond: 100 mm × 3 mm × 31·75 mm (nominal 4 in. × $\frac{1}{8}$ in. × $1\frac{1}{4}$ in.)

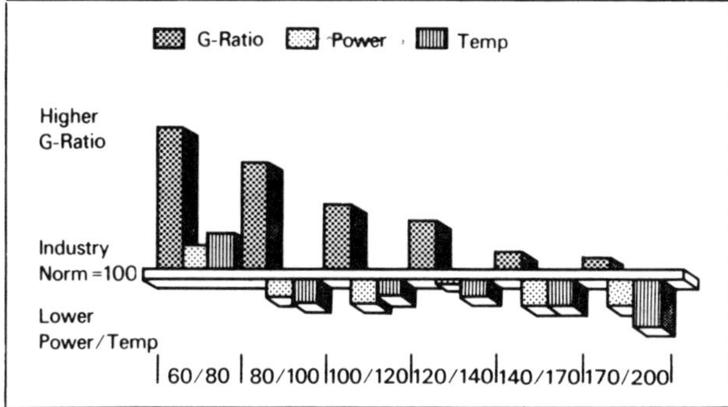

Fig 14 *CDA—relative performance in dry grinding P20 and K10 tungsten carbide. 60/80 to 170/200 US mesh*

Concentration:	75
Grit size:	Various
Grit type:	CDA, CDA55N and CDA50C 'Industry Norm'

In wet grinding, performance was measured in G-ratio and power drawn on the spindle motor. In dry grinding, the measurement of temperature on the wheel matrix face is also included, and has been measured by an infra-red pyrometer.

Fig 14 shows the performance of CDA in dry tool and cutter grinding ISO P20 and K10 tungsten carbides, using grit sizes 60/80 to 170/200 US mesh. On average the power and temperature are below the 'Industry Norm', and in only one case are they slightly above, which is caused by the relatively high G-ratio. CDA therefore makes a free-cutting wheel which is recommended for difficult-to-grind carbides. It is also very economical as can be seen from the relatively high G-ratios.

CDA55N performs extremely well in wet grinding tungsten carbide, and it is logical to establish its relative level of performance in dry grinding ISO P20 and K10 tungsten carbide with various grit sizes.

As shown in Fig 15 the power and temperature levels of the CDA55N wheels are lower, whilst most of the G-ratio levels are above the 'Industry Norm'.

CDA50C is specifically designed for dry grinding tungsten carbide since the copper coating is a particularly good 'heat sink'. Its success does depend to an extent on the use of a suitably developed bond, but the results achieved make that well worth while. Fig 16 shows the results for grit sizes 120/140 to 230/270 US mesh when grinding ISO P20 and K10 tungsten carbide. The trend of the G-ratios is to increase as the grit size decreases; the bonds used for the finer mesh grit wheels may have had an influence here, i.e. these bonds were perhaps more suitable for carrying a copper coated grit.

All CDA50C wheels were compared to 'Industry Norm' wheels which contained copper coated as well as nickel coated grits. The choice of grit type for the 'Industry Norm' wheels was left to the wheel manufacturer and more nickel coated grits were used in comparison to CDA50C in these finer sizes.

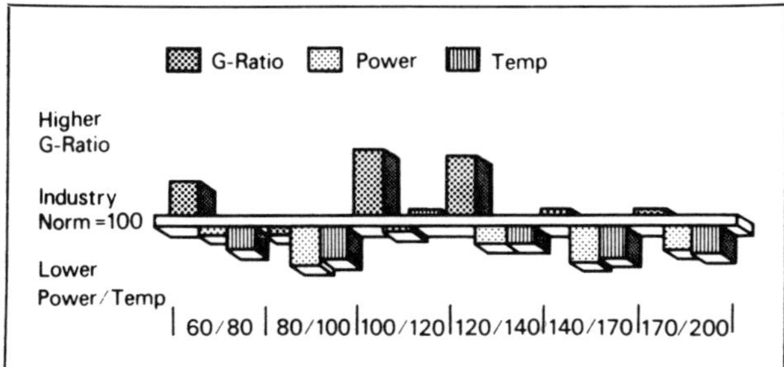

Fig 15 *CDA55N—relative performance in dry grinding P20 and K10 tungsten carbide. 60/80 to 170/200 US mesh*

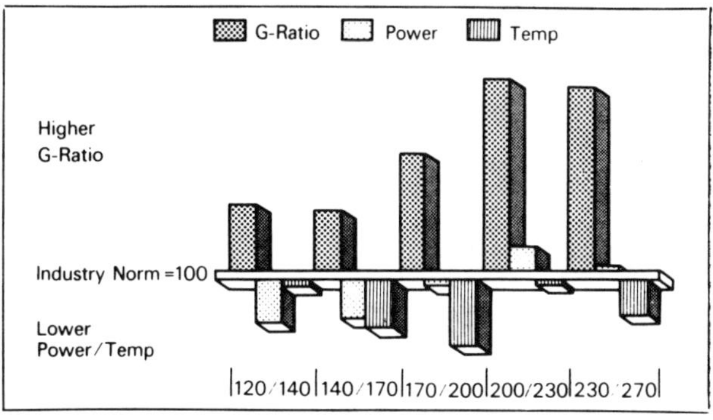

Fig 16 *CDA50C—relative performance in dry grinding P20 and K10 tungsten carbide. 120/140 to 230/270 US mesh*

TABLE 3

Dry tool and cutter grinding—CDA, CDA55N and CDA50C

Test conditions – 11V9 wheels 100 mm × 3 mm × 31·75 mm

Machine	Tacchella 6AP tool & cutter grinder
Spindle speeds	3500 r.p.m. for K10 2800 r.p.m. for P20
Wheel peripheral speeds	18·3m/s for K10 14·7 m/s for P20
Infeed	0·030 mm
Total infeed	2 mm
Table speed	2 m/min
Specimens	P20 and K10 tungsten carbide
Specimen face size	12·7 mm × 6·35 mm
Number of specimens	24 in an indexing head

Table 3 lists the parameters used for the three grit types, CDA, CDA55N and CDA50C, during the programmes of dry grinding tungsten carbide.

It should be pointed out that two carbide types, ISO P20 and K10, were involved, and each was ground at a different wheel peripheral speed. This was mainly due to discussions with diamond tool and carbide manufacturers who claimed that more ISO 'K' grades are being ground dry with wheel peripheral speeds around 18 m/s (3600 s.f.p.m.), whereas less 'P' grades are being ground dry. The results shown in Fig 16 were averages from both carbide grades. If CDA50C were to be related to the 'Industry Norm' on 'K' grades alone, the advantage would be even greater.

Conclusions

CDA is a completely new diamond abrasive from De Beers. It has been exhaustively tested and the results discussed in this report show that it outperforms the 'Industry Norm' in the grinding of cemented tungsten carbides:

—It produces consistently higher G-ratios;
—It draws less power from the spindle motor;
—In dry grinding it generates less heat at the surface of the wheel matrix;
—It is more free-cutting;
—It promotes more regular breakdown of the wheel matrix;
—It produces a greater number of new cutting edges as matrix breakdown progresses;
—It displays improved bond retention;
—Its grinding sound is smoother;
—It generates a superior surface finish; and
—It is more economical in use.

The distinct economic advantages, added to the impressive range of technical benefits to be gained by using CDA, must surely make it the most successful carbide grinding abrasive avaliable today ◊

Reprinted from The Carbide And Tool Journal, March-April 1979

Dry Grinding of Carbides

by

Roger J. Loecy
President
CITCO, Inc.

A new way to dry grind cemented tungsten carbide is with DeBeers CDA-M diamond in 5SD resin bond wheels by CITCO, Inc.

Industry's present demand for highly specialized diamond types designed for specific applications and operating conditions has been brought about by the progressive sophistication of grinding technology and continuing diversification by both wheelmaker and end-user.

A prime example of this specialization is a new diamond product CDA-M (Carbide Diamond Abrasive-Multigrain)(1) developed by DeBeers specifically for the dry grinding of cemented tungsten carbides and other ultrahard metals with resin bond wheels.

The distinguishing feature of CDA-M is that each individual particle is actually composed of much finer diamond grains which have been cemented together in a metal matrix. As a result, irregular shapes and sharp cutting points are evident in every particle.

The linkage between the grains is achieved by means of an eutectic alloy which is known to produce a powerful chemical bond with diamond. A thin layer of the alloy is formed around each single grain and the structure of the integrated particle is then consolidated by fusion of the metallic surfaces.

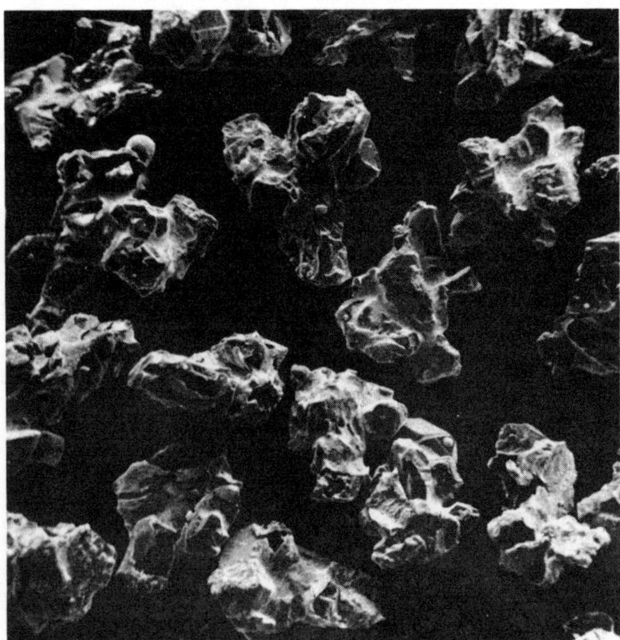

Figure 1. Scanning electron micrograph showing irregular shape and sharp cutting points in every particle of CDA-M.

Figure 2 shows a scanning electron micrograph of a single particle of CDA-M. The individual grains, of which the particle is formed, can clearly be seen protruding from the surface.

On the resin bond grit size scale, the composite CDA-M particles are relatively coarse, centering around 80 U.S. mesh. The constituent grains are much smaller, however, and cover a broad range of sizes of approximately 140 U.S. mesh and finer. The arrangement and distribution of grain sizes within this bond have been chosen after careful experimentation to yield an abrasive which is able to exhibit stock removal characteristics consistent with its coarse granular texture, while, at the same time, producing a surface finish typical of the fine diamond of which it is made. This unique particle structure and its distinctive blend of properties eliminates the need for a range of grit sizes. Therefore, CDA-M is manufactured in only one size.

An important property of CDA-M, and one which contributes significantly to its high performance during grinding, is the crystallographic character of the diamond grains. The required qualities are induced during synthesis by a process especially developed for this purpose. To obtain the maximum cutting efficiency and utilization of each particle, the small crystallites must ideally be blocky and irregular in shape and be equipped with sharp angular edges and points. They must also possess a certain degree of friability which will enable them to perform a firm and strong abrasive action under normal operating conditions yet fracture by small amounts during sudden surges in grinding

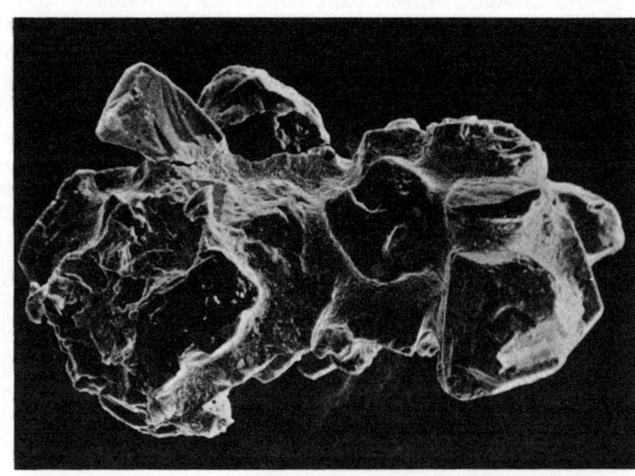

Figure 2. Scanning electron micrograph of a single particle of CDA-M.

stresses. In this respect, the friability of the constituent diamond grains and the strength of the metal matrix in which they are bonded, are complementary to each other. If the bonding is too weak, whole grains, even under low grinding forces, will rapidly be lost from the composite particle. Conversely, very friable diamond crystallites in conjunction with a high strength matrix will be prone to total disintegration before doing any work.

The ratio of metal to diamond in CDA-M is similar to that of standard metal clad resin bond grits. The alloy used to bond together the small diamond crystallites accounts for 55% ± 2% by weight of the whole particle. However, unlike a conventional metal coated grit, where the cladding covers the entire surface of the particle, in CDA-M the metal mainly occupies the spaces between the constituent diamond grains. In some cases the exposed surfaces of the individual grains may not be entirely covered with metal.

Technical considerations in the development of CDA-M are controlled fragmentation, bond retention, heat dissipation and impact shock.

(a.) **Controlled Fragmentation**

The frequent occurrence of particle fracture during the grinding of tungsten carbide and the subsequent generation of new cutting points is an indispensable requirement of a good resinoid grit. The development of wear flats, which can sometimes occur if the diamond is too strong, will inevitably result in higher grinding power, high wheel temperatures and undesirable residual stresses in the workpiece. Gross fracture of whole particles, on the other hand, will take a heavy toll on wheel life, due to inefficient diamond utilization. Fig. 3 illustrates some of the ways in which diamond wear can take place.

The most acceptable type of wear when grinding tungsten carbide is by a micro-chipping process, whereby each particle loses only a small fragment near the cutting surface when the grinding forces approach the critical bonding strength between the diamond and the resin matrix (Fig. 3.3). Not all particles, however, are able to wear in this idealized manner and while some may suffer catastrophic breakage in the bond (Fig. 3.2) others may be torn completely out of the matrix before being able to relieve the stress by fracture (Fig. 3.1). Thus, the manner of breakdown and the degree of control of its fracture properties are distinguishing features in the quality of a resinoid grit.

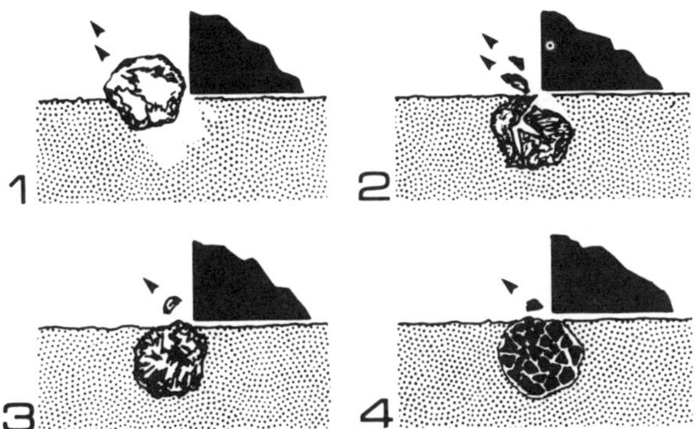

Figure 3. Modes of wear for different particle types in a resin bond.

CDA-M is a grit which has been specifically manufactured to comply with the philosophy of controlled grit breakdown. Each particle has, essentially, an in-built micro-chipping mechanism in that it is already composed of discrete diamond chips. Only the grains near the surface of the integrated particle are lost when the grinding forces become excessive and since this occurs across the junction of the binding alloy, the strength and structure of the remaining composite remain intact (Fig. 3.4). Consequently, the more effective and lower rate of wear of this diamond type is expected not only to increase wheel efficiency, but, because of its less erratic breakdown characteristics to result in much smoother grinding conditions.

(b.) **Bond Retention**

In contrast to the more rigid metal bonds, resin bonds have limitations with respect to particle retention. However, by their resilience, controllable wear rates, and contribution to a free-cutting action in the high temperature grinding of ultra-hard materials, resins offer considerable technical advantages.

To assist the keying in the bond and maintain a free cutting action, particles must be irregular in shape and endowed with many sharp cutting points. Fig. 1 shows a scanning electron micrograph which, highlights these features in several particles. In the grinding of very high modulus materials, additional support of the particle is required and is obtained by increasing the surface area of the diamond in the form of multiple re-entrant angle and corrugated features. In this latter respect, the irregular formation of CDA-M is unique, for it provides an almost ideal anchorage in the bond. The tenacious grip of the matrix on the particle by virtue of its extremely large contact surfaces enables the portion of the grit exposed to the workpiece to effect the necessary stock removal unhindered by any sudden loosening movement of the whole particle. This is an important consideration, since the wear characteristics of both the wheel matrix and the diamond depend largely on the degree of available support of the particle in the bond.

(c.) **Heat Dissipation**

The introduction of a metal cladding to diamond abrasive grits in 1966 represented a major advance in diamond wheel technology. In the wet grinding of cemented carbides, wheel efficiency was increased by more than two-fold compared with wheels containing similar, but uncoated grits. Dry grinding efficiency too, was substantially improved and although not so dramatic as the former case, this development did, nevertheless, mark a significant step forward in this area.

The general improvement in grinding efficiency with metal coated grits is attributable to a number of causes. Since metal to resin adhesion is superior to that of diamond to resin, a stronger retention of the particle in the bond is provided, while the cladding, which totally encapsulates the particle, offers support to the diamond and helps prevent premature release of fragments after fracture. The metal coating also absorbs some of the impact shock which, with unclad grits, is normally accepted totally by the resin. However, the most notable consequence of the cladding in dry grinding, is the part it plays in heat dissipation and in minimizing thermal damage to the resin.

Diamond is an excellent heat conductor, while the resin, in which it is embedded, is a very good insulator. Consequently, heat generated on contact with the workpiece is ordinarily conducted extremely quickly to the resin-diamond interface, and, depending upon the grinding conditions prevailing, may be sufficient to cause thermal degradation of the matrix in that vicinity. The interposition of a metallic layer between the diamond and the resin provides a thermal barrier, as far as the matrix is concerned, since some of the heat is absorbed within the metal sheath. Immediately after contact with the workpiece, and during the idling period, some of this heat is conducted back through the diamond and dissipated to the surroundings.

In the case of wet grinding, the coolant, which is in intimate contact with the wheel, is able to remove sufficient of this heat within one revolution to produce an acceptable terminal matrix temperature. Dry grinding, on the other hand, involves short workpiece contact and a relatively long idle period, during which cooling takes place.

The Multigrain grit provides numerous protective heat sinks between each grain. The diamond/metal interfacial area is very much greater, permitting the immediate contact heat to be more effectively absorbed. Consequently, the resin surrounding the composite particle feels less of the sudden influx of heat, because of the longer time taken for the heat to pass between the grains and their metallic interfaces. Thus, the matrix temperature with CDA-M will generally be much lower than with normal metal coated grits and is a significant factor in longer wheel life.

(d.) Impact Shock

During the course of each revolution of the grinding wheel, there are critical stages in the life of every abrasive particle. The particle is first subjected to a violent shock loading as it strikes, at high velocity, the leading edge of the workpiece. This is immediately followed by a relatively long period of direct contact with the workpiece, when the particle must endure a certain amount of attrition in compliance with the fluctuating grinding forces. Lastly, as it leaves the workpiece, the particle suddenly experiences only the reactive forces from the stored potential energy of the resin and which are intent upon its expulsion from the matrix.

Although all of these phases are heavily interdependent, the most significant and singularly dramatic event is that of impact shock. In this short time span, the normal and tangential forces on the diamond surge to probably several times those of the usual and relatively quiescent levels which are obtained during the continuous abrasive contact with the workpiece. Some particles, will, of course, be lost completely in this interval, since the peak value of the forces will far exceed the bonding strength of the diamond and the matrix. However, provided that the speed of impact and impulsive force are sufficiently great, it is more likely that the particle will, instead, be driven with high acceleration into the bond. The degree to which the bond is able to absorb this sudden incursion and soften the impact shock, will affect the future structural strength of the diamond and thus influence the course of events in the ensuing stages. Alternatively, some of the stress can be relieved by the loss of a small fragment near the point of impact, which will, in turn, considerably lessen the deformation of the bond as the latter brings the particles to rest. Certainly the strong single crystal will be at a far greater disadvantage in these circumstances than the complex crystal comprised of a multitude of grain boundaries.

One of the most notable characteristics of CDA-M is its ability to resist the detrimental effects of impact with the leading edge of the carbide workpiece in this very crucial first stage. The metallic interfaces between each grain act as numerous miniature shock absorbers, which cushion the effects of the collision and permit a large proportion of the shock to be dispersed within the composite particle. Thus, there will be less deformation of the matrix and the resilience of the resin will not be impaired. Although the bonding between the grains is strong, any high build-up of stress in the particle or the matrix will, most certainly, be additionally alleviated by the release of one or two grains at the surface. Of most importance, in this latter respect, is that this form of stress relief can be undertaken smoothly and without any fear of inducing cleavage cracks throughout the whole particle as would often be the case with a single crystal.

The CDA-M particle is, therefore, better able to work efficiently, with full support of the bond, during the second or frictional cutting phase. The particle is also better prepared for the sudden release of the compressive forces occasioned by contact with the carbide, as it passes out of the workpiece zone. During the non-cutting period of the cycle, the wheel surface is in a considerable state of agitation, and some particles will be striving to regain their stability in the bond when they again contact the workpiece.(2) Such instances will be less common with the CDA-M particle, where, because of its still intact structure and greater security in the bond after leaving the workpiece, a return to equilibrium conditions will occur much more rapidly.

Grinding Performance of CDA-M

So far, this paper has taken an "in-depth" look at the theory of how and why CDA-M is a successful carbide dry grinding grit. Some attention has also been devoted to the effect of this grit and other grit types on tungsten carbide, and the advantages of CDA-M have been highlighted.

As with any new DeBeers grit type, following the theoretical work largely carried out at the Diamond Research Laboratory in Johannesburg, the product was tested under industrially accepted conditions at the DeBeers Technical Service Centre at Charters, UK.

The grinding test programs were very carefully laid down, in order fully to explore the capabilities of CDA-M and to find out where its strengths lay. Such programs covered various types of carbide and widely differing test parameters.

The results of all the tests showed conclusively that in CDA-M, DeBeers had produced a grit which outperforms all other existing grit types in the dry grinding of tungsten carbide. Unlike other grit types intended solely for dry grinding and which generally require suitably developed resin bonds, CDA-M can be successfully used in a standard phenolic resin bond requiring no special development.

Fig. 4-6 show comparative results achieved by wheels containing CDA-M and wheels containing a standard nickel clad grit of size 140/170 U.S. mesh. Two types of workpiece fixture were used in each case; one being a conventional indexing head carrying 24 pieces of tungsten carbide and the other, a static device carrying four pieces held laterally.

The test programs also covered the use of both phenolic-aluminum hubs and bakelite hubs, in order to see

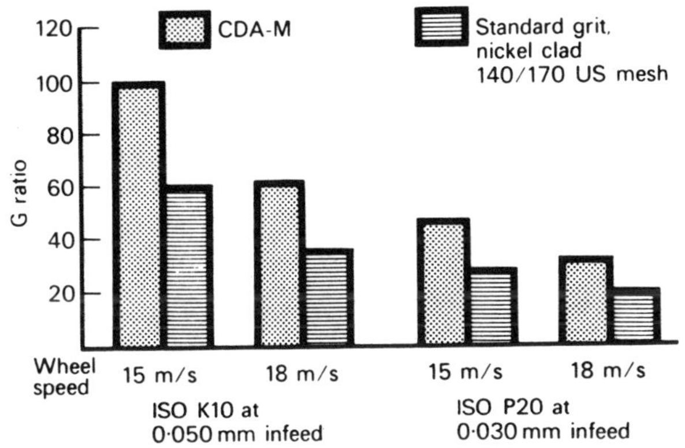

Figure 4. Relative performance of CDA-M and a standard nickel clad grit in dry tool and cutter grinding.

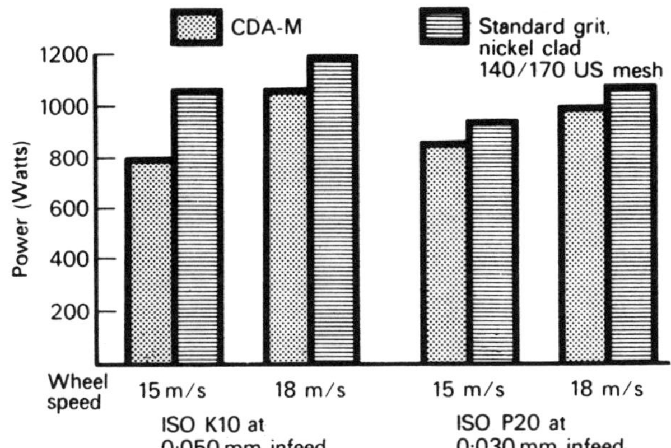

Figure 5. Relative levels of power drawn on the spindle motor for CDA-M and a standard nickel clad grit in dry tool and cutter grinding.

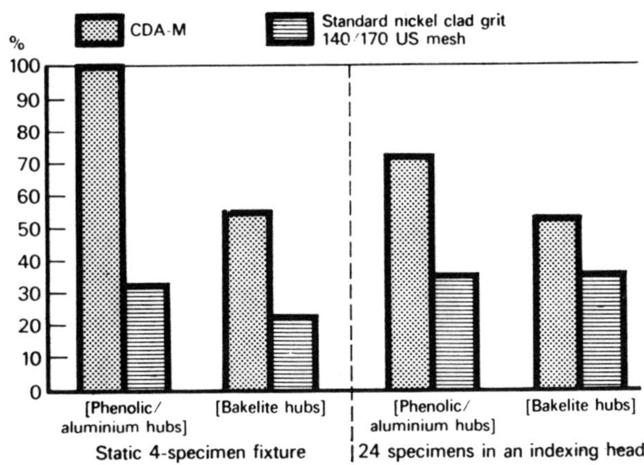

Figure 6. Grinding performance expressed as a percentage relationship between CDA-M and a standard nickel clad grit in phenolic/aluminium and bakelite hubs.

whether different hub materials would affect the results achieved by CDA-M. In general, the results with CDA-M in wheels with phenolic-aluminum hubs were better than those with bakelite hubs. However, it should be remembered that the bond formulation used with both hub types was always the same and it is, therefore, quite possible that a change in bond type would alter the results achieved with both bakelite and phenolic/aluminum hubs.

One interesting point which emerged during testing on various carbide grades was that, as a result of its unique structure, CDA-M is able to dry grind some of the really heat sensitive types, such as ISO P10. In industry these grades tend to be ground under "wet" conditions in order to avoid the problems associated with grinding heat.

TABLE 1

Machine	Tacchella 6AP tool & cutter grinder
Wheel peripheral speed	P20-15 m/s (2955 sf/min) and K10-18 m/s (3545 sf/min)
Infeed	P20-0.030mm (0.0012 in.) and K10-0.050mm (0.002 in.)
Table speed	2m/min (6.6 f/min.)
Specimens	ISO P20 tungsten carbide: (WC-75%; TiC-12%; Ta(Nb) C-5%; Co-8%) ISO K10 tungsten carbide: (WC-85.5%; TiC-7.5%; Ta (Nb) C-1%; Co-6%)
Specimen face size	12.7mm (0.5 in.) x 6.35mm (0.25 in.)
No. of specimens	24 in an indexing head; 4 in a static fixture

The parameters shown in Table 1 are based on those which have been successfully used with existing standard grit products and which are acceptable to industry. In fact CDA-M does not require the use of "special" speeds and feeds to achieve its good results.

The wheels used in the test programs were 11V9; the dimensions were 100mm diameter x 3mm matrix width and the grit concentration 75.

Tests on both ISO P20 and K10 tungsten carbide were carried out with wheel peripheral speeds typical of those used by industry for tool and cutter grinding. Fig. 7 shows the G-ratios achieved in these tests. Clearly CDA-M yields higher G-ratios in dry grinding than the standard nickel clad grit regardless of carbide

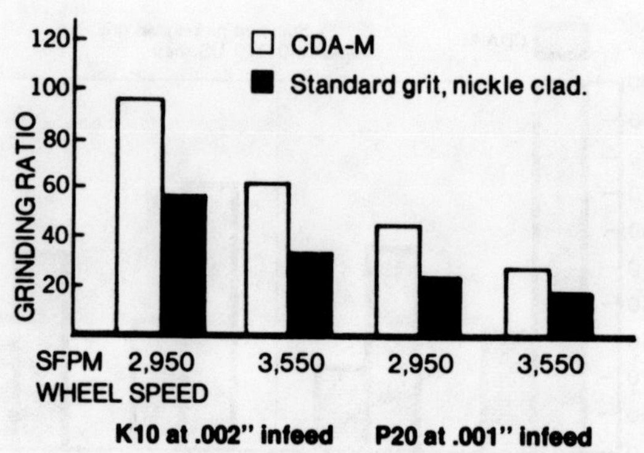

Figure 7.

type or wheel peripheral speed. ISO P20 and K10 grades were chosen since between them they cover the greatest amount of carbide grinding in industry.

In addition to achieving the higher G-ratios shown in Fig. 7, CDA-M also drew less power on the spindle motor in every case. This demonstrates that the wheel matrix surfaces containing CDA-M remain open and very free cutting. Quite apart from drawing less power, these features will ensure that less grinding heat is generated at the point of contact with the workpiece.

As mentioned previously, the relative performance of CDA-M when used with phenolic-aluminum and bakelite hubs was also investigated. Fig. 6 shows the grinding efficiency of CDA-M compared with a standard nickel clad grit on both these hub types using a static 4-piece specimen fixture and a 24-specimen indexing head arrangement.

Of particular note is that CDA-M gave its best result (G ratio) in the phenolic-aluminum hub when grinding four pieces of tungsten carbide in a static fixture (this result has, therefore, been expressed as 100%). When grinding 24 workpiece specimens in an indexing head, CDA-M performed at a lower level, although regardless of which hub system was used it was always better than the standard nickel clad grit. The lower level of performance with the 24-station indexing head could possibly be explained by the harmful vibrations built up when using this type of arrangement.

A second point of interest is that whereas the standard nickel clad grit gave much the same performance whether in wheels with bakelite or phenolic-aluminum hubs, CDA-M performed markedly better in a phenolic-aluminum hub.

CONCLUSIONS

CDA-M offers a solution to the problems associated with the dry grinding of tungsten carbide which were not entirely covered by existing products.

Other operations, such as wet grinding of tungsten carbide or the grinding of combinations of steel and tungsten carbide, will require grit types with physical properties different from these of CDA-M, in order to obtain the highest efficiency. Indeed, it is not expected that CDA-M would perform at all well in grinding steel, for example, where the mode of metal removal differs substantially from that of tungsten carbide. For that, and other types of grinding operations, abrasive products, with quite different physical properties particularly suited to the specific application, have been developed.

The evolution of CDA-M for the dry grinding of tungsten carbide was achieved by the integration of diamond synthesis technology and grinding theory. The results that have been presented demonstrate the potential of CDA-M and show the advantages of using the correct grit type for a specific application.

REFERENCES

1. TOMLINSON, P. N. and PENNY, A. L. CDA-M: A new dry grinding abrasive for tungsten carbide. DeBeers Industrial Diamond Technical Service Centre, Charters, Ascot, UK (1978)
2. HUGHES, F. H. "Diamond Grinding of Metals," Industrial Diamond Information Bureau, Charters, Ascot, UK (1975).

Reprinted from The Carbide Journal, September-October 1976

High-Performance Carbide Grinding

Presented at the
1975 Diamond Wheel Manufacturers' Institute Symposium
Chicago, Ill.
by

Laszlo M. Zsolnay
Manager of Technical Department
Diamond Branch, The Carborundum Company

Wheels designed for fast, low-cost carbide grinding should use low driving power in addition to yielding high grinding ratios at fast-stock removal rates.

Using finer grit, hybrid, resin-bonded diamond wheels, four times faster stock removal rates are recommended in 1975 than a decade ago.

Grinding Costs

The goal of high-performance carbide grinding is to cut the grinding costs by fast stock removal. The cost of grinding is conveniently calculated as the sum of the wheel cost and the cost of time. J. E. Niesse[1] established the rule of thumb that the total cost is at minimum when the cost of the diamond wheel approximately equals the cost of labor plus overhead.

By this rule, the stock removal rate would have to be slowed down if the cost of wheel usage were substantially higher than the cost of time. At slower stock removal rates, higher grinding ratios are obtained and, therefore, the wheel usage per cubic inch of carbide removed is decreased.

In recent years, polyimide resins and new hybrid-modified phenolic resins plus the use of metal-coated diamonds greatly increased the grinding ratios. To take full advantage of the available wheels, faster stock removal rates should be used.

But, unless very thin wheels are used, quite often the power requirement sets limits to the stock removal rates. Excessive power generates excessive heat, or more often it exceeds the mechanical rigidity of the grinder or its fixtures. Sometimes it simply stalls the available motor.

Therefore, it is not enough to measure the grinding ratios at various stock removal rates; but, one should also measure the power requirements and select wheels which draw less power.

Test Methods

The standard horizontal spindle surface grinding test published by the Diamond Wheel Manufacturers' Institute[2] yields results which can be directly related not only to horizontal spindle surface grinding, but also to cam grinding with peripheral-coated wheels, to centerless and to cylindrical grinding. The similarity of these production grinding methods with the grinding test consists of having a small arc of contact between the diamond wheel and the workpiece, and having ample flood coolant.

For present purposes, the standard test has been modified as follows:

1. The tests were conducted at two levels of severity; the second, four times as severe as the first.
2. The power was recorded.
3. The finish was measured.
4. The number of passes was greatly increased for better accuracy in measuring small wheel wear.
5. The number of test pieces ground in each test was increased from 10 to 90 to obtain better average grindability for every test.

The details of the test method are listed in Table 1.

TABLE 1

Details of Grinding Test Method

Grinding Machine	Gallmeyer & Livingston Horizontal Spindle Surface Grinder	
Wheel Type	1A1	
Wheel Diameter	6″	
Wheel Thickness	¼″	
Wheel Speed	3,450 RPM = 5,420 SFPM	
Coolant	3.6 gallons per minute CIMCOOL 1:40	
Material Ground	78B CARBOLOY	
Size of Carbide Block	90 pieces, ⅜″ x ⅝″ x 1″, assembled to a 5⅝″ x 3¾″ x 1″ block	
Table Speed	40 FPM	
Feed Rate	3.5 cu. in./hour/in.	13.8 cu. in./hour/in.
Down-Feed per Pass	.001″	.002″
Cross-Feed	.030″	.060″
Number of Passes per Test	750	100

TABLE 2
Characterization of Selected Test Wheels

Wheel Grading	Bond Type	Diamond Type	Diamond Grit Size	Diamond Concentration
MD120-R100-B	Phenolic Resin	Synthetic, not coated	120	100
MD180-R100-B	Phenolic Resin	Synthetic, not coated	180	100
MD120-L100-M	Soft Bronze	Metal Type, Synthetic	120	100
MD180-L100-M	Soft Bronze	Metal Type, Synthetic	180	100
CMD120-100-BX	Polyimide	Synthetic, Nickel coated	120	100
CMD180-100-BX	Polyimide	Synthetic, Nickel coated	180	100
CMD120-T100-B1	Hybrid Resin	Synthetic, Nickel coated	120	100
CMD180-T100-B1	Hybrid Resin	Synthetic, Nickel coated	180	100

Selection of Test Wheels

The test wheels were selected to represent the best available in their respective class. They are tabulated in Table 2.

The first formula represents the best phenolic resin bond wheels available in 1965. The second formula is a good metal bond wheel, a grading much used for carbide grinding in 1965. The third formula is made with the best polyimide available in 1975, and contains nickel-coated diamonds. The last formula is made with a hybrid combination of resins designed for fast stock removal, at lower power.

Discussion of Results

Table 3 contains the results of the grinding tests. Table 4 summarizes 10 years of progress, going diagonally from the upper left-hand corner to the lower right-hand corner of Table 3. The data show that it is possible to obtain high stock removal rates without large increases in power requirements.

They also show that, with both polyimide and with hybrid resins, 180 grit provides better performance than 120 grit. The use of 180 grit should be considered even if the finish obtained with 120 grit is satisfactory.

The dollar figures of Table 4 are examples only; the wheel cost will vary with the size of the wheel used. Larger wheels are cheaper per cubic inch of abrasive.

The cost of time of contact will vary not only with labor and overheat rates, but also with the ratio of contact time to idle time.

The example shows that, because of improvement in wheel performance, the grinding cost per cubic inch or per piece may have decreased in the last decade, despite increasing labor rates.

Definition of Terms

The grinding ratio is defined as the volume of carbide removed divided by the volume of wheel loss.

The feed rate is calculated in cubic inches fed during the time of contact between wheel and workpiece. The idle time at the end of passes is excluded. The rate is related to the width of the wheel face. It is expressed as cubic inches per hour per inch of wheel face.

The actual stock removal was measured on the carbide. Except for a small correction for wheel wear, it is almost equal to the total infeed.

The finish was measured without any spark-out, by interrupting the test at the end of a pass. It is measured in microinches and calculated by the root of the squares of the deviation from the mean surface.

TABLE 3
TEST RESULTS
Wet Surface Grinding Test
For composition of test wheels, refer to Table 2.

	Feed Rates					
	3.5 cu.in./hour/in.			13.8 cuin./hour/in.		
Wheel Grading	Grinding Ratio	Power HP/in.	Finish M.I. RMS	Grinding Ratio	Power HP/in.	Finish M.I. RMS
MD120-R100-B	126	3.6	20	31	5.9	40
MD180-R100-B	82	3.1	15	25	5.1	30
MD120-L100-M	512	7.1	30	Over 16 HP/in. Wheel smeared—test stopped.		
MD180-L100-M	Over 16 HP/in. Wheel smeared—test stopped.			Not tested.		
CMD120-100-BX	289	5.2	30	79	11.5	40
CMD180-100-BX	391	5.3	15	155	9.4	25
CMD120-T100-B1	424	4.2	20	98	6.3	30
CMD180-T100-B1	537	3.0	12	114	5.4	20

TABLE 4

CHANGE IN RECOMMENDED TECHNOLOGY IN LAST DECADE
Performance Figures from Table 3, Dollar Values Examples Only.

Year	**1965**	**1975**
Resin	Phenolic	Hybrid
Diamond	Synthetic, not coated	Synthetic, nickel coated
Grit Size	120	180
Feed Rate	3.5 cu.in./hour/in.	13.8 cu.in./hour/in.
Power Used	3.6 HP/in.	5.4 HP/in.
Finish	20 M.I.	20 M.I.
Grinding Ratio	126	114
Wheel Cost/cu.in. Abrasive	$665	$698
Wheel Cost/cu.in. Carbide	$5.28	$6.12
Time Cost/Hour Contact	$30	$60
Time Cost/cu.in. Carbide	$8.57	$4.35
Total Cost/cu.in. Carbide	$13.85	$10.47

The power was continuously recorded and graphically averaged for the contact time. The idling time, at the end of the passes, was excluded. The actual HP was multiplied by a factor of 4, to obtain power per inch of wheel face values from the HP measured with ¼″ wide wheels.

At the calculation of the cost of time, it was assumed that the wheel is in contact ½ of the time, while the other half is for idling out of contact, for set time and other time lost.

References

1. J. E. Niesse: Diamond Grinding of Carbides, January-February 1968, Cutting Tool Engineering. Reprinted in DWMI 105, Diamond Wheel Manufacturers' Institute.

2. J. R. Thompson: Test Procedures for Grinding Carbides Wet, DWMI 102, Diamond Wheel Manufacturers' Institute.

Reprinted from Tooling & Production, June 1977

COST-EFFECTIVE SURFACE GRINDING

of hardened steels and tough superalloys

by **Paul Grieb**
Program Engineer
Specialty Materials Dept
General Electric Co
Worthington, OH

Hardened steels, abrasion-resistant cast irons and tough superalloys are all notoriously difficult to grind. Nevertheless, hundreds of companies are surface-grinding them successfully in the toolroom and in production using standard horizontal-spindle machines.

These plants use General Electric Borazon® CBN (cubic boron nitride) wheels for surface grinding operations. Among the materials that are being successfully—and profitably—ground with Borazon CBN wheels are M-2, M-4, M-15, T-1, T-9, T-15, A-7 and D-7 steels hardened to 58 Rc and above; carbon and alloy steels hardened to 50 Rc and higher; difficult-to-grind cast irons; and Stellite® alloys and superalloys with hardnesses over 35 Rc.

Benefits of better wheels

Some typical benefits received when aluminum oxide wheels are replaced with CBN wheels are fast stock removal with less wheel wear, more consistent finishes with fewer wheel reconditionings (dressings), and cooler grinding, with less possibility of burning the workpiece.

The cooler grinding characteristic of CBN wheels—a consequence of their ability to remain sharp and free-cutting when other wheels dull—is often reason enough for adopting CBN wheels. For example, excessive grinding heat from dull wheels can result in accelerated fatigue cracking of superalloy parts in service. Grinding with CBN wheels improves part reliability in this case. Cool grinding also helps to prevent part distortion.

In most cases, however, CBN wheels are selected simply because they are the most cost-effective tools available for surface grinding hard, abrasion-resistant materials.

These wheels cost more than aluminum oxide wheels. But they generally improve productivity enough to pay a big return on that investment. A typical grinding cost analysis is seen in **Figure 1**.

Keep in mind that the most cost-effective surface grinding applications of CBN wheels are on difficult-to-grind materials. Aluminum oxide wheels can still be used for easier-to-grind steels, cast irons and superalloys.

Selecting the right machine

The performance of a CBN wheel—like that of any grinding wheel—depends on the capabilities of the machine.

Before replacing an aluminum oxide wheel with a CBN wheel, the grinding machine should be checked to make sure you can take advantage of the productive potential of the CBN wheel.

Here are some things to consider:

1) Is the machine in good condition?

Generally, material is ground at faster rates with CBN wheels than with aluminum oxide wheels. At high material-removal rates, grinding forces can be high. To contain high grinding forces, overall machine rigidity should be good. To minimize chatter—which can shorten wheel life and impair surface finishes—worn spindle bearings should be replaced. Feed mechanisms should be examined to make sure they operate smoothly and precisely.

2) Is the coolant system adequate?

For optimum wheel life at high material-removal rates, CBN wheels require an adequate flow of coolant, applied directly to the wheel-workpiece interface. If the machine doesn't have a suitable coolant system, one should be installed.

3) Is spindle speed adequate?

With CBN wheels, excellent results are obtained by wet grinding at wheel speeds of 5500 to 6500 sfm (28 to 33 m/sec) and higher. The spindle speed of the machine should permit grinding at surface speeds in this range.

4) Is spindle horsepower adequate?

It takes plenty of spindle horsepower to grind hard, abrasion-resistant materials at high material-removal rates.

Minimum and preferred spindle horsepowers for grinding with CBN wheels of different diameters are shown in **Table 1**. With horsepower lower than the minimum values shown, the spindle may slow down, reducing material-removal rates, while adversely affecting wheel life and surface finishes.

5) Does the machine have automatic incremental downfeed?

While it's not mandatory that a surface grinding machine have it to operate successfully with CBN wheels, such systems help the operator produce more accurate work faster.

Because CBN wheels wear so slowly, the operator doesn't have to make allowances for wheel wear when setting downfeeds. For all practical purposes, the amount he feeds down is the amount of material that will be removed from the work.

Digital readout systems are useful accessories for surface grinding machines equipped with CBN wheels. Before grinding, the operator can set the readout so the magnetic chuck represents "zero." During grinding, he can check the dimension of the part being ground by simply observing the digital readout. Because of the extremely low wear rate of CBN wheels, there is no need to provide accommodation for wheel loss due to dressing or wheel wear. He doesn't have to bother with handwheel graduations or zero slip-ring adjustments, so he gets more work done in less time.

Wheel selection guidelines

Successful surface grinding with Borazon CBN wheels depends, to a large extent, on selecting the right wheel for the job. Here are some guidelines:

- For most jobs, use resin-bond wheels made with CBN Type II abrasive. They provide the most cost-effective combination of fast material-removal capability and long life for the majority of surface grinding operations. Type II abrasive is metal coated for improved retention in resin-bonds.
- For grinding complex, deep forms, use electroplated wheels. Wheels made with CBN Type I or Type 500 abrasive are recommended. These extra-tough abrasive types are designed for good retention and long life. Generally, form accuracies as close as ±0.0001" (0.0025 mm) can be attained with them. Finishes are usually slightly rougher than those produced by resin-bond wheels of the same mesh size.
- Specify normal wheel diameters and widths. When an aluminum oxide wheel is replaced with a CBN wheel, it's usually best to specify a CBN wheel of the same diameter and width as the aluminum oxide wheel. Specifying a smaller CBN wheel to reduce initial wheel cost can be a false economy. Undersized wheels wear much faster than normal-sized wheels. Due to this accelerated wheel wear, wheel cost per ground piece may be greatly increased. Also, larger-sized wheels grind cooler when operated at the same surface speeds.

If, because of a machine horsepower or rigidity limitation, it's necessary to specify an undersized CBN wheel, it's better to reduce the width of the wheel rather than its diameter. If wheel diameter is reduced, surface speeds may be lowered to the point where the CBN wheel cannot perform effectively. Wheel speeds below 5500 sfm (28 m/sec) may not give satisfactory results in wet grinding.

When aluminum oxide wheel sizes are 12" x 2" (305 mm x 50 mm) and larger, it's sometimes possible to successfully apply a narrower CBN wheel. For example, when grinding cobalt-type high-speed steels, a 12" x 1" (305 mm x 25

1. Cost comparison of aluminum oxide and CBN wheels.

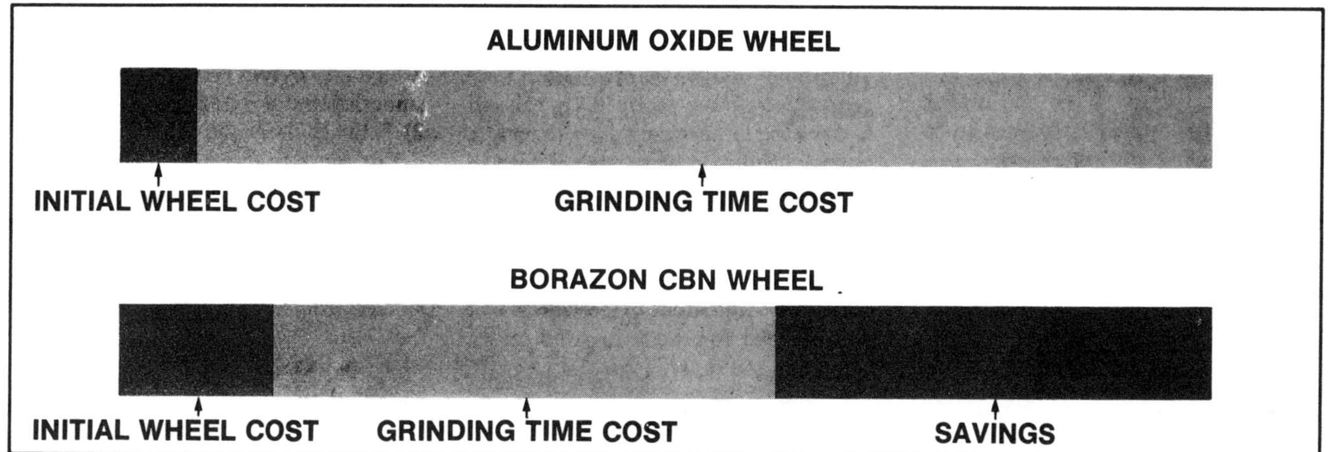

mm) CBN wheel will ordinarily remove stock faster than a 12" x 2" (305 mm x 50 mm) aluminum oxide wheel and still deliver good wheel life.

• Choose the largest abrasive mesh size that gives the desired finish. For a given set of operating conditions, a wheel containing coarse (large mesh size) CBN abrasive will have longer life than a wheel containing fine (small mesh size) abrasive. However, the wheel with fine abrasive may produce smoother finishes if downfeed rates are lowered. For grinding superalloys, a wheel with fine abrasive may be specified to help reduce grinding heat.

Table II suggests what abrasive mesh sizes to select when replacing an aluminum oxide wheel with a CBN wheel.

• Specify wheels with the correct abrasive concentration. The material-removal efficiency and life of a resin-bond CBN wheel are strongly influenced by the abrasive concentration—the amount of abrasive per unit volume of wheel rim. Wheels with high abrasive concentrations cost more than wheels with low abrasive concentrations but—because they remove stock faster and last longer, **Figure 2**—they're usually more cost-effective.

For most surface grinding operations, resin-bond wheels with 100-concentration CBN abrasive are recommended, provided that spindle horsepower is not lower than the minimum shown in **Table I**. A 100-concentration wheel usually contains approximately 72 carats of abrasive per cubic inch (4.4 carats per cubic cm) of wheel rim material.

In some cases, a 75-concentration wheel provides a good compromise between initial wheel cost and long wheel life. Also, 75-concentration wheels are effective for sidewheel-grinding and slot-grinding operations.

Nominal Wheel Diameter inches	mm	Horsepower per Inch (25.4 mm) of Wheel Width Minimum	Preferred
6-10	152-254	2	3
10-12	254-306	3	5
14-16	356-406	5	7½
20-24	508-610	7½	10
30	762	10	12½

Table I
Horsepower Requirements for CBN Wheels

Wheel mounting, trueing and conditioning

Unless correctly mounted, CBN wheels cannot perform well. After mounting, all CBN wheels except electroplated wheels must be correctly trued and conditioned to insure good grinding performance.

In trueing, the wheel rim is cut or abraded with a tool to develop perfect roundness, concentricity with the spindle and the desired profile. In conditioning, bond material is removed to expose the sharp cutting edges of the CBN abrasive crystals.

The following guidelines are general. For more specific information on trueing and conditioning, consult the instructions that come with the wheel.

Secure wheel mounting is absolutely essential when CBN wheels are used to remove material fast, as they usually are. Grinding forces at high material-removal rates can be high enough to make the wheel shift on its mount, causing chatter, vibration and reduced wheel life.

For correct mounting of a CBN wheel, the mounting flanges and faces must be in good condition. To reduce the amount of wheel rim material that has to be removed in trueing, the wheel should be mounted as true as possible. With the aid of a dial indicator, it's possible to position the wheel so as to minimize runout.

2. The relations of grinding ratios to material removal rates. Using a 100-concentration CBN wheel results in the lowest total grinding cost per piece for most surface grinding jobs.

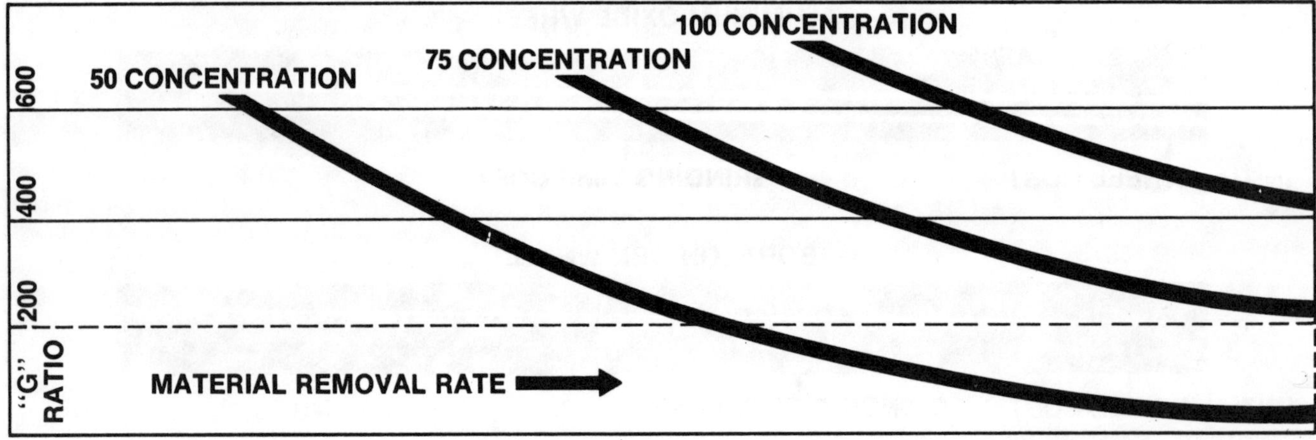

If the wheel hole and wheel mount are close to the same diameter, the wheel should be rotated on the mount to the minimum runout position.

Paper washers, blotters and metal shims should not be used when mounting a CBN wheel. They make it difficult for the wheel to be held securely on the mount.

Accurate trueing is important. An out-of-round wheel will pound the workpiece, wear fast and produce poor finishes. The wheel should not be over-trued, however. Excessive trueing simply wastes wheel material and shortens wheel life.

Excessive trueing can be avoided by marking the wheel periphery with a wax crayon. When the crayon mark disappears from the wheel periphery, trueing should be stopped immediately. Use of other types of marking materials such as dyes or layout blue can result in over-trueing, and a loss of 5 percent or more wheel life.

CBN wheels are always trued wet, with a flood of coolant applied as directly as possible to the tool-wheel interface.

Single-point and cluster type diamond tools should not be used for trueing CBN wheels as they may damage the wheel.

One recommended trueing tool for CBN wheels is a metal-bond nib containing 40-150 mesh diamond abrasive. With this type of tool, trueing increments should be in the 0.0005″ to 0.001″ (0.013 mm to 0.025 mm) range.

An alternative to using a diamond nib for trueing is to use a 46 to 60 grit silicon carbide wheel in a brake-type trueing device. Vitrified-bond wheels with bond hardnesses in the J through M range give best results.

Table II
Comparative Abrasive Mesh Sizes

Aluminum Oxide Wheels	BORAZON CBN Wheels
46-60	80-140
80-120	170-325
150-220	325-400

CBN wheels with diameters of 14″ (355 mm) and above may require dynamic balancing. This will aid in improving surface finishes and wheel life.

Trueing glazes CBN wheels. *They will not grind when glazed.* Before grinding, they must be conditioned to make them free-cutting.

CBN wheels with diameters of 12″ (305 mm) or less can be effectively conditioned by abrading the wheel surface with a stick containing fine (220 grit) aluminum oxide abrasive in a G to K grade vitrified bond.

The stick should be pushed into the wheel surface aggressively. If the wheel is badly glazed, it may be necessary to use an entire stick to obtain sufficient abrasive crystal exposure. Typically, a 7″ x ½″ (178 mm x 13 mm) CBN wheel may consume up to 3″ (76 mm) of a ⅝″ (16 mm) square stick during conditioning. And a 12″ x 1″ (305 mm x 25 mm) wheel may consume 3″ to 5″ (76 mm to 127 mm) or more of a 1″ (25.4 mm) square stick.

When conditioning a wheel with an aluminum oxide stick, only enough coolant is used to create an abrasive slurry on the wheel surface. This minimizes dust and promotes good conditioning action.

CBN wheels of 12″ (305 mm) diameter and larger can also be quickly conditioned with a water-soluble polyethylene glycol wax stick containing 220-grit aluminum oxide abrasive.

For conditioning, the wheel is brought into light contact with the workpiece or a suitable piece of scrap metal. The wax stick is then applied to the wheel surface at the "4 o'clock" position (approximately), using light pressure.

The wax melts and coats the wheel surface with an abrasive slurry. As the wheel rotates, it carries this slurry into the wheel-workpiece interface.

Because there is zero clearance between the wheel and the work, the aluminum oxide abrasive in the slurry is forced into the wheel bond. It quickly abrades away bond material and, to a lesser extent, the workpiece material. As conditioning proceeds, the wheel is fed straight down just enough to maintain zero clearance.

When conditioning a CBN wheel with an aluminum oxide or wax stick, normal grinding safety precautions should be observed. The wheel guard should be in place. And the stick should always be long enough so there's no possibility of a hand coming into contact with the wheel.

Once grinding is underway, CBN wheels tend to condition themselves further. After working for 10 to 15 min, the wheel should reach steady-state operating performance levels.

Since electroplated wheels have only a single layer of CBN crystals, they should not be trued or conditioned. Special care must be used to keep runout to a minimum when mounting electroplated wheels; most have indicating rings for this purpose.

The best way to eliminate excessive runout is to have the wheel manufacturer machine the wheel core integrally with its wheel mount. Radial and lateral runout will be close to zero, resulting in good wheel performance.

Putting the wheel to work

With the right CBN wheel for the job, installed on the right machine, and correctly trued and conditioned, the stage is set for efficient grinding. To obtain the full benefits of CBN wheels, however, it's necessary to establish the right operating conditions.

When CBN wheels are used to grind difficult-to-grind materials at high material-removal rates, *a continuous flow of coolant* is required to reduce grinding friction and carry away heat.

Because of the possibility of grinding heat damaging the workpiece, dry grinding is not recommended. If it's absolutely necessary to surface-grind dry, feeds, table speeds and spindle rpm should be reduced to minimize the possibility of burning the workpiece.

CBN wheels remove metal by making chips. Grinding pressures can be high and coolants should have good lubricity to prevent chips from sticking to the abrasive crystals.

Good results can often be obtained with a 5 percent solution of light-duty soluble oil. *But for truly superior results and longest wheel life, a 5 to 10 percent solution of heavy-duty soluble oil is recommended,* **Figure 3**. Use of straight water (or water with rust inhibitors or detergents) is not recommended.

When grinding superalloys, sulfurized or sulfochlorinated mineral oils may be required to minimize grinding friction, since these metals generally cannot be cut as cleanly as hardened steels and hard cast irons.

Coolant nozzles should be parallel to the direction of table reciprocation. Nozzle spouts should be within ⅛" to ¼" (3 mm to 6 mm) of the work surface. To keep the coolant flowing where it's needed, supplemental shielding can be installed on both sides of the spout, creating a gutter.

It's also a good idea to provide extra shielding below the wheel guard to contain spray generated by the wheel. The shielding can cover the sides of the wheel down to within ⅛" (3 mm) of the machine table (or whatever the wheel rim thickness is), since there's no need to allow for wear of the CBN wheel.

Also recommended is the use of a dummy block with short workpieces to distribute coolant effectively, **Figure 4**.

CBN wheels perform best when wet grinding wheel surface speeds are in the 5500 to 6500 sfm (28 to 33 m/sec) range. At speeds within this range, longest wheel life and finest surface finishes are obtained.

Optimum performance can be obtained from CBN wheels when high table speeds—up to 150 sfm (0.76 m/sec)—are used. At high table speeds, material-removal rates are high, increasing the number of pieces ground per hour, and reducing labor and overhead cost per piece.

For normal grinding to size, crossfeed increments for CBN wheels should be from one-fourth to one-half of wheel width. Smaller increments can be used for finishing passes.

Optimum downfeed rates for surface grinding with CBN wheels range from 0.001" to 0.002" (0.025 mm to 0.051 mm) per pass for relatively easy-to-grind hardened steels such as M-2 and T-1, to 0.0005" to 0.001" (0.013 mm to 0.025 mm) per pass for more difficult-to-grind hardened steels such as M-4 and T-15.

When grinding a number of parts to the same dimension, a positive handwheel stop should be used, if available. Because of the low wear rates of CBN wheels, parts ground over fairly long periods of time will be the same size within ±0.0001" (0.0025 mm) or so. This consistent size control is one of the most important benefits obtained by using CBN wheels.

There's no need to recondition CBN wheels before making finishing passes.

To obtain good finishes, it is only necessary to reduce the last increment of downfeed to 0.0002" to 0.0005" (0.005 mm to 0.013 mm) per pass and reduce the crossfeed increment to about one-eighth of wheel width—¹⁄₁₆" (1.6 mm) for a ½" (12.7 mm) wide wheel, for example.

If it's impossible to attain the required finish with a given CBN wheel, it may be necessary to change to a wheel with a different CBN abrasive mesh size. In most cases, changing to a wheel with abrasive three mesh sizes smaller or larger than the abrasive in the original wheel will be necessary to achieve a substantial change in surface finish.

Surface finishes produced by CBN wheels often look less polished than finishes produced by aluminum oxide wheels. The reason is that CBN abrasive crystals stay sharp and cut—rather than burnish—the workpiece surface.

When examined with a profilometer, surface finishes produced by CBN wheels are usually better than they appear to be to the eye. And surface finishes produced by CBN wheels are usually more consistent than those obtained when grinding with aluminum oxide wheels.

Sidewheel and slot grinding

The consistent sharpness and low wear rates of CBN wheels make them extremely effective in sidewheel and slot-grinding applications.

The first step in preparing a CBN wheel for sidewheel or slot grinding is to undercut the wheel core about 0.010" (0.25 mm) to prevent the core from rubbing against the workpiece.

To avoid wobble, secure mounting is an absolute necessity for sidewheel and slotting operations. Once the wheel is securely mounted, the side of the wheel rim is trued and conditioned, using the techniques already described.

Normal wheel speeds in the 5500 to 6500 sfm (28 to 33 m/sec) range are usually desirable for sidewheel and slot grinding operations with CBN wheels. The recommended table speed is in the 30 to 50 fpm (0.15 to 0.25 m/sec) range for fast grinding to size. It should be reduced to 12 to 20 fpm (0.06 to 0.10 m/sec) for finishing passes. When sidewheel grinding, crossfeed increments in the 0.0005" to 0.001" (0.013 mm to 0.026 mm) range are appropriate.

When side wheel and slot grinding with CBN wheels, wheel performance can be optimized by following the general guidelines in this article. The most important of these is to use a 5 to 10 percent heavy-duty soluble oil coolant, properly distributed.

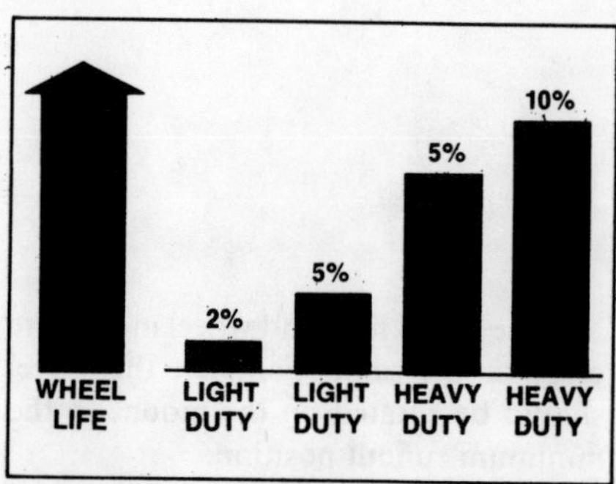

3. Relation of oil concentration to wheel life.

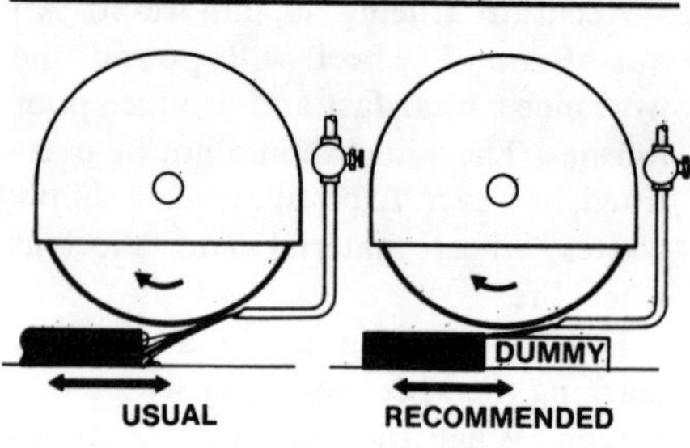

4. Use of a dummy block increases the ability of the coolant to get to the wheel-work interface.

Presented at the SME/GE Abrasives Workshop, March 1975

High Precision Grinding With Superabrasives

By Hank Deutschendorf
Moore Special Tool Co., Inc.

The utilization of BorazonT and Diamond grinding wheels for precision grinding of tools and dies have greatly enhanced the capability of the jig grinder and surface grinder.

Using the jig grinder, featuring numerical control, complex curviliniar, internal and external forms can easily be generated using standard, formed and special shaped grinding wheels.

Using formed wheels manufactured with Borazon or Diamond abrasive, the capability of the surface grinder to accurately grind complex forms is greatly increased.

This paper also discusses various grinding techniques utilizing different types of wheels employing superabrasives. Significant advantages are realized when grinding complex forms to high precision accuracies.

INTRODUCTION

Since the grinding wheel is the basic cutting tool used to form or generate the dimensional detail of a workpiece, its proper selection is important to the performance of the grinding operation.

This paper deals primarily in the area of applying various types of grinding wheels manufactured with the superabrasives Diamond or Borazon. When using this type of wheel to accurately form grind or generate a particular contour in tool and die manufacturing, its advantage is easily realized because of minimal wheel wear.

The accuracy achieved and the increased material removal rate can be related to the cutting tool employed.

This paper also discusses some unusual applications of Diamond and Borazon wheels utilizing the jig grinder and surface grinder. Many of these grinding jobs could not have been performed or would not be economically feasible if conventional abrasives were used.

Also covered in this paper are manufacturing techniques of various types of Fiamond or Borazon wheels.

BACKGROUND

Two machines essential to the tool and die industry are the precision jig grinder and surface grinder.

T Trademark of the General Electric Co.

Both of these machines are used to establish the functional and dimensional detail of die components after hardening, thus eliminating dimensional inaccuracies resulting from distortion during heat treatment.

The jig grinder with various accessories can produce interchangeable die components for press tools. With the use of cutting tools manufactured with the superabrasives Borazon or Diamond and the accurate dimensional control of a tape controlled jig grinder, hand fitting punch and die components are eliminated resulting in cost savings in their fabrication. High production life and product accuracy is assured. (Fig. 1)

Conventional surface grinders may also be employed providing they have an accurate downfeed and a rigid spindle.

WHEEL TYPES

Conventional Abrasives

Before the advent of superabrasives, conventional abrasives such as aluminum oxide or silicon carbide wheels were used in various types of grinding machines. These wheels are easily shaped and trued using a single point diamond tool. The major disadvantage of these conventional abrasive wheels is the constant need for redressing and relative poor "Grinding Ratio" when compared with Borazon or Diamond wheels. This is especially apparent when grinding parts manufactured from high wear resistant materials or super alloys.

Impregnated Wheels

With the development of carbide, the need for the Diamond wheel became obvious. A formed Diamond wheel had a particular need in the tool and die industry. One of the first multi-formed Diamond wheels to be used was an aluminum or copper impregnated wheel.

The desired form is turned into the aluminum or copper rim of the surface grinding wheel using a pantograph or other types of form generating attachments with the aid of a template.

A single layer of diamond is impregnated or forced into the wheel using a matching female form ground into a hardened roller or bar. (Fig. 2) (Ref. 1)

The rotating crusher roll and wheel are brought into mating contact. The forcing together of the wheel and roller embeds the diamond particles into the rim of the wheel. (Fig. 3)

Although this method proved to be efficient and relatively easy to manufacture, diamond pull-out would occur, making the wheel inefficient in heavy stock removal operations.

With the development of Borazon, the same wheel manufacturing techniques can be performed.

Resin Bonded Wheels

With the increased use of carbide and the efficient grinding results obtained from a resin bonded diamond wheel, the use of these wheels increased dramatically.

When straight type peripheral wheels were used, truing and dressing of these wheels could be done fairly easily. Employing a brake truing type of device, the bond of the wheel would be eroded to bring the wheel into uniformity and run concentric.

If this type of wheel had to be used in form grinding operation, the form had to be ground in a prefabricated wheel, using aluminum or silicon carbide wheels. This time consuming process would increase the wheel cost tremendously. It also had limitations in the complexity of forms which could be generated.

When resin bonded Borazon wheels are used in steel grinding, form generating is achieved more easily. A diamond wheel can be used to grind or generate forms into the wheel. A wheel having multi-formed configuration can be ground on a separate machine. Low R.P.M. of the wheel to be formed is employed. Using a metal bond or plated diamond wheel which follows the desired form path from a template, the form is ground into the wheel. Care must be taken to have the wheel run concentric after transferring it to the machine used to do the grinding operation.

Plated Wheels

With the development of electroplated wheels using Diamond or Borazon abrasives, a new type of grinding technique has opened up.

Standard shank type mandrels or points are widely used in jig grinding and internal grinding application. These types of wheels do not require any truing or dressing of the abrasive, but it is important that the wheel is manufactured geometrically accurate. (Fig. 4)

This cutting tool has a single layer of Diamond or Borazon abrasive mechanically retained on the surface of a steel or aluminum substrate with a nickel matrix. A common Watts type plating bath is used to electrolytically deposit the abrasive.

The primary advantage of this type of wheel is its free cutting ability and high stock removal rate.

Formed plated wheels for surface or jig grinding applications, using steel as the core material, are ground using conventional abrasives dressed to the proper forms. At times when multi-formed surface grinding wheels are required, generating the appropriate form prior to plating is time consuming.

Using aluminum as the core material, formed plated wheels are manufactured much more easily. The ability to generate the desired form directly into the wheel core using a carbide tool, makes this particular wheel less costly to manufacture.

Using a pantograph type of dresser, the desired form is cut into the aluminum wheel blank with the aid of a template. (Fig. 5)

Since aluminum core wheels have less mass when compared to a steel wheel, the vibration is reduced.

Aluminum - Epoxy Wheels *

Another type of wheelmaking process involves the bonding of Borazon or Diamond abrasive to the pre-shaped aluminum wheel core, using an aluminum type of epoxy.

In this process the female form desired is ground into a roll. A conventional abrasive wheel is used, its shape having been generated from a template. The unit holding the finished master roll is motorized, a feature that enables the roll to rotate and drive the wheel being manufactured.

Removing the conventional wheel from its spindle, and replacing it with an aluminum wheel blank is the next step. The objective at this point is to impart the shape ground into the roll to the aluminum wheel.

Prior to the actual manufacturing of the wheel, the aluminum wheel blank has to be machined to the approximate desired configuration. This procedure is very similar to dressing the form on the conventional abrasive wheel which was used to grind the master roll. After the aluminum wheel blank is coated with an epoxy mixture, the master roll is brought into contact with the wheel and the desired form is rolled into the epoxy coated surface.

Next, the Diamond or Borazon abrasive is applied to the still soft epoxy surface which is again rolled to ensure penetration

* Manufactured under license to Toolmasters, Ltd., England, U.S. Patent No. 3,551,125.

of the abrasive and bring the wheel into conformance with the master roll. (Fig. 6)

APPLICATIONS

Jig Grinding

All commonly available abrasive wheels are used in the jig grinder, each accommodating the requirements of the various materials being ground. Because by nature of its function much of the work on the part is in restricted areas such as holes or enclosed contours, a wide range of wheel sizes are required. It is important that the user have the right wheel for the right job available.

Although die work will remain the primary pre-occupation of the jig grinder, many other areas are being explored to make use of this machine and various types of wheels manufactured with the superabrasives Diamond and Borazon. In die work a large increase in total production is realized from a tool which requires fewer sharpenings since there is uniform clearance between punch and die. Also, when these tools are manufactured from steel and ground using Borazon, tool life is increased due to the cool grinding ability which minimizes the danger of burning or annealing the surface. (Fig. 7)

In mold cavity work, the use of preformed plated or formed resinoid bond type wheels manufactured with Borazon or Diamond allows us to efficiently grind parts without constantly redressing the grinding wheel. (Fig. 8)

Ball socket generating has opened another new area for Borazon wheels. Minimal wheel wear assures precision form control. (Fig. 9)

Another application is precision grinding of glass lenses having aspheric shapes. Wheels containing Diamond abrasive are used to generate these forms to high degree of precision.

The jig grinder and a wide selection of cutting tools, manufactured with the superabrasives Borazon and Diamond, is providing the solution to difficult machining operations.

Surface and Form Grinding

The surface grinder or a precision form grinder are two machines used in grinding application where larger wheels are employed.

Many different types of abrasive wheels are commonly used. Again, when these machines are used in precision form grinding applications, the constant need for redressing is very time consuming.

When machining difficult to grind materials, the availability of various types of wheels manufactured with Borazon or Diamond makes the operation much easier.

Work has been performed in evaluating deep form or creep feed grinding. Tests have shown that plated formed wheels containing Diamond or Borazon abrasive performed well in this type of grinding mode. Some of the latest testing being performed is slow wheel R.P.M. and slow table feed. Further testing is being conducted in this area.

CONCLUSION

Certainly with all the different methods of wheel manufacture, their application and usage in various grinding machines, more research and development has to be performed to answer all the questions regarding the application of superabrasive wheels. One thing is certain, that they can provide performance in precision grinding which exceeds that of conventional abrasive tools.

The proper machine and wheel selection enables one to generate many complex forms heretofore very difficult to obtain. One has to be sure that when using Borazon or Diamond wheels, it is the right application.

RECOMMENDATION

Here are some basic guidelines to use when confronted with the proper wheel selection.

1. A plated wheel should be used when dry grinding and fast stock removal are the prime criteria. In many cases they are used for roughing operation.

2. Resinoid or epoxy wheel for finishing operation where tight tolerance and good surface finish is required.

3. Make sure the proper wheel is selected to machine the part most efficiently, as well as proper machine selection.

4. All wheel types must run with minimum vibration. Proper preparation of the wheel, such as proper truing and dressing, is essential to wheel life and performance in high precision grinding.

REFERENCES

1. "Holes, Contours and Surfaces" by Richard F. Moore and Fred C. Victory, 1955, Moore Special Tool Co., Inc.

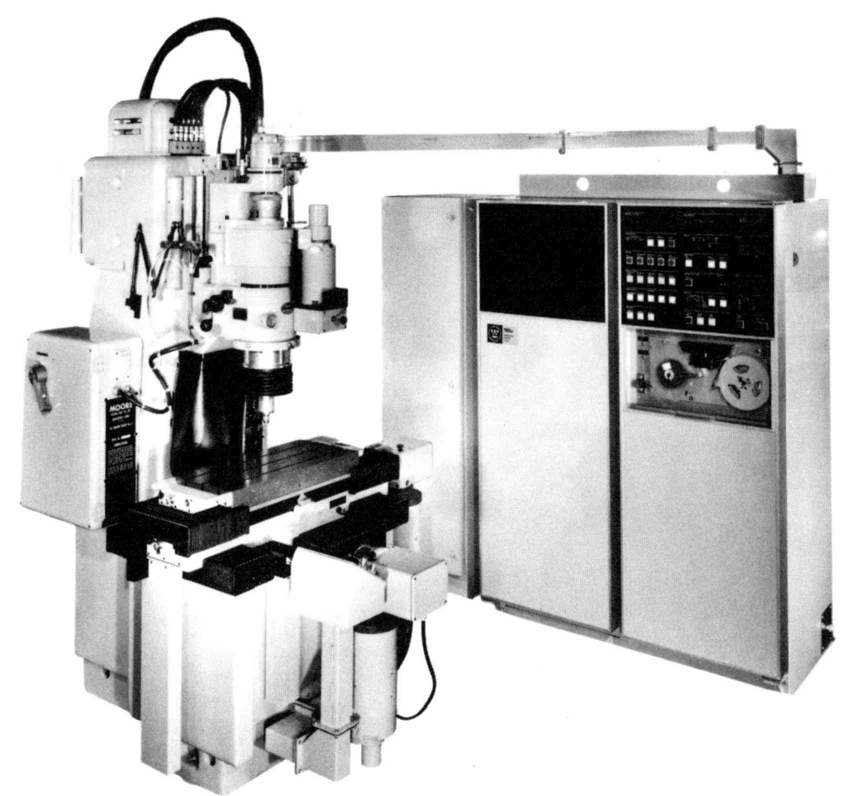

Fig. 1 Tape Controlled Jig Grinder

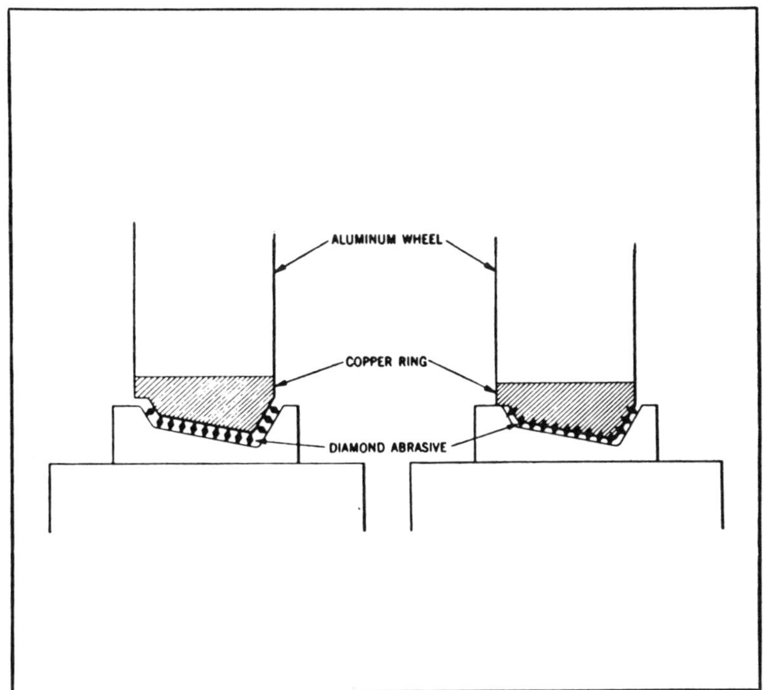

Fig. 2 Diamond particles project from copper by amount allowed for them in relating the form on the crusher roll or bar to that turned on the rim of the wheel.

Fig. 3 Form crusher roll in contact with wheel blank.

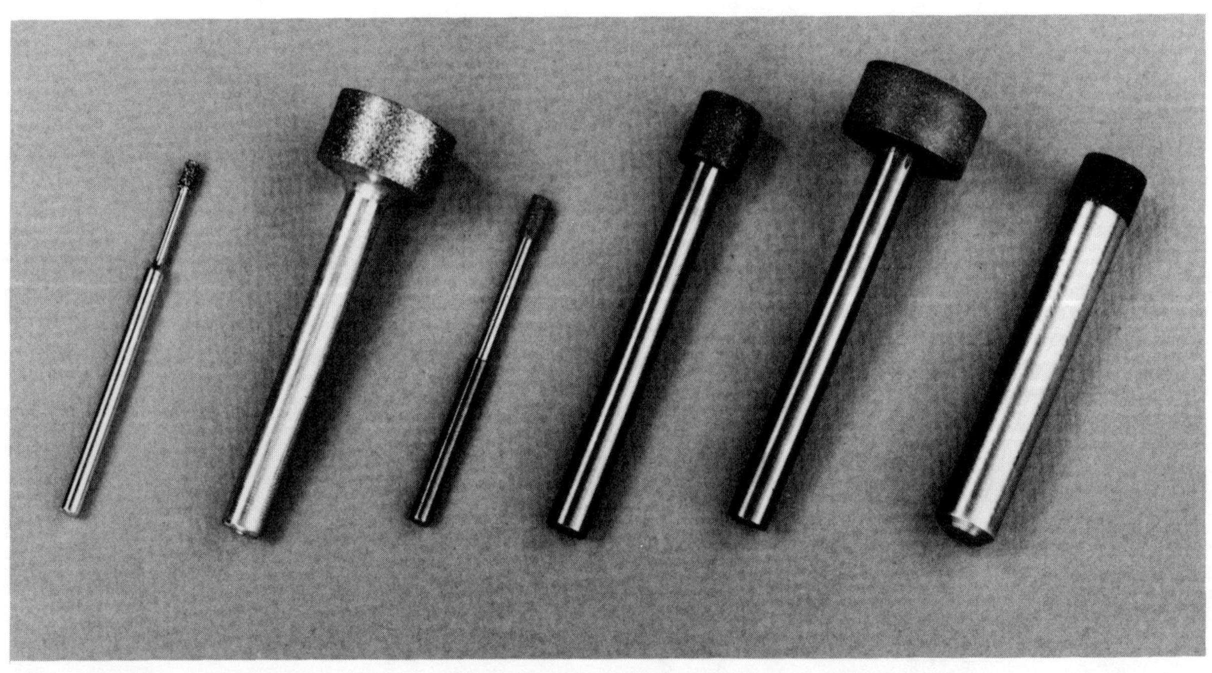

Fig. 4 Shank type mandrels containing Borazon or Diamond abrasive.

Fig. 5 Desired form is turned into aluminum wheel blank from a template. Allowance is made for grit size.

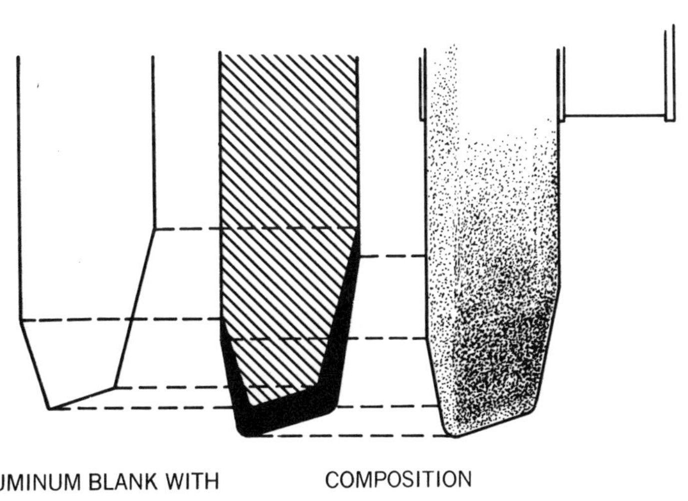

Fig. 6 Formed Diamond or Borazon wheel manufactured on its own spindle.

Fig. 7 Intricate punch & die shown being ground on tape controlled jig grinder.

Fig. 8 Typical forms ground with preformed plated wheels.

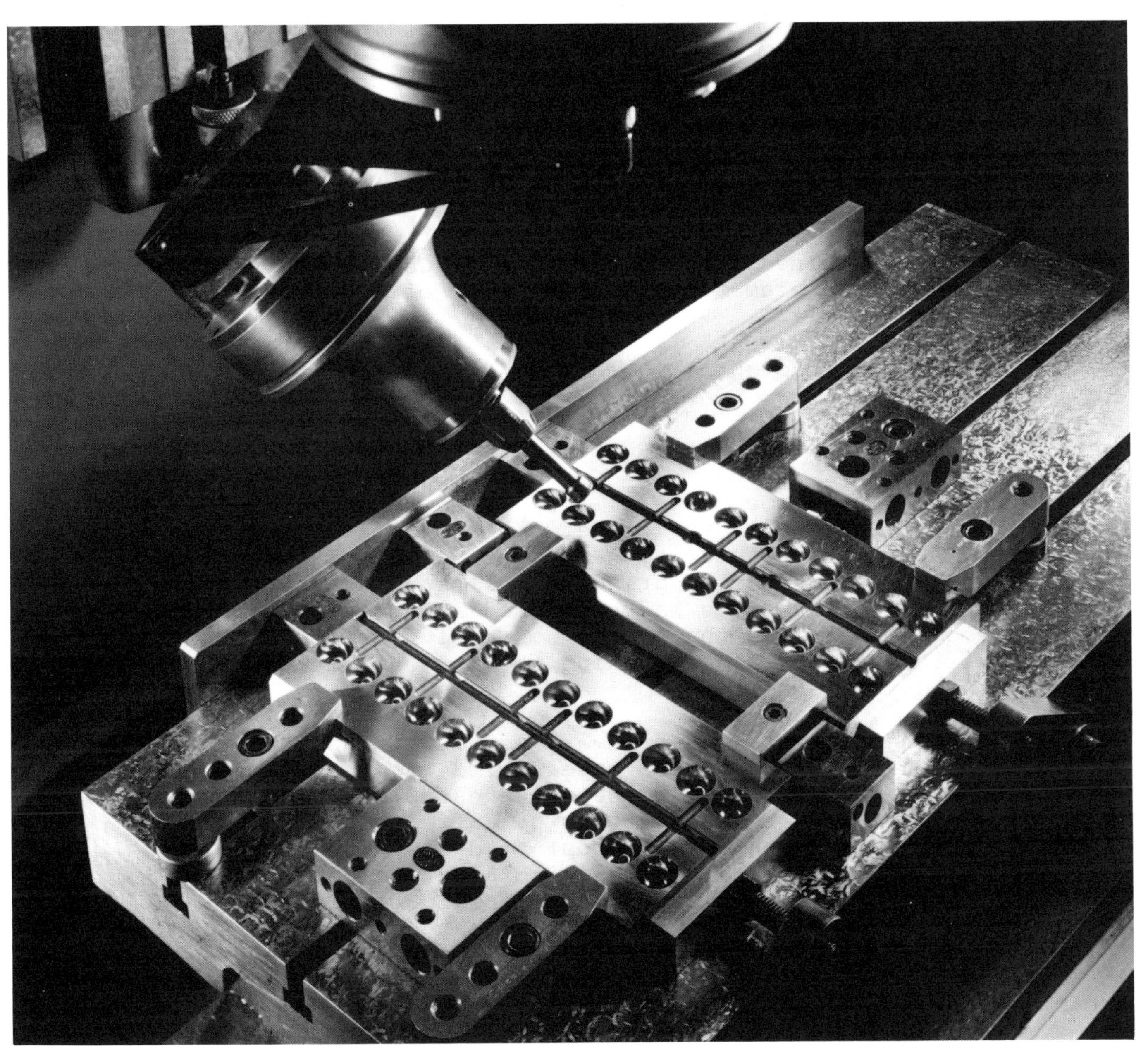

Fig. 9 Mold cavity being ground on Jig Grinder with special Borazon wheel.

Presented at SME's South Atlantic Conference, February 1977

Grinding High Speed Tool Steels

By R. A. Moir
And W. N. Ault
Norton Company

The high speed tool steels, AISI Types M & T, are the most highly used steels for cutting tools.

The more modern M & T Types demand more sophisticated grinding methods. It has been demonstrated that cubic boron nitride (CBN) is most efficient for the grinding of the more difficult of the high speed steels.

Separate and distinct procedures for the use of CBN wheels must be used.

G Ratios obtained from comparisons of CBN to conventional abrasives, prove dramatically, the value of CBN to the tool steel user.

CBN has been proven no less than 50 to 1 superior, by G Ratio comparison.

INTRODUCTION

The high speed tool steels, AISI Types M and T, are the most highly utilized steels for cutters, hobs, drills, and other tools. With this popularity has come the demand for new and more efficient methods of grinding. Although, historically, aluminum oxide abrasives had been used in most applications, it has been demonstrated that cubic boron nitride (CBN) is the most efficient method of grinding the more difficult of the high speed tool steels in some applications.

The high speed tool steels vary as to grindability. In order to distinguish between steels of different grindability, a set of grindability groups has been established. These groups distinguish between steels of different grindability indexes (better known as G ratios)[1] and the size of the area being ground. In specifying grindability groups, the G ratio used is that for the optimum conventional abrasive wheel on a given job. The molybdenum and tungsten types of high speed tool steels are in groups 2, 3, 4 and 5, which means the G ratio obtainable with the optimum conventional abrasive specification for all these steels is less than 12.

The best wheel specification for grinding these steels varies not only due to the material being ground and the size of the area to be ground, but also due to the coolant, if any, used and the type of grinding operation or mode. Figures 1 and

[1]G ratio is defined as the cubic inches of material removed per cubic inch of wheel used.

2 show the standard marking systems for conventional and CBN abrasive grinding wheels, respectively.

ALUMINUM OXIDE WHEELS

Aluminum oxide abrasives are normally the first choice for grinding the high speed tool steels. In fact, aluminum oxide can be used to grind any of these steels in any grinding mode. There are some modes, however, where silicon carbide or CBN abrasives show a marked superiority.

High speed tool steels are usually ground using semi-friable aluminum oxide abrasives. These abrasives fracture under grinding pressure which creates new sharp cutting surfaces and helps dissipate the heat of the grind. While the high speed tool steels are not generally as susceptible to cracking due to heat as the D series cold work tool steels, the M40 series and the high carbon high vanadium types--particularly at high hardness levels--are more sensitive to cracking due to grinding. Actually, grinding burn, even if not accompanied by cracking, can cause serious metallurgical damage to the ground surface. In some operation such as knife grinding, where heat is a major grinding problem, a friable (white) aluminum oxide abrasive is often used.

SILICON CARBIDE WHEELS

Silicon carbide abrasives are used to grind high speed tool steels in very few special applications. Silicon carbide abrasives are by nature harder than aluminum oxide, and the crystal shape is jagged and sharp. Also silicon carbide vitrified wheels have a higher coefficient of thermal conductivity than similar aluminum oxide wheels and are therefore cooler cutting. However, high production grinding of high speed tool steels with silicon carbide is not as efficient as grinding with aluminum oxide. In other words, G ratios with aluminum oxide are usually higher than with silicon carbide.

There are several special applications where the properties of silicon carbide are utilized. These include fine pitch thread and form grinding with vitrified wheels, lapping, and honing. All of these operations utilize 240 and finer grit silicon carbide vitrified wheels. The wheels hold form well due to the hardness of the abrasive and do not generate as much heat while grinding. All three operations are normally done with oil or water-soluable oil coolants, in order to provide lubricity.

COATED DIAMOND WHEELS

Before the advent of CBN abrasives, much work was done developing Armored[2] diamond crystals which could be used in

[2]Armored is a trademark of the General Electric Company.

grinding steels similar to the high speed tool steels. The hope was that by using coated diamond it would be possible to grind these steels at a very high G ratio without the carbon content of the steels chemically interacting with the diamond. These wheels met with some small success, but CBN abrasives are used in an identical fashion with superior results (See Figure 3).

CBN WHEELS

CBN (Borazon)[3] abrasive was introduced commercially by General Electric in 1969. Since that time, it has proven to be a highly effective abrasive in the grinding of hardened high speed tool steels; especially those with low grindability indexes. Much of this success can be attributed to the ability of CBN grinding wheels to retain form and remove material rapidly with low heat generation. The hardness of the abrasive (shown in Figure 4) is the major reason for these characteristics. The chips produced are not unlike those produced in a machining operation. This helps to maintain the metallurgical integrity of the material being ground. Since the introduction of CBN abrasive, the manufacture and resharpening of high speed steel tools have been among the most successful CBN applications.

There are several criteria to be considered when attempting to select a CBN wheel for a particular material:

(a) Degree of difficulty to grind with conventional abrasives.
(b) Hardness (Rc 55 or harder).
(c) Need for a burn-free grind.

Experience indicates that the chance of successfully grinding high speed tool steels also depends on the grinding mode utilized. For example: internal, tool and cutter, and honing have relatively high success rates, while cylindrical, centerless and vertical spindle modes have a lesser chance of success. New CBN crystal types and bond materials are being developed in efforts to increase success rate in these modes.

The choice of grit size plays an important role in the surface finish obtained. The conventional abrasives produce a predictable finish by grit size based on many years of history. The sharpness of CBN dictates that it is necessary to use a finer grit size--than with conventional abrasives--to obtain a like surface finish. Extremely fine surface finishes are difficult to obtain with CBN abrasive wheels, especially in the line-contact type operations such as cylindrical grinding. The wide area of contact modes such as vertical spindle-rotary table surfacing, however can produce fine finishes.

---

[3]Borazon is a trademark of the General Electric Company.

When using CBN abrasive wheels, machine condition is another important consideration. A machine in need of maintenance will influence finish and wheel life to a greater degree with CBN wheels than it will with aluminum oxide or silicon carbide wheels. Machine vibrations tend to be hidden or dampened by rapid conventional wheel breakdown, while CBN "non-forgiving" wheels will chatter under identical conditions.

It is most important to properly condition the face of a CBN wheel prior to attempting to grind with it. Truing and dressing are both required.

TRUING

Resin-bonded CBN wheels may be trued to good roundness and concentricity by several techniques. For best results the following types of tools are recommended:

(a) Diamond truing tool (metal bond):
There are five standard tools, which are illustrated in Figure 5. Recommendations for use are given in Table 1.

(b) Brake Controlled Truing Device:
The brake controlled truing device (Figure 6) is designed for truing straight wheels on horizontal spindles as well as cylinder and cup wheels on vertical spindle surface grinders. The truing device should be mounted so that its truing wheel spindle is parallel to that of the peripheral type wheel, or perpendicular to that of a cylinder or cup-type wheel. This truing device must be used dry.

(c) Rotary diamond dressers or diamond form blocks:
These specialized tools are usually custom engineered for individual applications by truing tool manufacturers (such as Koebel Diamond Tool, Detroit).

DRESSING

After truing, all CBN wheels must be dressed to open up the glazed surface left after truing (see Figure 7).

The wheel is slowed down by jogging[4] and aggressively scrubbed with an abrasive stick as it slows to a stop. The hand-held stick is applied to the center of the wheel and rolled to the edges. A holding device can be used.

A small amount of coolant should be used to create a paste or slurry that rolls like grains of sand between the wheel and the stick. This abrades out bonding material which would otherwise prevent the wheel from cutting freely.

Before dressing, the wheel will feel smooth; a properly dressed wheel will have a rough texture.

---

[4]Jogging is turning the power off and on, as the Borazon wheel should only be dressed while it is coasting to a stop.

Rapid wear of the dressing stick is a good indicator that the wheel is being opened.

If workpiece burning or poor cutting action is experienced under normal grinding conditions, this indicates a need for additional dressing.

Three examples of grinding operations, where CBN abrasive wheels have replaced aluminum oxide wheels, are listed in Table II.

Additional comparisons in grinding performance and efficiency between CBN abrasive wheels and aluminum oxide wheels--as applied to high speed tool steels T15, M3, M2 and M43--are graphically shown in Figure 8.

GUIDELINES FOR USING CBN WHEELS

Be sure spindles and backplates run true, and spindle is not worn undersize.

When possible, mount on adapter to facilitate wheel changes.

When changing wheels, remove CBN wheel and adapter as a unit (helps keep the wheel in running truth).

Mount cup wheels to run true on the face within 0.001" indicator reading; straight wheels within 0.0005" by "tapping" lightly and/or "truing".

After the wheel is running true, it is important to dress the wheel face with a 37C220-K8V silicon carbide stick to assure sharp wheel face condition. A rubber abrasive stick (A120-B2RR) can also be sued to clean the wheel face.

CBN wheels seldom need dressing during use.

When required, use same silicon carbide stick mentioned above or equivalent aluminum oxide stick.

Maintain correct wheel speed. For best results, resin bonded CBN wheels should be run with a flood coolant at 3500 to 8500 sfpm.

If grinding dry, speeds in the 3000-5000 sfpm range are recommended.

Avoid excessive grinding pressure or feeds on precision grinding jobs, the depth of cut per pass should not exceed:

(a) 0.004" for 60 and 80 grit wheels
(b) 0.002" for 100 and 120 grit wheels
(c) 0.001" for 150 to 220 grit wheels

(d) 0.0005" for wheels 240 grit and finer

On final finishing passes, reduce the feed to a fraction of these amounts and let the cut "spark out" to get the best possible finish.

Grind wet whenever possible. Using a flood water-soluable grinding coolant reduces grinding costs.

FIGURE 1--Symbols for Designation of Composition of Aluminum Oxide and Silicon Carbide Grinding Wheels. Example is 32A46-H8VBE. Alundum and Crystolon are registered trademarks of Norton Company.

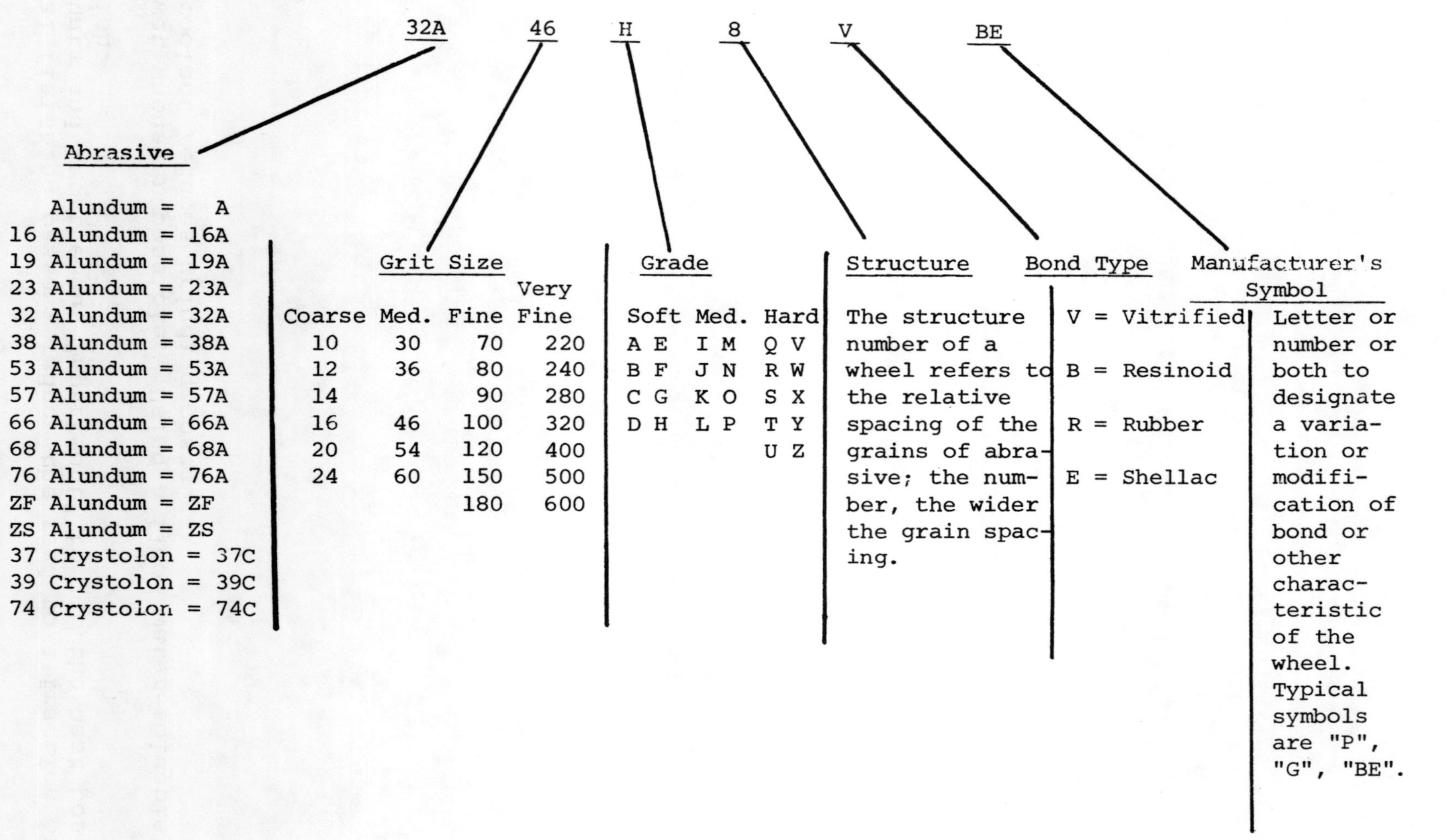

FIGURE 2 -- Symbols for Designation of Composition of Borazon Grinding Wheels. Example is CB100-WBA 1/16.

CB ABRASIVE	100 GRIT SIZE		W GRADE	BA BOND TYPE	1/16 CB DEPTH
Cubic	60	180	W -- durable	BA -- wet	1/16
Boron	80	220	T -- intermediate	BB -- dry	3/32
Nitride	100	240	Q -- mild	BC -- wet	1/8
	120	320	N -- very mild	BD -- dry	
	150	400		BE -- dry	

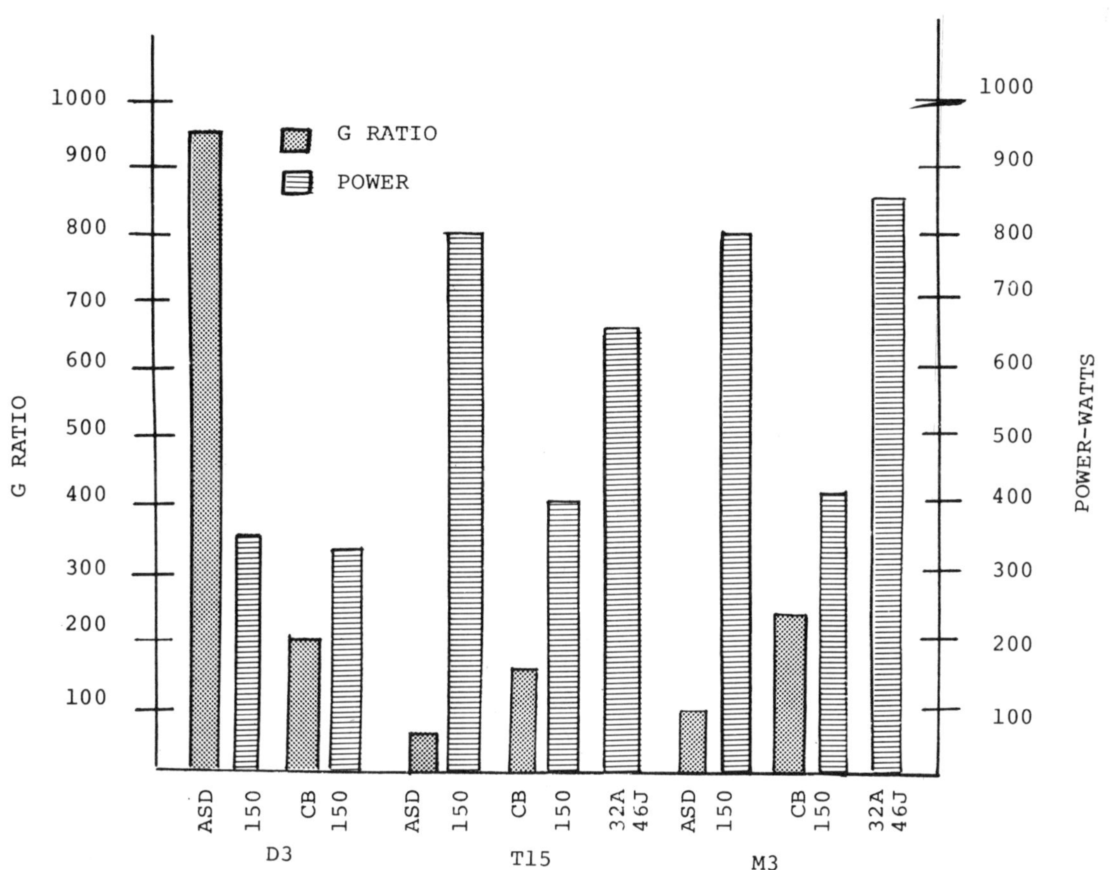

FIGURE 3 -Comparison of G Ratio and Power Requirements for Wet Surface Grinding High Speed Tool Steels Types D3, T15 and M3, with Armored Synthetic Diamond vs. Borazon Wheels. Conditions -- 5000 sfpm, 0.001" downfeed. G Ratios for aluminum oxide wheel too low to indicate on chart.

FIGURE 4 -- Relative Hardness of Selected Materials, Including Abrasives.

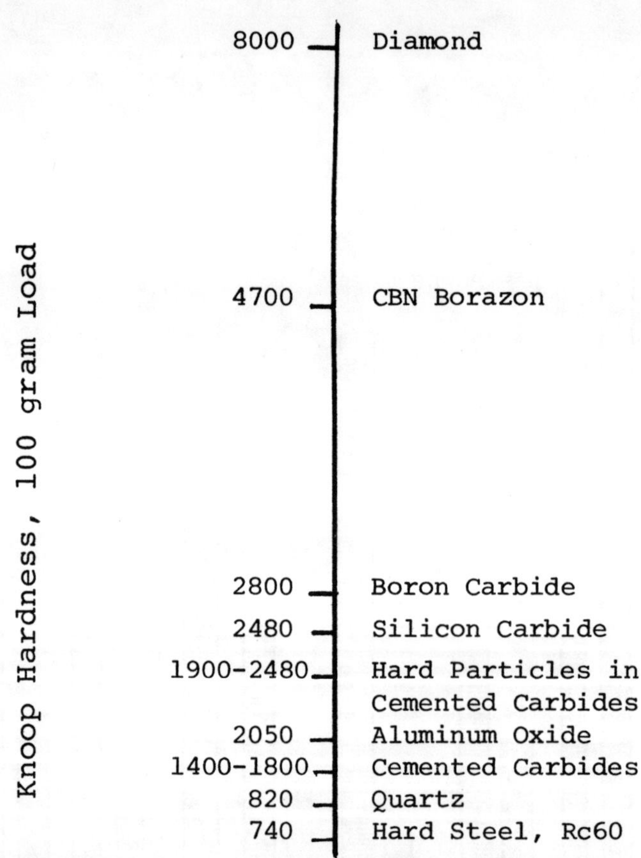

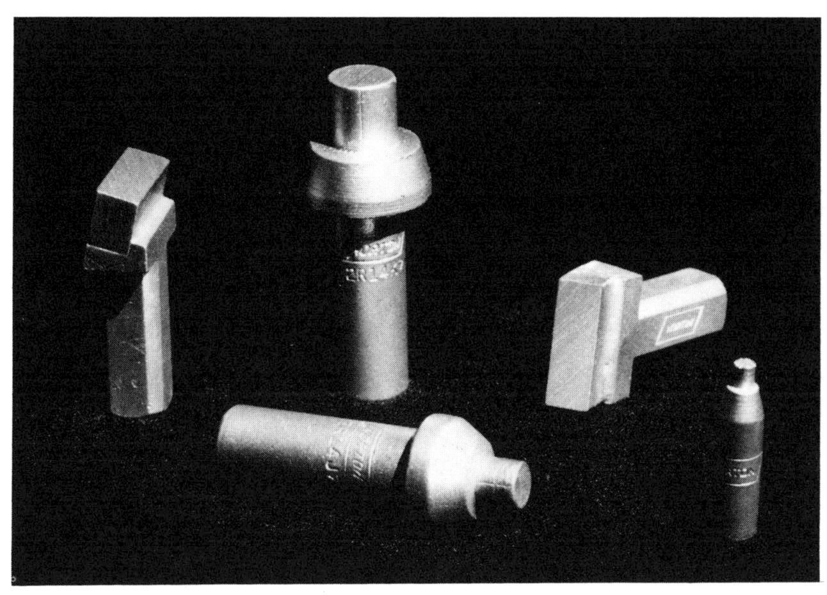

Figure 5 - Diamond Tools for Truing Resin Bonded Borazon Grinding Wheels

Figure 6 - Brake Controlled Truing Device.

Glazed Surface after Truing

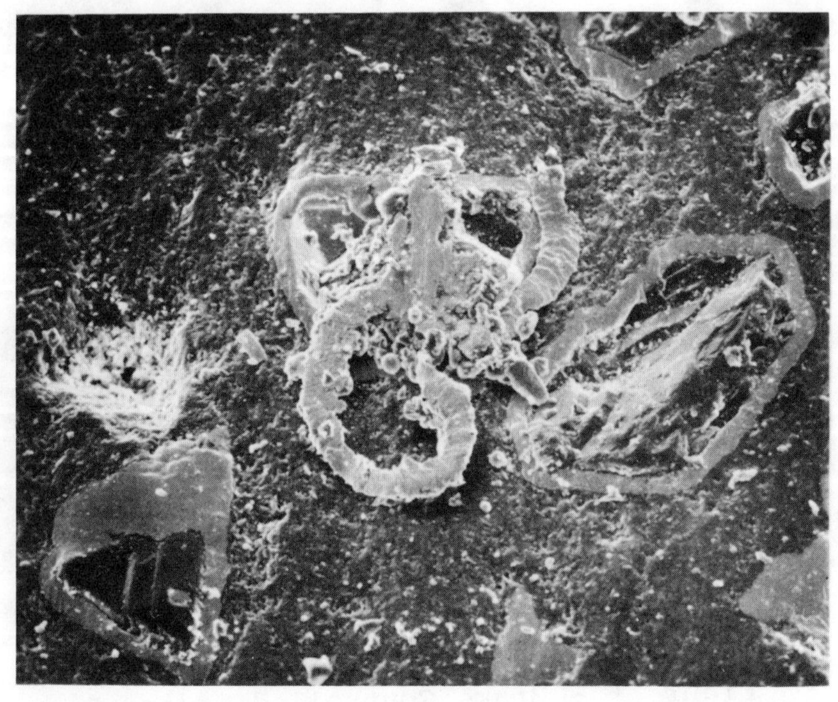

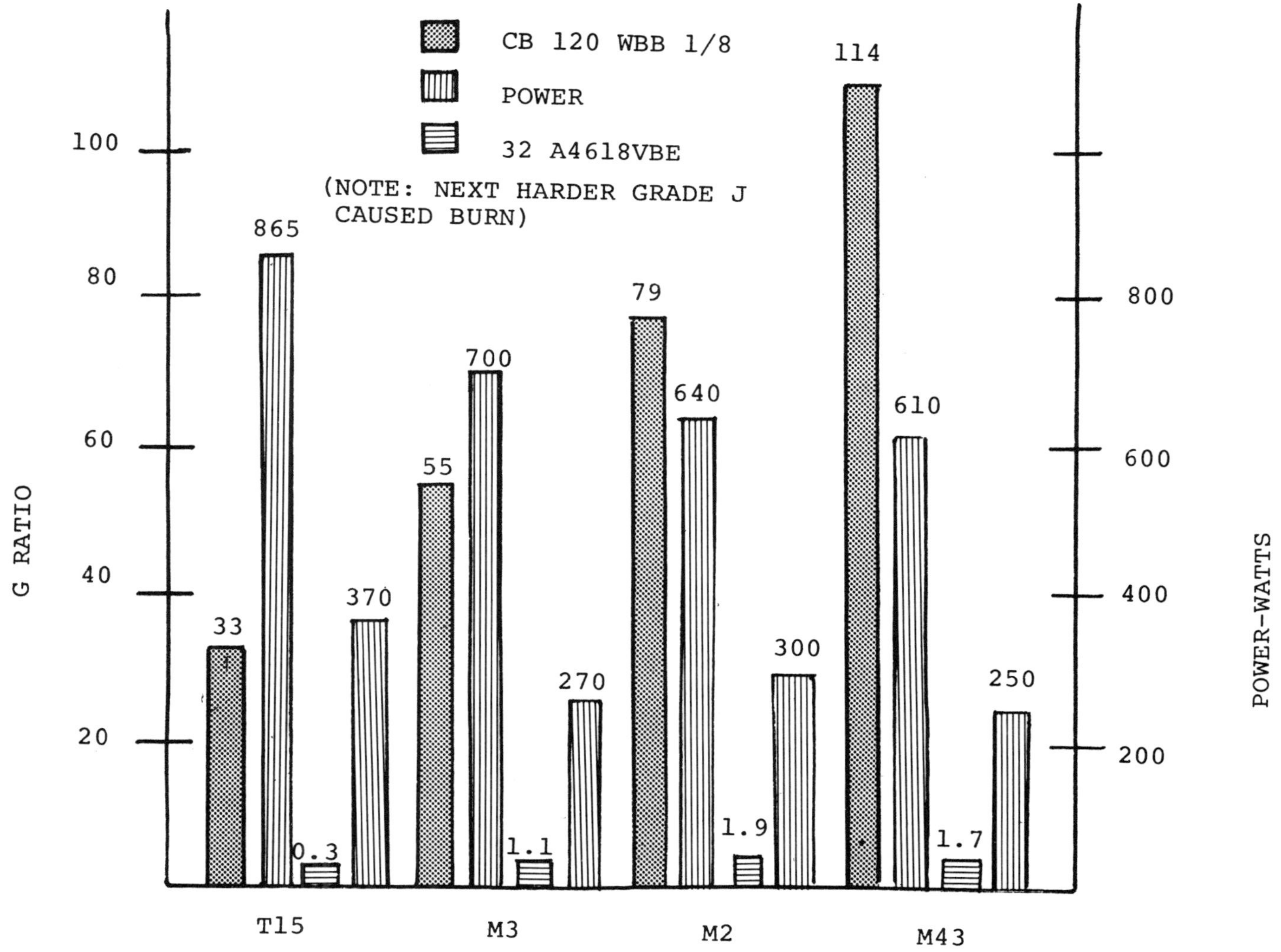

FIGURE 8 -Comparison of G Ratio and Power Requirements for Dry Tool and Cutter Grinding High Speed Tool Steels Types T15, M3, M2, and M43, with Borazon and Aluminum Oxide Wheels. Conditions -- 3500 sfpm, 0.002" infeed.

Table I
Truing Resin-Bonded CBN Wheels with Diamond Tools

Machine Type	CBN Wheel Size	Truing Post Angle	Tool Specification*
Centerless & Cylindrial	Any	0°	2R14K7
		15	1A14C7
		30	2B14C7
Surface	6" and Under	0	1R14J7
	6" and Over	0	2R14K7
Tool Room	General	0	2R14K7
Internal	Mounted Points		
	¼" and less	0	3R14L4
	Over ¼"	0	1R14J7

*Special Shapes and Sizes upon Request.

Table II
Grinding High Speed Tool Steels
with CBN Abrasive Wheels

EXAMPLE 1 -- Hob Grinding

Part Ground -- Hob 2" long -- 1" diameter
Type M2 (Rc 61-63_
Remove 0.040" sharpening
Standard Wheel -- Alunimum oxide -- vitrified
CBN Wheel -- 3x5/64x7/16" -- 180 grit -- 100 conc
Results -- Grinding time per piece reduced by over 50%. Burning and poor geometry produced with aluminum oxide was eliminated

EXAMPLE 2 -- Cutter Sharpening

Part Ground -- 12" diameter Inserted Milling Cutter
24 inserts 1 3/8x0.450"; Type T15 (Rc62)
Remove 0.003 to 0.005"/insert
Standard Wheel -- 46 grit-vitrified aluminum oxide
CBN Wheel -- 3 3/4" 11V9 -- 100 grit -- 75 conc
Results -- Change from aluminum oxide to CBN wheel reduced grinding time per cutter from 2 ½ hours to 30 minutes with virtually no wheel wear.

EXAMPLE 3 -- Relief Grinding

Parts Ground -- Various Type T5 Reamers and milling cutters (Rc 62-64)
Dry operation with varying removal
Standard Wheel -- 4" diameter 11V9 -- 100 grit -- 100 conc
Finish -- Commercial (approx. 20 Mi AA)
Results -- The CBN wheel produced 13 parts/hour compared to 8/hour with aluminum oxide.

Reprinted from Stanki I Instrument, Volume 48, Issue 2, 1977

APPLICATION OF SUPERHARD MATERIALS IN THREAD GRINDING

A. D. POKHOROVSKII
A. M. BELIKOV

Test results of introducing Elbor (c.b.n.) wheels for grinding hardened leadscrews of thread pitch 1mm and less and length exceeding 250mm are described. Use of Elbor provides means of manufacturing threads to 0-1 class accuracy. Data are given on the use of Elbor wheels in grinding threads on ball transmission nuts. Application of diamond rollers to dressing aluminium oxide grinding wheels in preliminary and finish grinding of trapezoidal threads and finish grinding of acute angled locking threads is considered.

Grinding the threads of precision, hardened leadscrews of small pitch and the corresponding hardened nuts involves many technological difficulties because of the rapid spalling of the tips of the abrasive wheel and losses in wheel profile accuracy. Accordingly, threads of pitch less than 1mm and length exceeding 250mm on unhardened leadscrews are usually produced by cutting. Special difficulties arise in grinding internal threads, since small siameter abrasive wheels show rapid wear.

One of the most effective trends in improving thread grinding techniques is the introduction of Elbor (c.b.n.) wheels, which can be explained mainly by their long life, greater by a factor of almost 50-100 than that of alpha alumina wheels. Furthermore, with hardened leadscrew threads R_a = 0,32-0,16μm, and the accuracy is Class 0-1. Metallographic and X-ray analysis shows that there are no structural changes characteristic of conventional abrasive machining in the surface layers of KhVG steel after thread grinding.

At MZKRS, the grinding of small pitch threads on long leadscrews has been studied, using a Model 5822B3 machine in grinding over the whole of the screw, using screws of diameter 28-30mm and length 800-1100mm in KhVG steel (HRC 58-60). The technical requirements for leadscrews (in accordance with OST 2N33-2-74) are as follows: thread pitch error within the limits of one pitch 2μm; deviation of thread pitch 2μm; cumulative pitch error 2, 3, 4, 5, 6, 7, and 10μm on lengths of 25, 50, 100, 200 and 300, 400, 600, and 1000mm. The permissible deviations in profile on each side of the thread is 6μm. The surface roughness of the ground profile should not exceed 0,25μm.

The wheels used are Type LM20ST1K, 100%, and LM28T1K, 100%, diameter 400mm (dressing with Type N2 diamond pencil) with a coolant consisting of industrial oil 20 with addition of 20% 'Volzhsk-100'. In grinding screws with a thread pitch of 0,5-1mm with monocorundum or electrocorundum wheels, even during the first 150-200mm of thread length the internal diameter of the thread increases because of wheel wear, and grinding burns arise on the surface which eliminate the possibility of further machining without dressing the grinding wheel.

In grinding the same screws with Elbor wheels this does not occur. It should also be noted that the long life of Elbor wheels enables correction devices to be exploited to the full thus providing additional correcting axial displacements (1-2μm) of the workpiece relative to the abrasive wheel during grinding.

The experimental conditions are shown in Table 1, and the magnitude of the transverse feeds in grinding the screws in Table 2.

Elbor and aluminium oxide wheels were also compared in machining the internal threads of anti-friction screw-nut pairs, using a Model MV-28 internal thread grinding machine, and L6ST3K and L8ST2K wheels, both 100%. An automatic cycle with a special diamond dressing tool (0,3carat) in a mandrel is used for finish dressing of the wheel along the radius. The grinding and dressing conditions are given in Table 1.

The experimental results show that the machining output in using Elbor wheels is directly related to the cutting speed. The highest metal removal rate is obtained at a wheel speed of 40-50m/s; after 80-100 passes, no changes in wheel profile geometry can be observed. The depth of cut is 0,2mm. The surface of roughness scarcely changes on varying the grinding depth, and there is no elastic springback of the wheel.

Two main methods are currently used for profiling multi-ribbed grinding wheels, which satisfy the accuracy and surface finish requirements of the components. These include dressing with a diamond tool (to a copy or with the aid of a dressing device) and dressing with steel rollers (rolling) of corresponding profile. A disadvantage of the first method is the disturbance of the grinding wheel profile when the diamond becomes blunted, the complexity and high cost of the dressing devices, and the necessity of using highly skilled labour.

Rolling with steel rollers is widely used in profiling and dressing multiple grinding wheels in batch machining of components with acute angled threads, but because of the specific features of the method the wheel life is lower than one profiled with a diamond tool. Furthermore, this method involves additional costs in the manufacture of a reduction gear for decreasing the wheel speed during dressing, and the necessity of increasing the stiffness of the machine spindle bearings. Furthermore, the manufacturing cost of the rolling rollers is high, and they can be used for no more than 3-5 dressings.

Thread grinding machines have recently been equipped with instruments for dressing the wheels with diamond rollers, which decreases the dressing time and significantly decreases the consumption of abrasive tool, because of its increased life. At MZKRS the dressing of multiple grinding wheels with diamond rollers in preliminary and semi-finish grinding of trapezoidal threads has been introduced, and also the finish grinding of acute angled locking threads.

It is established that down rotation of the roller provides the most favourable cutting conditions in thread grinding. The factors affecting the wheel life are the roller feed and speed per wheel revolution, and sparking out during dressing on down rotation of the roller. The recommended dressing conditions are: roller speed 25-27m/s, roller feed 0,5μm per wheel revolution, and sparking out over not more than 1s.

The test results were applied to designing a special automatic thread grinding machine, Model MV-145. This is equipped with diamond multi-thread rollers, and has been successfully used at KamAZ, BAZ, the Minsk Tractor Factory, and various other factories.

TABLE I

Component	Component dimensions (mm): Total Length	Component dimensions (mm): Thread Length	Component dimensions (mm): Thread diameter X pitch	Thread Type	Wheel material	Grinding conditions: Wheel speed (m/s)	Grinding conditions: Component speed (rpm)	Grinding conditions: No. of passes to obtain complete profile	Dressing conditions: Diamond feed (mm per pass)	Dressing conditions: No. of passes	Dressing conditions: Dressing time (min)	Wheel life (in number of passes)
Screw	1102	960	30X0.942	Trapezoidal	Elbor	30-35	6-8	3-4	0,005-0,01	2	-	85
Screw	802	635	28X0.628	Trapezoidal	Elbor	30-35	5-6	3-4	0,005-0,01	2	-	85
Nut	70	70	50X10	Round r = 3,62mm	Elbor Aluminium-oxide	30-50 16-18	5-10 2-4	4-6 12-15	0,005-0,01 0,03-0,04	1-2 3-4	2 3	85-100 2-3
Nut	30	30	80X12	Round r = 3,62mm	Ebor Aluminium-oxide	30-50 16-18	3-5 1-2	6-7 15-20	0,005-0,01 0,03-0,04	1-2 3-4	2 3	70-80 2-3

TABLE 2

Type of pass	Magnitude of transverse feed (mm per pass) for lead-screws of pitch (mm): 0,628	0,942
Rough grinding		
First pass	0,3	0,40
Second pass	0,1	0,10
Third pass	0,05	0,10
Finish Grinding	0,05	0,05

Machining turbine parts—problem solving with diamond and CBN tooling

BY ING GRAD **H HUJER** KLÖCKNER HUMBOLDT DEUTZ AG
OBERURSEL W GERMANY

Reprinted from Industrial Diamond Review, December 1978

In turbine construction, machining problems are often encountered which, because of the materials involved, can be solved only by the use of high-performance grinding and cutting tools. Because the machining operations very frequently have to be performed on new materials, or on new combinations of materials, for which reliable tooling information is not available, the choice of tool is often very difficult. This article describes a number of solutions which have been found in the manufacture of turbine components by the use of diamond and CBN tooling

ings of the axial compressors of jet propulsion units. With a height of 400 mm, the housing measures 650 mm in diameter and accommodates six rows of guide blades. To facilitate mounting on the propulsion unit, the housing is divided into two half-shells which are then bolted together. In this condition, the entire machining of the inside profile is carried out on a vertical turret lathe. Machining has to be performed on layers of carbon/epoxy resin which are sprayed cold on to the bore between the grooves accommodating the guide blades, and then hardened at ≈ 470 K. The requirements here are a surface finish (R_z) of ≦ 12·5 μm and IT8 tolerances on the bore diameter. Early trials with conventional lathe tools resulted in tearing of the surface, in addition to which the life of such tools was excessively short. Tests carried out with diamond tools produced very good results as regards both tool life and surface finish, so that it was possible to make a start on optimizing the machining parameters. At the present time, diamond tools are in use at our works for normal production at cutting speeds of 200 m/min, cutting depths up to 0·5 mm and a tool feed of 0·1 to 0·12 mm. Machining is carried out dry. The machining debris takes the form of powder and no chips in the conventional sense are produced. Fig 3 above shows the bore of an axial compressor housing which has been diamond machined.

Initially, only single-point natural diamond tools were used, but for some time now use has also been made, under the same operating conditions, of tools incorporating synthetic polycrystalline diamond inserts, eg SYNDITE. The total amount of work performed during the life of these tools, i.e. ten housings, was found to be equivalent to that achieved with natural diamonds. At eight, the number of possible regrinds was also the same. A cutting distance of about 70,000 m is therefore attained for each tool. Commercial arguments in favour of the diamond composites are the price of the tools and the lower cost of regrinding. In principle, however, both types of tool are suitable for performing this kind of machining operation.

The machining of boron-treated components

For some components of the fuel control system of a newly developed small-scale gas turbine, very strict requirements were imposed on the wear characteristics of frictional surfaces in relative movement. The boron-treatment of these parts was therefore built into the manufacturing sequence right from the design stage. Diffusion of boron into the workpiece surface is carried out by the 'powder pack' process at a temperature of about 1175° K maintained for about 1 hour. The borided layer, 40 to 60 μm deep, was found to have an HV 0·3 hardness of 17,000 to 18,000 N/mm².

Published information indicated that, if at all, borided surfaces could be machined only with diamond tools.

The first thing to establish was whether machining was required at all, i.e. given the required tolerances and surface finishes which had already been achieved prior to the boriding process, did the latter process provoke any appreciable deterioration in quality. It was established that the boriding process did provoke some swelling of the material and a certain amount of roughening of the surface, and therefore both these factors were allowed for in the preliminary design. For instance, for those dimensions of the components which were subject to tight tolerances (see Fig 2), allowance was made for a 15 μm growth to the material due to the boriding process. Similarly, to arrive at the specified surface finish of $R_z \leq 1$ μm, machining prior to the boron treatment was designed to produce a finish of $R_z \leq 0{\cdot}63$ μm. Measurements carried out on the first production batch after boriding showed that, despite all this, finish machining would still be needed. In the case of the washer-shaped austenitic chrome-nickel steel thrust ring (Fig 2a), finish lapping of the flat face was necessary to achieve the requisite finish of $R_z = 1$ μm. This operation was performed by hand using a grey cast iron plate and diamond compound with a 0–0·5 μm particle size. In view of the small number of components, mechanization of this process was unnecessary.

In the case of rotationally symmetrical bearing supports (see Fig 2b), made of Nimonic 75 a nickel alloy possessing high heat resistance, bores of diameter 9H6 and 10H6 no longer had the requisite dimensional accuracy after boriding. Although the relevant literature says that diamond tools are the most suitable for machining borided surfaces, we started off by using an available, metal-bonded CBN grinding stick measuring 8 × 10 × 6 mm and containing D91 (170/200 US mesh) size grit. As the machining allowance on the bore was a mere 0·01 mm, alignment before grinding was a fairly intricate operation. The work was carried out at a cutting speed of 13·5 m/s on an Overbeck internal grinder. Contrary to our expectations, the finish machining of the bores with the CBN tool gave rise to no problems (Fig 1). With a second manufacturing batch, the bores were 0·05 to 0·06 mm narrower after boriding. This means that, in the present state of boriding techniques, because of the varying degree to which the components swell, finish grinding is necessary in order to maintain the requisite bore tolerances. Because of the costs involved, we have not carried out any further tests, e.g. with diamond internal grinding tools ◊

Fig 1 Above *A bearing support (left) after machining with a CBN grinding tool*

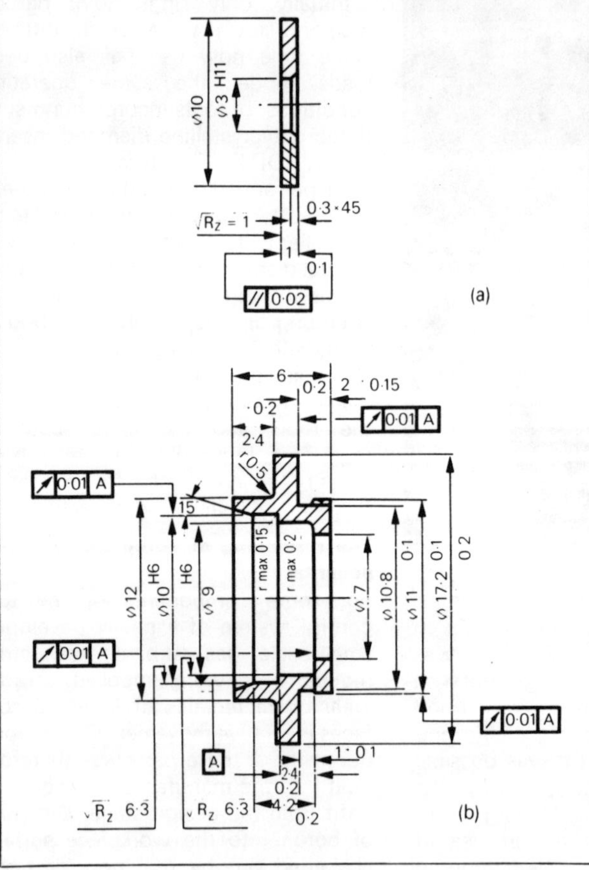

Fig 2 *Dimensions and tolerances of the borided components to be machined: (a) Thrust ring; (b) Bearing support*

Fig 3 Opposite page *Bore of an axial compressor housing coated with carbon/epoxy resin after machining with a diamond lathe tool*

Grinding hardenable steel bonded carbide alloys with diamond wheels

BY **F FREHN** THYSSEN EDELSTAHLWERKE AG FERRO-TITANIT DIVISION KREFELD W GERMANY
AND **A T NOTTER** DE BEERS INDUSTRIAL DIAMOND DIVISION TECHNICAL SERVICE CENTRE CHARTERS ASCOT UK

Reprinted from Industrial Diamond Review, June 1979

After a description of the metallurgical differences between tool steel containing Cr_3C_2, WC-Co sintered carbide, and the FERRO-TITANIT range of hardenable steel bonded carbide alloys, some practical advice is given on correct heat treatment procedures for the latter materials. The technical and economic advantages of diamond over conventional abrasives for grinding FERRO-TITANIT in the hardened state are then described and illustrated. The report concludes with a description of some surface and cylindrical grinding tests carried out at De Beers Technical Service Centre, giving results for resin bond wheels containing CDA55N and DXDAMC metal clad synthetic diamond grits, which can be applied in industry

1. Introduction

The progress achieved over the last 50 years in mechanization and automation of production processes in manufacturing industry derives not least from continued research into tool development. For example, the higher output capacity and/or improved life of the tooling enabled manufacturers to repeat machining operations within tight limits of accuracy and finish, a precondition for ensuring the interchangeability of standard components produced in long manufacturing runs. In the endeavour to improve the wear resistance or the load-bearing capacity of the most widely varying products, it has been necessary to increase the hardness and mechanical strength of both metallic and non-metallic workpiece materials and to develop new materials, which could be machined only with difficulty. As far as steels are concerned, manufacturers have tried to meet the demands of the market with an ever higher proportion of high-strength alloys. The technical limits encountered in this direction have been overcome by the development of very hard sintered carbides, in which WC, TiC and TaC are metallurgically incorporated in a metal binder phase (cobalt). The most recent advance in this field is represented by hard metals in the production of which powder metallurgy is employed to incorporate metal carbides into a hardenable steel bond.

It is well known that powder metallurgy processes eliminate all the negative aspects inherent in fusion metallurgy, including fibre flow, segregation, irregular crystal size, deleterious inclusions, etc. Among other things, this makes it possible to increase the carbide content of these steel-bonded materials to twice that of steel without causing any substantial deterioration of the machining properties. For this reason, the TiC content of high-chromium, hardness-retaining hard metal alloys can be raised to as much as 35% by weight (about 50% by volume)[1].

The hard materials described in this report, known by the trade name FERRO–TITANIT*, are characterised by a carbide content which in terms of volume is twice as great as that of high speed steels. Mostly, the very hard and light titanium carbide (TiC) is employed, which cannot be used in steelmaking. Depending on the intended application of the hard alloy, the TiC is incorporated—up to the proportion by volume or weight mentioned above—into a matrix made up chiefly of chromium-molybdenum-carbon steel (Fig 1). In the as-delivered condition (annealed) with a Rockwell C hardness of 42 to 50, this hard material can be worked by the normal machining techniques. When hardened to 54 to 71 HRC according to composition or grade (see Table 1), FERRO–TITANIT can be economically machined using diamond wheels and can be used to solve the most wide-ranging wear problems[2].

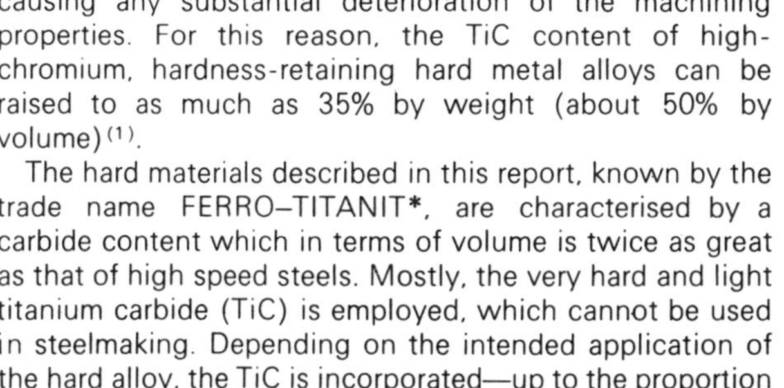

Fig 1 *Structure of a grade of FERRO-TITANIT showing titanium carbides incorporated in a steel matrix*

Correct heat treatment of the binder phase, or matrix, is vital to obtaining the maximum benefit from FERRO–TITANIT hard alloys. Steel-bonded hard alloys, which are also known as 'maximum-carbide' steels, possess a number of properties which greatly facilitate their heat treatment. To ensure the maximum durability of tools and components subject to wear, particular attention needs to be paid to the following[3]:

Neither over-long application of the austenitizing temperature nor overheating will produce a coarse crystalline structure. By virtue of their high melting point and low coefficient of thermal expansion, titanium carbides have

*Sole manufacturer: Thyssen Edelstahlwerke AG, FERRO-TITANIT Division, Krefeld, W. Germany.

TABLE 1
The most important grades of FERRO-TITANIT

Grade	*Harden-able to HRC*	*Matrix structure*	*Properties and applications*
C–Special	69–71	Martensitic	Cutting and stamping tools of all kinds, deep drawing dies, cold forming tools, parts subject to wear up to about 200°C
WFN	69–71	Martensitic	Cutting and stamping tools of all kinds, parts subject to wear, cold and hot forming tools up to 690°C
S	66–67	Martensitic	Parts subject to both wear and corrosion, pump valves, bearings, measuring tools
NIKRO 128/292	62–69	Nickel/ martensite	Cold and hot forming tools, parts subject to wear, tools for plastics processing up to about 750°C
UNI	54–56	Austenitic	Not magnetizable, highly resistant to corrosion—used for pressing dies, gauges, measuring instruments, guide rolls. Highly resistant to corrosion and sea-water—used for valves, bearings and rollers
GU 30	65	Martensite + micrographite	Self-lubricating with resistance to seizure—used for bearings, slides and rollers

an inhibiting effect on growth of the matrix. However, too low hardening temperatures and too brief holding times must be avoided.

Tools and components subject to wear can be annealed, machined and re-hardened any number of times without running the risk of an increase in volume due to crystal growth, such as can occur with steel after repeated treatments.

The change in volume following hardening is less and more uniform than in the case of steels, and warpage is greatly reduced. It is for this reason that, in contrast to steels, all hardenable steel-bonded carbides are given allowances of only 0·03 mm (0·001 in.) maximum per side prior to hardening, which considerably reduces the machining costs after hardening, if indeed machining is required.

In view of the small volume changes and minimal warpage, heat treatments should be performed wherever possible in neutral atmospheres (vacuum-hardening being preferred), as otherwise the negative areas of influence on

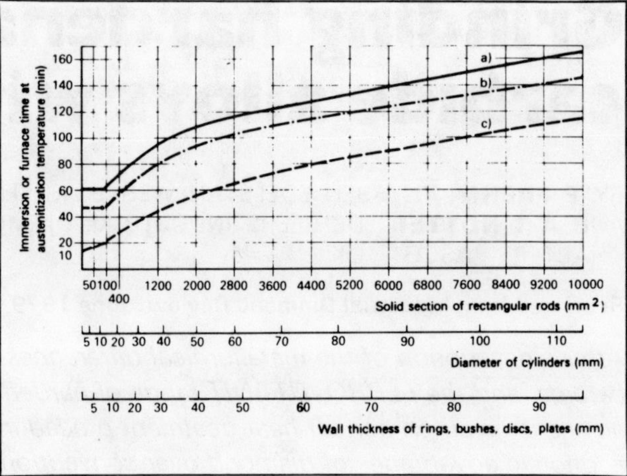

Fig 2 *Guidelines for the furnace and saline bath immersion times required to bring about the complete austenitization of FERRO-TITANIT: (a) shielding material, e.g. DESOTIC with minimum thickness of 20 mm per side; (b) vacuum furnace; (c) saline bath. The times indicated are minima. With FERRO-TITANIT, longer immersion and furnace times do not lead to deterioration*

the tools would exceed the machining allowances.

Because of their lower thermal conductivity compared with steels due to the high carbide content, all grades of FERRO–TITANIT require for the transformation of their crystal structure at least twice the holding time needed for tool steels. This holding time is further increased by shielding materials and steel sheets. Thanks to the direct transfer of heat, shortest times are required with saline baths. Fig 2 indicates the essential austenitizing times in relation to section for the common treatment techniques.

2. Grinding

Before considering in greater detail the machining of FERRO–TITANIT, and in particulaı the diamond grinding of the hardened grades, let us compare the chief materials presently used for heavy-duty punch tools[4]. Fig 3 shows polished sections of typical tool materials, including:

(a) 12% tool steel with about 22% by volume of chromium carbide, in which, owing to the fusion-metallurgical production technique employed, the Cr_3C_2 is irregular in shape, size and distribution. The maximum hardness which can be achieved with this material is about 62 HRC.

(b) The machinable hard alloy with its steel matrix shown in Fig 3(b) contains about 50% by volume titanium carbide which, thanks to the powder-metallurgical manufacturing process, is homogeneously distributed and made up of crystal sizes which guarantee the optimum mechanical characteristics. The maximum hardness which can

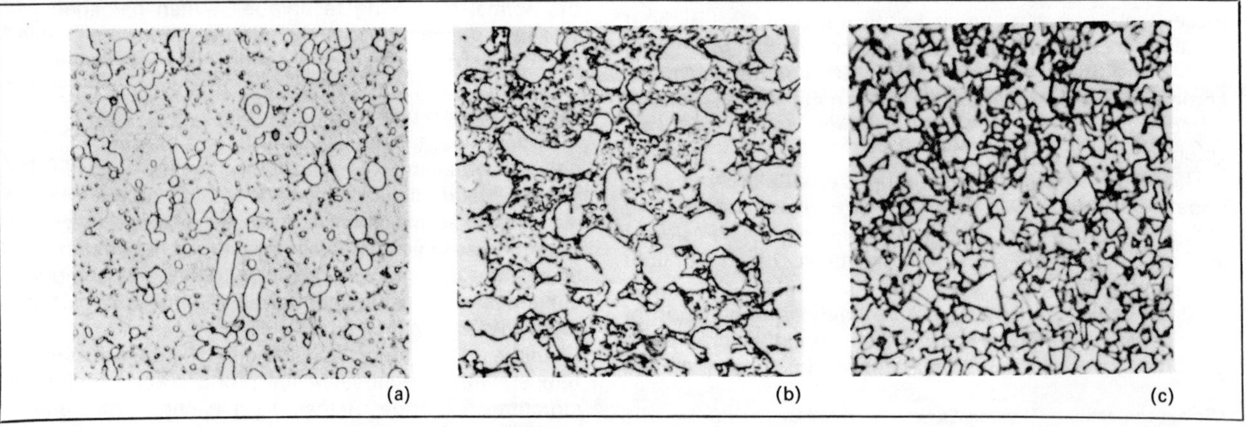

Fig 3 *Polished sections of various hard alloys: (a) 12% chromium steel; (b) FERRO-TITANIT; (c) G30 carbide*

be attained is 71 HRC. The exceptionally hard and light titanium carbide is not used in fusion metallurgy since it is virtually insoluble in the melt. This also explains its marked resistance to welding when used in tools and components subject to wear.

(c) G 30 sintered carbide, also produced by powder metallurgy, contains about 70% by volume tungsten carbide, in a cobalt binder phase. In contrast to the rounded titanium carbide of the hardenable steel-bonded alloy, the carbide particles are angular and sharp. The hardness of this naturally-hard carbide alloy is also around 71 HRC.

The fact that, despite their different carbide contents, the hardened alloy and the naturally-hard carbide both possess identical hardness values is due to the superior hardness of the titanium carbide (TiC) and martensitic steel bond of the former, as compared with the tungsten carbide (WC) and cobalt of the latter.

In the present state of technology, the question is not so much to determine what machining operations are practicable upon given machines, but to ascertain the effect which these machining processes have on the surface of the material, i.e. on the boundary zones. The effects on the boundary zones which result from inexpert execution of the machining operation have a mostly adverse influence on the tool performance. In this context we should mention, for example, surface rehardening, apart from which material-related problems are also of frequent occurrence.

The point has already been made that, in order to increase their resistance to wear, materials used for high-performance tools contain a greater or lesser amount of carbides (Fig 3). The hardness of these carbides contained in the sintered carbide, hardenable alloy or steel (e.g. $Fe_3C \approx$ 1100 HV; $Cr_3C_2 \approx$ 1300 HV; $Mo_2C \approx$ 1500 HV; WC $\approx$ 2400 HV; VC $\approx$ 2800 HV; TiC $\approx$ 3200 HV) constitutes a particular criterion for the grinding process. As far as the hardenable materials are concerned, it is also important whether the carbides are incorporated in an annealed or hardened steel matrix, as the hardness values of these bonding constituents (ferrite $\approx$ 80 to 90 HV; perlite $\approx$ 210 HV; austenite $\approx$ 180 HV; martensite $\approx$ 900 HV, ledeburite $\approx$ 900 to 1000 HV) also significantly affect the removal of material during grinding. To cut such materials properly, especially when they are in the hardened condition, conventional grinding media like corundum, Al_2O_3, and silicon carbide, SiC, are inadequate since, at 2000 HV, the hardness of corundum is below that of all the carbides mentioned and SiC, though it is superior to Al_2O_3 in hardness, is brittle and cannot for that reason produce a uniform cut. What happens, rather, is that the carbides are plucked out of the binder phase with the result that, as the conventional grinding media engage the matrix or the carbides, the surface structure of the work acquires a more or less grooved profile dependent on the carbide content of the material.

This phenomenon is illustrated by Fig 4, a scanning electron micrograph of a FERRO–TITANIT surface which has been ground by conventional means. Fig 5 demonstrates how unfavourably such grinding can affect the later polishing operation. While it is true that, on account of its mirror finish appearance when examined with the naked eye, such a polish is frequently considered to be good by the men on the shop floor, the microscope picture here clearly reveals two adverse effects of grinding with conventional media:

—Pores, left by carbide particles which have been pulled out, and

—Remaining carbide particles which have been covered over by smearing of the steel matrix.

As far as tool surfaces are concerned, the consequence of these phenomena is that, after only a few operating cycles, binding occurs as the workpiece material welds on to the

Fig 4 *Scanning electron micrograph of a FERRO-TITANIT surface machined with conventional abrasive*

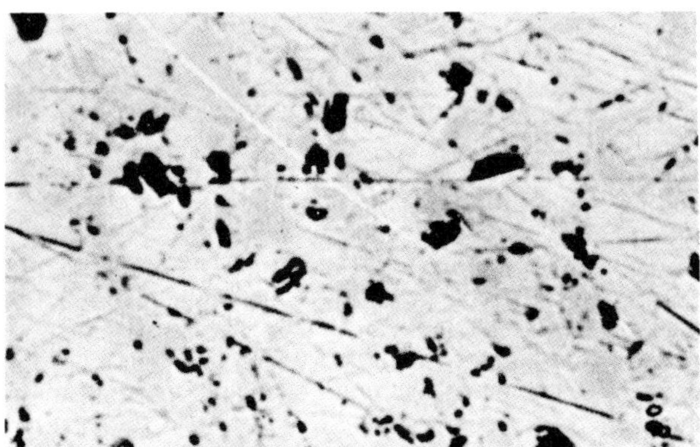

Fig 5 *Surface shown in Fig 4 after polishing*

smeared steel phase and keys on to the porous cavities (Fig 6). Admittedly, subsequent lapping with diamond compound would lay bare the carbides, but it would also cause diamond particles to become embedded in the pores, resulting in damaging 'comet's tails' on the tool surface (see Fig 7).

A reliable remedy is to be found only in the use of diamond grinding wheels, where the outstanding hardness of the diamond abrasive ensures the uniform removal of both bond and carbides. If, following grinding with a diamond wheel, the polishing operation is correctly executed, the surface shown in Fig 8 must be achieved without etching. The outlines of the titanium carbide particles can clearly be made out as 'islands', while the light portions represent the steel matrix. (In tool and high speed steels, the hard constituents, i.e. the carbides of chromium, vanadium, molybdenum, niobium and tungsten, would be shown up, whereas in the case of sintered carbides it would be chiefly tungsten and mixed carbides containing titanium and tantalum which would become visible.)

Fig 8 therefore no longer shows any trace of pores or of carbides which have been covered over by smeared metal. To the observer, the surface now appears rather matt, milky and grey and only partially reflecting. The differences in appearance illustrated respectively by Figs 5 and 8 are due to optical phenomena, i.e. in the latter polished specimen the structure at the surface is well formed, often indeed producing a relief effect, so that, to the observer, the incident rays of light are diffracted differently than by the smeared surface with pores.

The most meticulous attention should be paid to these differences in surfaces in maching shops, as the life of the tools and the appearance of the work are primarily determined by the quality of the polishing.

3. The present state of our knowledge of diamond grinding

3.1 Diamond wheels

On the basis of the foregoing remarks, the use of diamond

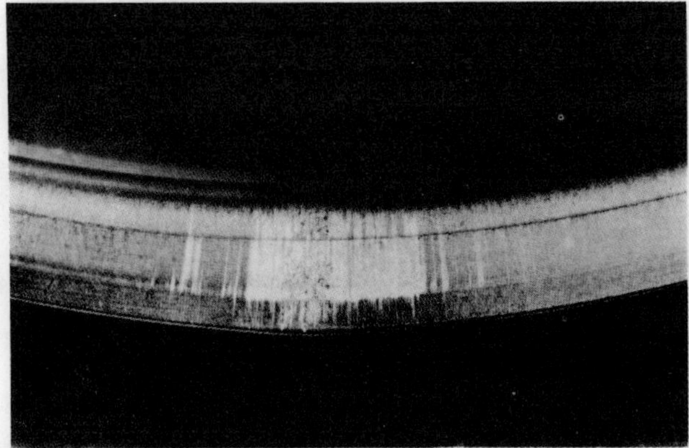

Fig 6 *Microscopic view of tool surface after improper machining as shown in Figs 4 and 5, revealing cold welding*

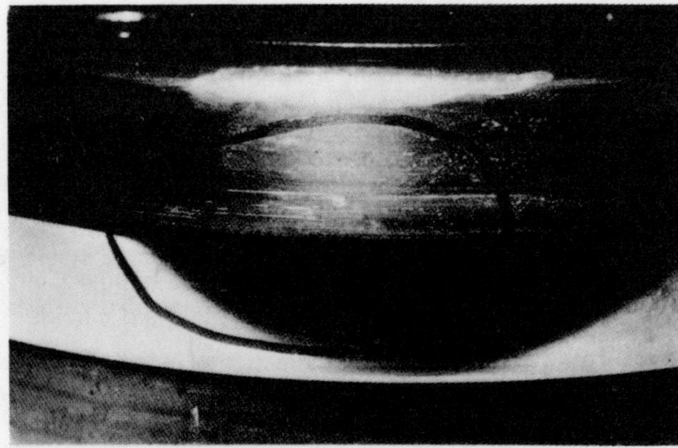

Fig 7 *Microscopic view of tool surface shown in Fig 6 after additional diamond polishing (comet's tails)*

wheels is strongly recommended for grinding hardenable FERRO–TITANIT materials in both the annealed and the hardened condition, if the maximum performance by tools and parts subject to wear is to be achieved. It is primarily resin bond wheels containing metal-coated synthetic diamond grit in sizes 100/120 US mesh (D151) to 140/170 US mesh (D107), at 75 to 100 concentration (≙ 3·3 to 4·4 ct per cm³ of abrasive volume), which are able to provide free-cutting properties which are 'kind' to the workpiece combined with an economic rate of metal removal.

Grinding machines in perfect working order are essential to efficient metal removal. They must be stable and constructed with the absolute minimum in the way of bearing play and fit clearances, failing which harmful vibrations will be likely to occur. The operation also benefits from adequate preparation of the diamond wheel by truing and dressing, to ensure maximum concentricity of the wheel and proper exposure of the diamond grits above the wheel matrix.

3.2 Heat in the grinding operation

The grinding result is also much affected by the proper and adequate supply of coolant to the cut. It is the job of the coolant to remove from the working area both the machining debris and the frictional heat, and thereby to contribute to the grinding performance and the quality of the work. Castrol's Syntilo 3 is particularly recommended as a coolant when grinding FERRO–TITANIT alloys.

The defects most frequently encountered in tools are thermal cracks caused by excessive grinding heat. This danger increases with decreasing thermal conductivity of the workpiece material, which in the case of metals generally means with an increasing carbide content. The use of diamond as an abrasive has the marked advantage over conventional grinding media in that diamond is known to be the best conductor of heat. In this context another decisive factor is the ratio of the wheel peripheral speed to the workpiece velocity. This ratio should not exceed 90 and, particularly in cylindrical grinding, this calls for appropriate adjustment to the r.p.m. of the wheel and work spindles.

Fig 9 shows a production sample which illustrates some typical consequences of heat generated during grinding. The profile has been ground into machinable hard metal with a steel matrix, of hardness 70–71 HRC without prior roughing following hardening. The chatter marks themselves (Fig 9(a)) already point to the poor quality of the machining. The heat generated by the grinding operation is shown up by the polished section (Fig 9(c)). The annealing caused by the grinding heat extends to a depth of 0·8 mm, and the hardness has been reduced to 62 HRC. Fig 9(b) reveals—starting from the top cutting edge—very severe erosion down to the hard 70–71 HRC core. What happens

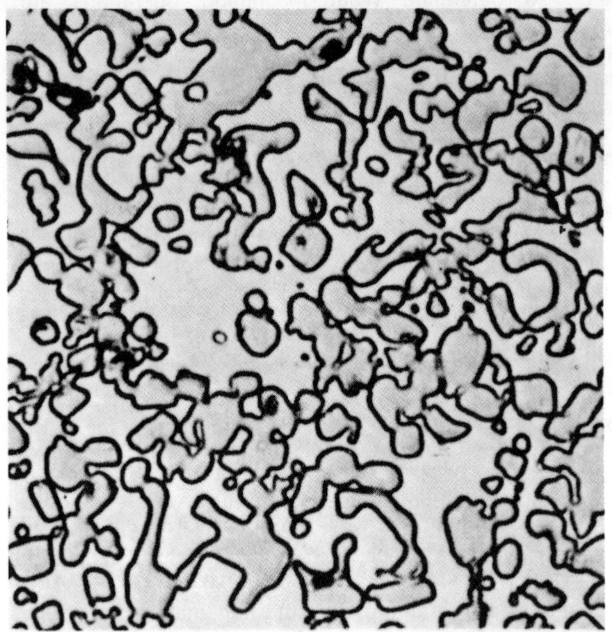

Fig 8 *FERRO-TITANIT surface structure attainable after diamond grinding ana correct polishing*

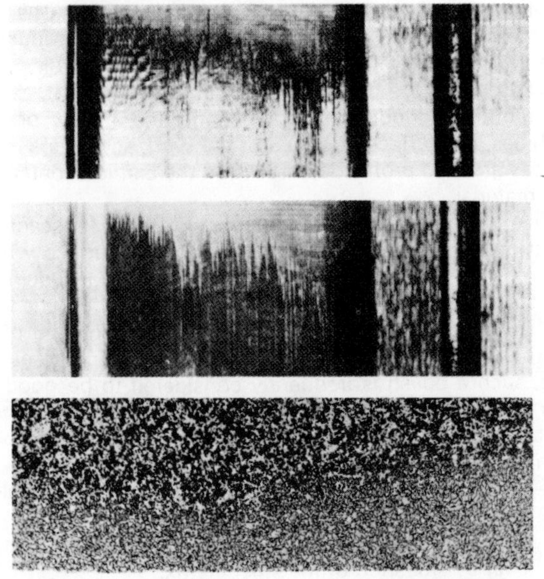

Fig 9 *The adverse effects of excessive grinding heat exhibited by a FERRO-TITANIT die segment. For explanation of (a), (b) and (c) refer to text*

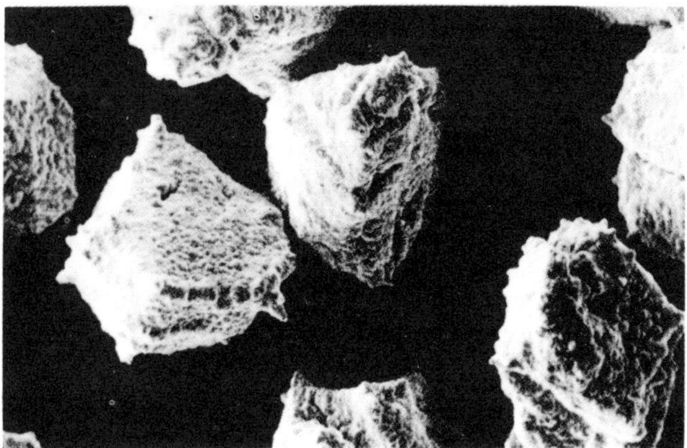

Fig 10 *CDA55N metal-coated synthetic diamond grit for the grinding of hard carbide alloys*

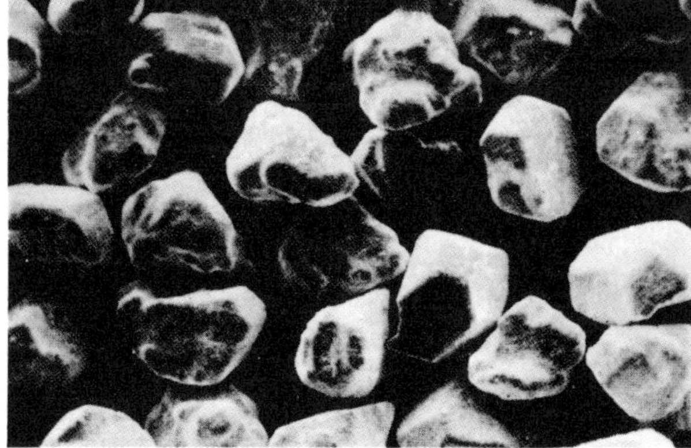

Fig 11 *DXDAMC metal-coated synthetic diamond grit for the grinding of steel/carbide combinations*

here is that the surface becomes rough, welding occurs, and the enlarged cutting gap causes prominent burrs on on the workpiece. The tool quickly becomes unusable.

3.3 Surface finish

The life of steel, hardenable carbide alloy and carbide tools is vitally affected by the surface roughness of the working edges and faces. The general view is that the maximum life in terms of output is achieved when the carbides are freely exposed at the surface and the degree of roughness is minimal. However, the ability of the lubricant film to adhere to the surface, which also depends among other things on the surface roughness, should not be overlooked. In this respect, the most varied requirements can be satisfied by choosing diamond wheels with an appropriate grit size and concentration, and by selecting the correct machining parameters. Wherever possible, cutting edges should be machined with an entering wheel so as to avoid chipping and to minimize any raggedness of the edges.

3.4 Economics

Apart from the technical factors which determine surface finish and metal structure, machining costs constitute the most important economic criterion for assessing a grinding operation. In the case of diamond wheels, the greatest significance continues to attach to the G-ratio which characterizes the useful life of the abrasive layer. For the surface grinding of hardened FERRO–TITANIT, of C-Special type particularly, grinding ratios currently being achieved in practice are around 30 to 45. Assuming an average wheel price of DM 80 per cm^3 for a 100 concentration resin bond abrasive layer, this gives us tool costs of DM 2 to 3 per cm^3 of material removed. Given the high removal rates and the consequently low time-related costs which can be attained with diamond wheels, this means that total costs are significantly lower than with Al_2O_3 or SiC wheels. This economic advantage is, of course, reinforced by the aforementioned technical advantages which derive from the use of diamond wheels, and by the fact that the greatly superior wheel life provides better possibilities for maintaining the dimensional and geometrical tolerances of the components.

4. The latest test results in diamond grinding

4.1 Workpiece materials

The investigation described below was carried out in the grinding department of De Beers Technical Service Centre, Charters, at the instigation of Thyssen Edelstahlwerke AG's FERRO–TITANIT manufacturing plant in Krefeld. Building on the information provided ten years previously by similar tests[5,6], the purpose was to ascertain the present state of the art in the light of grit development work carried out in the meantime by De Beers, and taking into account also the development of FERRO–TITANIT grades.

Surface and cylindrical grinding tests were therefore performed with resin bond 1A1 wheels, 300 mm in diameter and 30 mm in width, containing De Beers CDA55N (Fig 10) and DXDAMC (Fig 11) metal coated grits. The tests were carried out on the following three grades of FERRO–TITANIT in both the annealed and the hardened condition:

Material	*Rockwell hardness (HRC)*	
	Annealed	Hardened
WFN	49	69
UNI	51	55
NIKRO 128	53	63

In respect of their properties and recommended areas of application (see Table 1), the above are representative grades from the FERRO–TITANIT range. The specific chemical composition and properties of these grades are summarized in Table 2 and in Figure 12.

4.2 The test programme

4.2.1. Wet surface grinding

The parameters used for these tests, together with the specifications of the wheels, are given in Table 3 As far as the G-ratios and power consumption of the wheels are concerned, the results are shown in the form of bar charts in Fig 13 and 14. The surface finishes which can be achieved are shown in Table 4.

Fig 13 shows the results achieved with annealed material. According to these results, the CDA55N grit is slightly superior for grinding grades WFN and NIKRO 128, whereas with the UNI grade, higher G-ratios were achieved with wheels containing DXDAMC.

Fig 14 shows a parallel set of results for the same materials in the hardened condition. In these tests considerable machining problems were encountered with CDA55N, especially on the hardened WFN grade. After every third test, corresponding to a total downfeed of 1·5 mm, this wheel had to be dressed, i.e. opened up, in order to reduce the increasing cutting forces and hence the power consumption. With the DXDAMC wheel these difficulties were not encountered. Nevertheless, in grinding the hardened grades WFN and UNI, the DXDAMC wheel achieved lower G-ratios compared with the CDA55N wheel, although on NIKRO 128 the G-ratio of the DXDAMC wheel was superior.

Both diamond grits produced almost identical surface finishes on the hardened materials, ranging from R_a 0·1 to 0·3 μm. On the annealed materials, on the other hand CDA55N produced better results (Table 4).

4.2.2. Wet cylindrical grinding

The relevant parameters and wheel specifications for

TABLE 2
Chemical composition and properties of the FERRO-TITANIT grades included in the tests

Grade	*Chemical composition (% approximate)*										
	Hard phase		*Matrix*								
WFN	TiC 33·0		C 0·65	Cr 14·0	Mo 3·0	Cu 0·8	V 0·5	Ni 0·4	Fe rest		
UNI	TiC 28·0		Cr 18·0	Mo 2·0	Cu 0·8	Nb 0·5	Al 1·0	Ti 2·0	Ni rest		
NIKRO 128/292	TiC 20..30	Ni 5·5..15	Co 10..15	Mo 5..15	Cr 0..14		Ti 0·2 ..1·0	Al 0..1·0	Cu 0·8	B 0·02	Fe rest

Grade	*Mechanical characteristics (in hardened condition)*						
	Spec. weight g/cm³	*Compressive strength N/mm²*	*Tensile strength N/mm²*	*Elastic modulus N/mm²*	*Coeff. of thermal expansion $10^{-6} \cdot K^{-1}$*	*ISO round notch JE AV (J)*	*Bending strength N/mm²*
WFN	6·5	3600	1800—2000	294 000	(20..700°C) 8·9..11·7	1..3	1800—2000
UNI	6·95	2200—2400	1400—1500	297 000	(20—600°C) 9·8..11·7	0·8	1500
NIKRO 128/292	6·8	3500	2200—2400	294 000	(20—540°C) 8·0..9·6	1..2	2000—2400

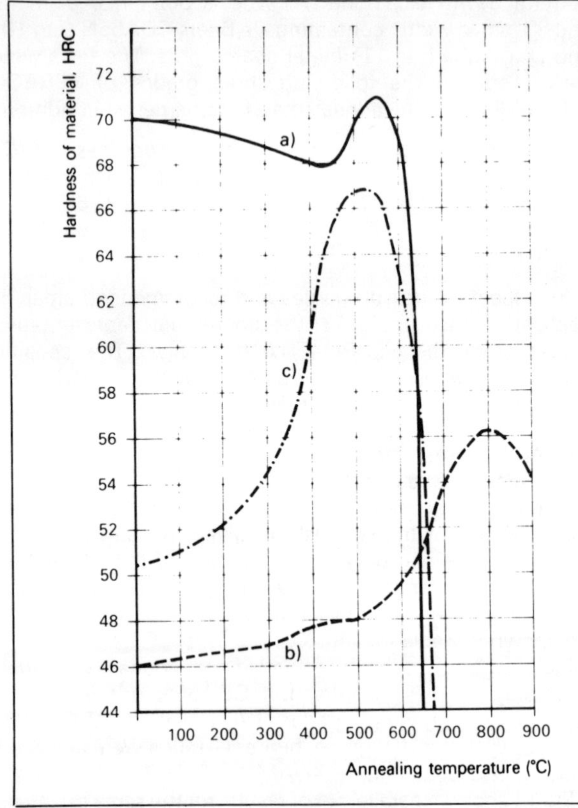

Fig 12 *Hardness curves for the grades of FERRO-TITANIT included in the grinding tests: (a) WFN, (b) UNI and (c) NIKRO 128*

TABLE 3
Grinding parameters and wheel specifications used for wet surface grinding tests

Machine	Mägerle F10VR: 11·9 kW
Wheel peripheral speed	28 m/s
Downfeed	0·008 mm
Total downfeed per test	0·5 mm
Crossfeed	2·5 mm
Table speed	16 m/min
Coolant	Syntilo 3, 1 : 80 in water; 16 l/min
Grinding wheel	1A1, 300 × 30 mm
Bond	Resin
Diamond grit	CDA55N 120/140 US mesh (D126); DXDAMC 100/120 US mesh (D151)
Concentration	100 (≙ 4·4 ct/cm³ abrasive volume)

TABLE 4
Surface finish after wet surface grinding
(All figures are averages of at least three measurements)

Diamond grit		*Surface finish—R_a centre line average height; R_t maximum peak-to-valley height (μm)*					
		Annealed			Hardened		
		UNI	WFN	NIKRO 128	UNI	WFN	NIKRO 128
CDA55N	R_a	0·17	0·30	0·30	0·20	0·10	0·17
	R_t	1·30	1·95	2·05	2·20	1·00	1·18
DXDAMC	R_a	0·37	0·38	0·42	0·30	0·15	0·22
	R_t	2·25	2·67	2·40	1·60	1·00	1·50

TABLE 5
Grinding parameters and wheel specifications used for wet cylindrical grinding

Machine	Karstens ASA 900; 5·5 kW
Wheel peripheral speed	16; 25 m/s
Workpiece speed	20 to 40 m/min
Infeed	0·01; 0·005 mm
Table speed	0·5 m/min
Coolant	Syntilo 3, 1 :80 in water; 20 l/min
Grinding wheel	1A1, 300 × 30 mm
Bond	Resin
Diamond grit	CDA55N 120/140 US mesh (D126); DXDAMC 100/120 US mesh (D151)
Concentration	100 (≙ 4·4 ct/cm³ abrasive volume)

these tests are shown in Table 5. As in the case of the wet surface grinding, the results have been summarized in the form of bar charts (Figs 15–18) and a table (Table 6).

Some difficulties arose from the relatively low static stability of the available cylindrical grinding machine (~ 20 N/mm as compared with 33 N/mm for the surface grinding machine). Given the high cutting forces required to grind the hardened FERRO–TITANIT grades, the relatively low rigidity of the machine resulted in elastic deformation of the machine-tool-workpiece system which reduced the amount of metal removed as compared with the downfeed increment. It was necessary, where appropriate, to reduce the downfeed so as to avoid deflections. As regards the surface finish, the primary aim was to avoid spiral marking of the work. Given constant table and crossfeed, the workpiece r.p.m. was found to be an important factor in this respect.

On the annealed materials, satisfactory results in terms of G-ratios, power consumption and surface finish were obtained with a single combination of parameters for all three grades of FERRO–TITANIT (Fig 15). The wheel life and grit performance results were analogous to those recorded in surface grinding, with CDA55N being superior to DXDAMC when grinding WFN and NIKRO 128, and the situation reversed for the UNI grade.

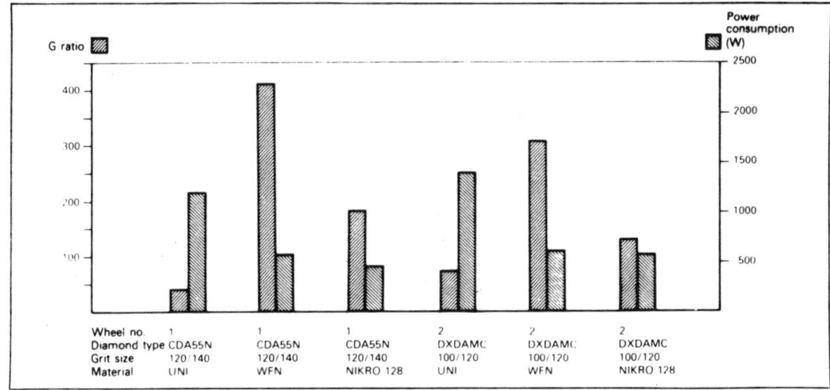

Fig 13 *G-ratios and power consumption recorded during the wet surface grinding of various grades of FERRO-TITANIT in the* annealed *condition*

wheel speed	*28 m/s*
downfeed	*0·008 mm*
crossfeed	*2·5 mm*
table speed	*16 m/min*

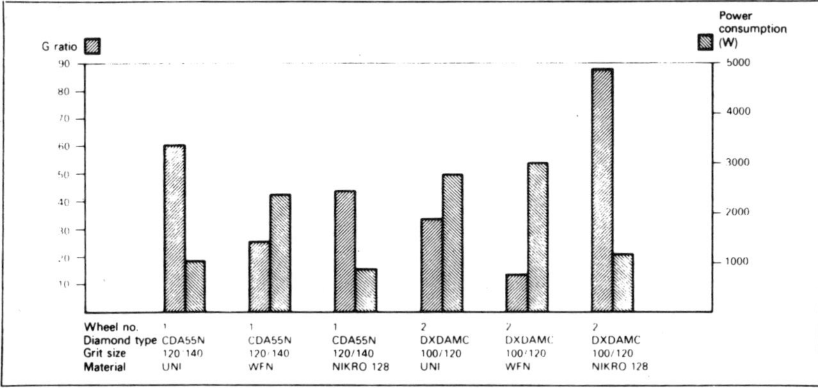

Fig 14 *G-ratios and power consumption recorded during the wet surface grinding of various grades of FERRO-TITANIT in the* hardened *condition*

wheel speed	*28 m/s*
downfeed	*0·008 mm*
crossfeed	*2·5 mm*
table speed	*16 m/min*

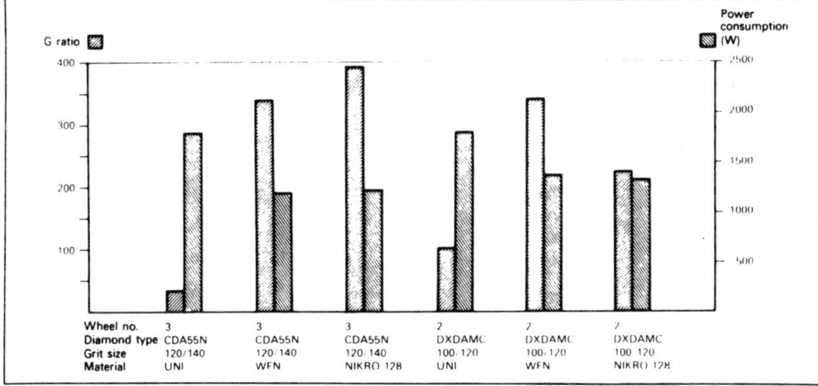

Fig 15 *G-ratios and power consumption recorded during the wet cylindrical grinding of various grades of FERRO-TITANIT in the* annealed *condition*

wheel speed	*25 m/s*
workpiece speed	*140 r.p.m.*
infeed	*0·01 mm*
table speed	*0·5 m/min*

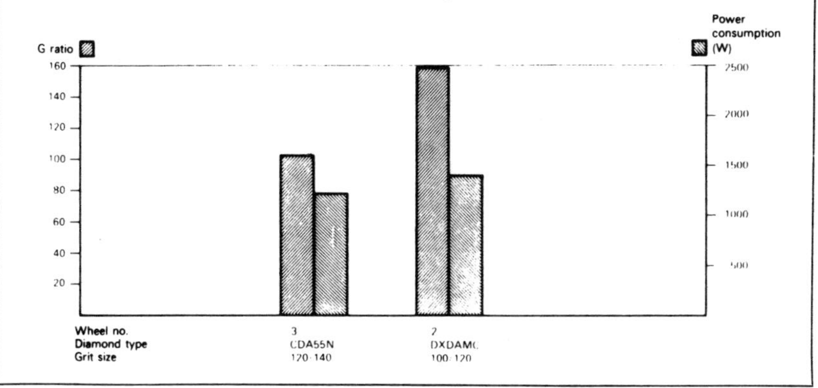

Fig 16 *G-ratios and power consumption recorded during the wet cylindrical grinding of FERRO-TITANIT grade UNI in the* annealed *condition and with reduced wheel peripheral speed*

wheel speed	*16 m/s*
workpiece r.p.m.	*140 r.p.m.*
infeed	*0·01 mm*
table speed	*0·5 m/min*

Since, in the annealed condition and with the parameters listed in Fig 15, the UNI grade had proved to be the most difficult to machine, an attempt was made to improve on the results by reducing the wheel peripheral speed to 16 m/s. As shown by Fig 16, it was possible in this way to improve the G-ratio of the CDA55N wheel—as compared with the results shown in Fig 15 for a wheel speed of 25 m/s—by a factor of almost 3, and for the DXDAMC wheel by about 50%.

In cylindrical grinding of the hardened grades, as shown in Fig 17, no significant difference was found between the two diamond grits as regards the G-ratios achieved. At the same time, in order to avoid elastic deformation of the machine resulting from extreme grinding forces, as well as spiral marking of the work surface, different machine parameters were required for the hardened grades. The r.p.m. of the workpiece was altered slightly, the infeed reduced from 0·01 mm to 0·005 mm for the hardened UNI

Fig 17 *G-ratios and power consumption recorded during the wet cylindrical grinding of various grades of FERRO-TITANIT in the* hardened *condition*

wheel speed	*16 m/s*
workpiece r.p.m.	*variable, see Fig 17*
infeed	*variable, see Fig 17*
table speed	*0·5 m/min*

G ratio
Power consumption (W)
180
160
140
120
100
80
60
40
20
2500
2000
1500
1000
500

Wheel no.	3	3	3	2	2	2
Diamond type	CDA55N	CDA55N	CDA55N	DXDAMC	DXDAMC	DXDAMC
Grit size	120/140	120/140	120/140	100/120	100/120	100/120
Material	UNI	WFN	NIKRO 128	UNI	WFN	NIKRO 128
Infeed	0,005	0,010	0,005	0,005	0,010	0,005
Workpiece rpm	100	140	120	100	140	120

Fig 18 *G-ratios and power consumption recorded during the wet cylindrical grinding of FERRO-TITANIT grade WFN in the* hardened *condition and with increased wheel peripheral speed*

wheel speed	*25 m/s*
workpiece r.p.m.	*120 r.p.m.*
infeed	*0·005 mm*
table speed	*0·5 m/min*

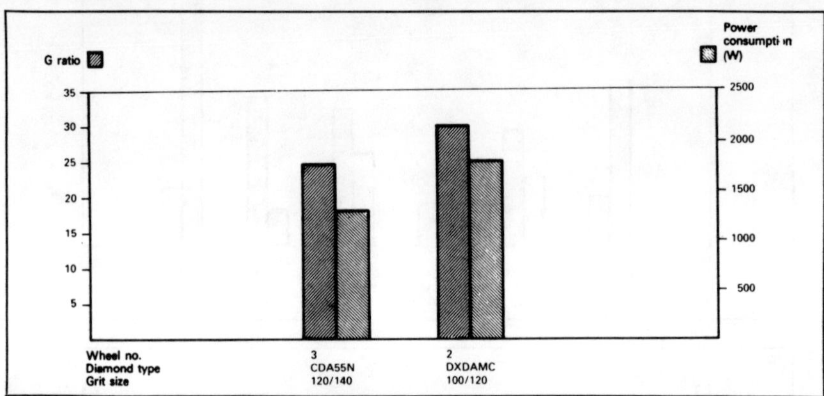

TABLE 6
Surface finish after wet cylindrical grinding

Diamond grit		Surface finish—R_a centre line average height; R_t maximum peak-to-valley height (μm)					
		Annealed			Hardened		
		UNI	WFN	NIKRO 128	UNI	WFN	NIKRO 128
CDA55N	R_a	0·42	0·28	0·42	0·44	0·47	0·42
	R_t	2·70	1·70	2·60	1·50	3·05	3·05
DXDAMC	R_a	0·40	0·18	0·48	—	0·63	0·58
	R_t	3·10	1·25	2·80	—	3·83	4·15

and NIKRO 128 grades, and the wheel speed reduced to 16 m/s (Fig 17). Tests performed on the hardened WFN grade at a wheel speed of 25 m/s failed to improve the G-ratio (Fig 18).

Table 6 shows the surface finishes achieved in cylindrical grinding. According to this data CDA55N seems to yield slightly better surface finishes than DXDAMC.

5. Conclusions

All in all, resin bond wheels containing CDA55N and DXDAMC diamond abrasives proved entirely suitable for grinding hardenable FERRO-TITANIT carbide alloys. The results provide fundamental confirmation of the earlier investigations [5,6] which showed that, with regard to both wheel performance and metal removal, metal-coated diamond grit is best suited to the economic and technically satisfactory machining of these materials. The choice between CDA55N and DXDAMC must be left to the wheel manufacturer, to be made in the knowledge both of his bond formulations and of the machining conditions applied by the end-user.

If, within the scope of the tests described above, DXDAMC produced, overall, somewhat more favourable results than CDA55N, it should not be overlooked that slightly different grit sizes were used (see Tables 3 and 5).

The following grinding parameters produced the best results:

Wet surface grinding	
Wheel peripheral speed	28 m/s
Downfeed	0·008 mm
Crossfeed	2·5 mm
Table speed	16 m/min
Wet cylindrical grinding	
Wheel peripheral speed	16 m/s
Workpiece speed	20—40 m/min
Infeed	0·01 mm
(for hardened UNI and NIKRO 128	0·005 mm)
Table speed	0·5 m/min

References

1. FREHN, F. Neuartige warmfeste Hartstofflegierung und ihre Anwendung (A new heat-resistant hard metal alloy and its application) *ZWF 70* (1975), No. 3, p 131 (In German)
2. Anon. FERRO-TITANIT zerspan- und härtbare Hartstoff-Legierungen lösen Ihre Verschleissprobleme (FERRO-TITANIT machinable and hardenable hard metal alloys solve your wear problems) Thyssen Edelstahlwerke AG, FERRO-TITANIT Div., publication No. 1153/3, Nov. 1977 edn. (In German)
3. FREHN, F. Die Wärmebehandlungen der bearbeitbaren und härtbaren Hartstoffe FERRO-TITANIT mit Fehlerkatalog (The heat treatment of machinable and hardenable FERRO-TITANIT hard metal alloys, with a list of defects). Unpublished report (In German)
4. FREHN, F. Richtige Auswahl und Handhabung der Bearbeitungsverfahren für Stanzwerkzeuge. Verwendung von Hartstoffen. (The correct choice and application of machining processes on punching tools. The use of hard metal alloys). Special reprint from 'VDI–Bildungswerk', Association of German Engineers, technical publication No. 6353/10 (In German)
5. FREHN, F. and DOBERITZSCH, H. Wirtschaftliches Schleifen mit Diamantscheiben der bearbeitbaren und härtbaren Hartstoffe FERRO-TIC (Economic grinding of FERRO-TIC machinable and hardenable carbide alloys with diamond wheels). *Industrie Diamanten Rundschau* (1968) No. 3, p 167 (In German)
6. Diamond grinding of FERRO-TIC machinable and hardenable hard alloys. 'Diamond Information L14', De Beers Industrial Diamond Division, England; Diamant-Information No. M 11, De Beers Industrial Diamond Division, Düsseldorf.

Diamond tools as a means of rationalising production in the optics industry

BY **E LOH** WILHELM LOH KG WETZLAR W GERMANY

Reprinted from Industrial Diamond Review, February 1977

The author reviews the present state of technology in the production of optical glass components. At the same time he indicates new possibilities for rationalisation, particularly in small-run and one-off production. The article shows how diamond tools have already contributed considerably to rationalisation and must, moreover, be considered as an essential part of any future moves in this direction

1. Introduction

The aim of this article is to survey new production methods for glass machining in the optical industry, particularly with reference to modern machinery, and thus show the latest stage of development in the manufacture of precision optical components.

The machining techniques involved will be considered in turn in the following sequence: milling, lapping, polishing and centring.

Recent years have seen particular advances in the use of diamond pellets for precision machining and also in the use of plastic in final polishing. There is adequate knowledge on the type of diamond, the grit size and bond to be used in diamond pellets for economically machining the broad spectrum of optical glasses. The development of suitable polishing plastics is also well advanced.

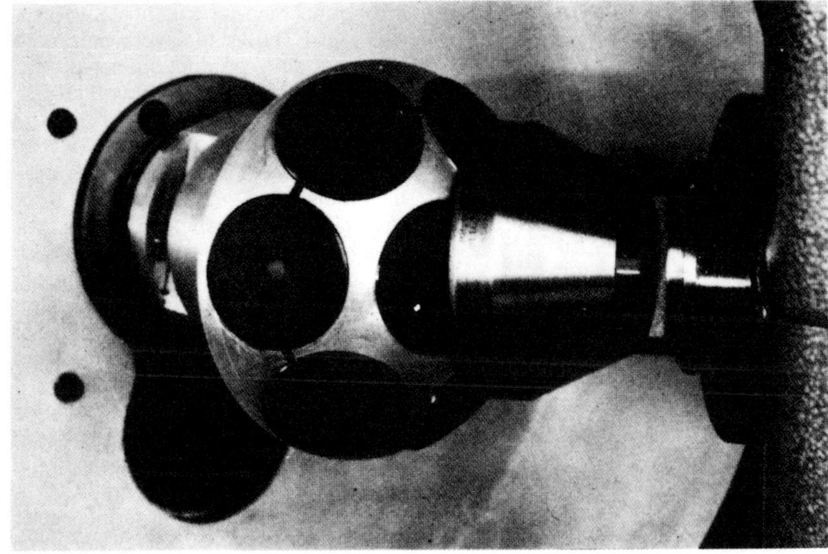

2. Machining operations

2.1 Milling

The rough shaping of optical components is done with diamond impregnated cup wheels and has been a standard production technique for many years (Fig 1). It must be mentioned here, however, that the production tolerances used earlier in milling have been considerably tightened for lapping to be successful. For these tolerances, although they must take account of the finish of the milled surface and the desired geometric form, have to be less than the removal depth in the subsequent process, lapping, which is between 0·08 and 0·1 mm.

To meet these tolerances in milling requires stable machines—stable primarily in relation to spindle alignment, since this controls the geometric precision of the surfaces. Fig 2 shows a machine for milling spherical and flat surfaces on single lenses and batch mounted lenses, which fulfils these requirements. Universal milling machines are used for milling prisms.

2.2 Lapping

Milling is followed by lapping. Where the motion of the machine in the first operation determined to a major degree the geometrical form of the workpieces, in lapping profiled diamond tools take over this role. These are set with sintered diamond pellets, which endow the tool's working surface with long life and good shape retention. As far as geometrical form is concerned, it is also necessary to have regular workpiece surfaces for subsequent polishing as these considerably reduce the time and effort required for this fine finishing stage.

In order to be able to make full use of the excellent life and form retention properties of these tools, certain requirements must be fulfilled as regards the kinematics of the lapping machines employed. The motion of such a machine should guarantee even wear for good retention of the tool's shape. This can only be achieved by a constant working pressure, i.e. the direction of the working pressure must always be towards the radius centre.

Fig 3 shows the work area of a lap

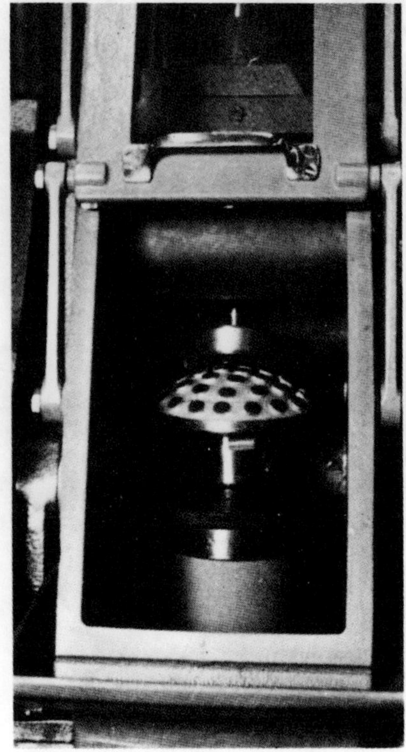

Fig 1 Opposite page *Milling spherical surfaces on optical glass with a diamond impregnated cup wheel*

Fig 2 Above *Radius milling machine for machining spherical and flat surfaces on single lenses or lenses batch-mounted on carriers*

Fig 3 Top *Work area of a lap/milling machine for finishing spherical surfaces of optical glass using a diamond pellet tool*

Fig 4 Top right *Flat or low curvature pellet tools are used for lapping flat workpieces or those with large radius curvatures*

milling machine for spherical surfaces, and Fig 4 shows the work area of a corresponding production set-up for lapping flat or large-radius workpieces.

2.3 Polishing

Polishing too is carried out with profiled tools faced with felt, pitch or resin to retain the polishing abrasive, for which various oxides, e.g. cerium or zirconium oxide, are used. Theoretically polishing is also effected by chip removal, the thickness of the chips of course being extremely small (between 0·005 and 0·002 mm). Nevertheless it is necessary for the finishing tolerance of the preceding lapping process to take account of the thickness of the chip.

The accuracy of optical surfaces only achieved previously by individual polishing is today obtained extremely effectively by lapping, and indeed to the tolerances prescribed for finished parts. The economic effect of this benefit, guaranteed by the correct use of lapping, can be seen primarily from the fact that the correction of the workpiece shape is raised from a process with minimal stock removal to a finishing stage with far greater material removal capability. Polishing thus has the task merely of rendering the surfaces of optical glass components transparent. Only when these requirements are met can one talk of 'high-speed polishing'. In order to achieve this, the kinematics of polishing machines must meet the same requirements as the lapping machines, namely the maintenance of constant polishing pressure. This point takes on all the more importance when one considers that the polishing tools with their resin work-faces cannot be compared with lapping tools as regards life and form retention. Development therefore has tended towards ever harder workfaces for polishing.

Fig 5 and 6 illustrate two designs of polishing machine, again for machining spherical, flat or curved surfaces.

2.4 Centring and facetting

The last process in the manufacture of a lens is finishing the edge to centre the lens. This includes both cylindrical grinding the edge and simultaneously facetting the lens.

Data on the methods used for setting the lens according to its optical axis and according to lens form, including the technique for centring using a laser beam, have been published elsewhere, and since an explanation of the principles involved would require a considerable amount of space, nothing more will be said on the subject here. Basically it may be remarked that very special requirements must be met by centring machines, because centring has more influence on the quality of an optical system than any other lens machining operation. These requirements include primarily distortion-free, rigid machine design, precise axial alignment of the centring spindles and precise rotation of these spindles.

The centring and facetting machines illustrated in Fig 7 and 8, for example, guarantee a tolerance on lens diameter of $\leqslant$ 0·01 mm. Axial deviation of the spindle is $\leqslant$ 0·005 μm and rotary run-out of the centring spindle is

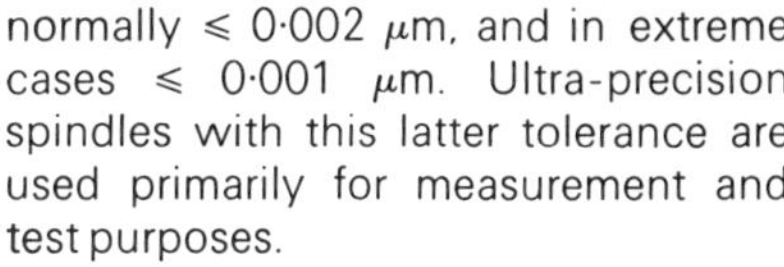

normally ≤ 0·002 μm, and in extreme cases ≤ 0·001 μm. Ultra-precision spindles with this latter tolerance are used primarily for measurement and test purposes.

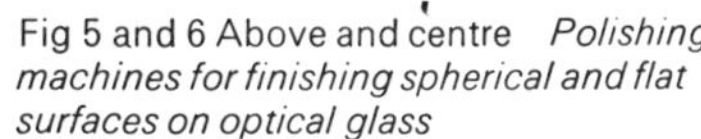

Fig 5 and 6 Above and centre *Polishing machines for finishing spherical and flat surfaces on optical glass*

3. New developments

Following the above review of the machining techniques used today in the production of optical components, let us briefly examine further developments which offer considerable future prospects of increased rationalisation in production.

For short series runs it is generally not worthwhile to obtain lapping tools for precision machining. It was therefore necessary to look for new ways to replace the wasteful and imprecise process using loose abrasives. This was not merely successful, but a complete machining operation was saved.

Fig 7 Right *Fully automatic centring and facetting machine for working lenses according to the 'clock' mounting technique*

Rough machining by milling, and precision machining with lapping or grinding are all today carried out in just one operation on a specially developed radius milling machine. Since at the end of this operation there only remains the polishing to be done, the process is referred to as 'fine milling'.

Ring tool

In contrast to the 'old-fashioned' lapping in which formed tools are used, in fine milling the geometrically precise surface is produced with a ring tool (Fig 9). As in lapping, in fine milling too a special tool is required for each radius, the face lip of which corresponds to the radius to be machined. In the context of this lip, the tool may thus also be regarded as formed. Since, however, it is actually a ring tool it is considerably cheaper than a pellet tool and it can be adapted more easily to a change in workpiece geometry.

The particular demands this process makes on the machine are, in addition to a special degree of stability, extremely high precision in the axial play between tool and workpiece spindle, and increased rotational accuracy of tool and workpiece. Moreover, an accurately controlled continuous infeed is just as essential a prerequisite as the possibility of setting the optimum cutting speeds related to the advance, i.e. to the rotational speed of the workpiece, for the different tool diameters.

All these requirements are to a very large degree met by the newly developed type of machine shown in Fig 10. Continuous infeed is provided by a hydraulic servo, the tool speed is steplessly adjustable between 4000 and 12,000 r.p.m. and the workpiece spindle speed may be set between 1 and 120 r.p.m. The diamond used in the tools is in practice between 20 and 40 μm in size and 30 to 50 in concentration. Depending on operational demands, natural and synthetic MICROGRIT SND and DA have

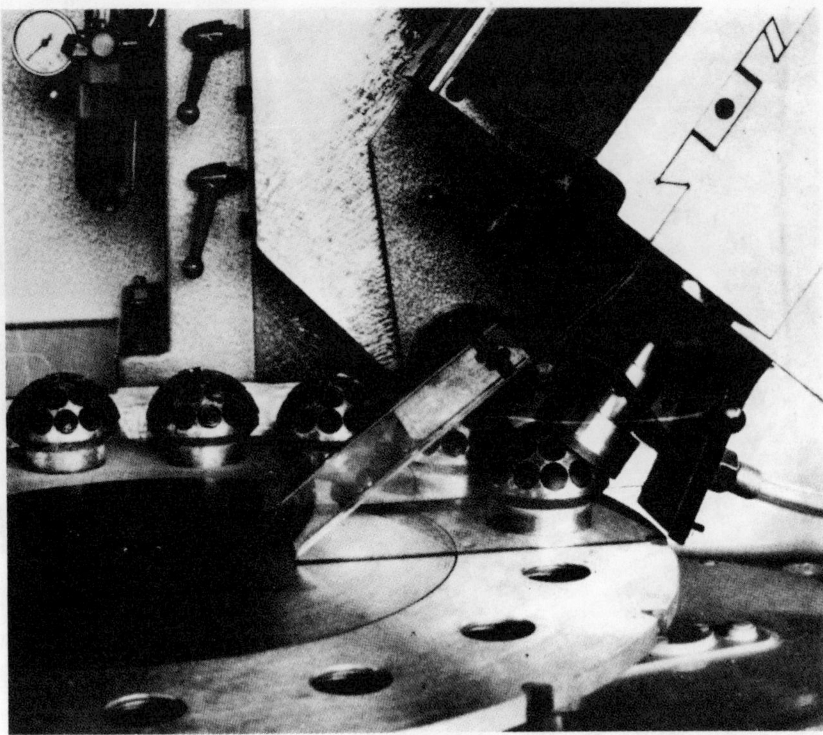

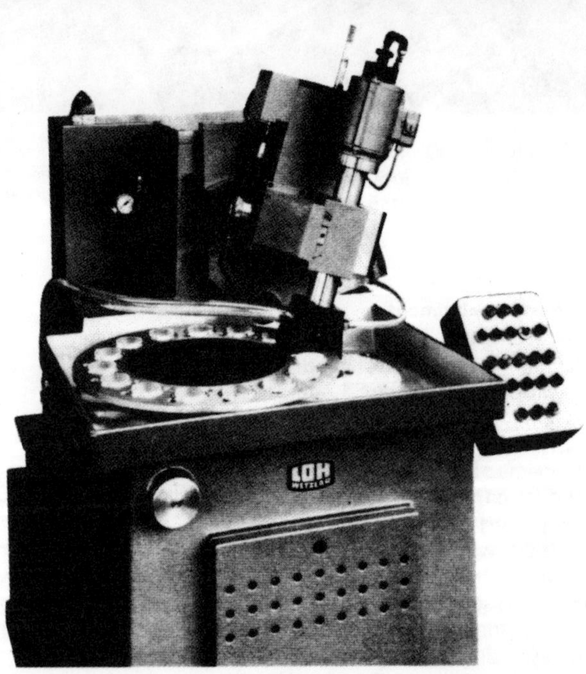

Fig 8 Above *Semi-automatic centring and facetting machine with laser centring device for machining lenses according to the 'clock' mounting technique*

Fig 9 Top right *Fine milling small numbers of optical components with a diamond impregnated cup wheel*

Fig 10 Right *Recently developed fine milling machine for efficient finishing of small series and single workpieces*

proved equally successful as abrasives.

The economic advantages of the new 'fine milling' process may be summarised as follows:

In comparison with previous techniques, it offers a single process, i.e. one process, either grinding with loose abrasive or lapping with diamond pellets, is eliminated. In addition, manual and subsidiary process times are reduced, equipment for the unnecessary process is no longer required and, as already explained, tool costs are reduced.

It is important to note that, with the application of fine milling and the cost advantages it guarantees, the expenditure, both in time and material, is no greater in the final polishing stage, and the increases in efficiency and rationalisation obtained without loss of workpiece quality, i.e. at the end of the production process, are attained in full.

Polishing

Similar demands, namely to rationalise production of small series of high quality components, existed for polishing too. Machine development has led to a basically new operating principle. The new generation of polishing machines work with constant work pressures which are easily matched to any radius or a flat surface ◊

Reprinted from Vestnik Mashinostroeniya, Volume 55, Issue 6, 1975

DIAMOND HONING OF HARD NON-METALLIC MATERIALS

V. R. KANGUN

R. Z. TSYPKIN

Components in very hard non-metallic materials such as glass, devitrified glass, and ceramics, the cylindrical holes in which require to be machined to high precision, are used in engineering and instrument construction. Internal grinding and lapping of holes in components made from these materials are very difficult operations.

At VNIIALMAZ the possibility of applying diamond honing to finishing cylindrical holes in bushes made from such very hard non-metallic materials has been investigated[1].

These tests involve the honing of bushes made from K8 optical glass and 22KhS ceramic, of internal diameter 40mm, wall thickness 3-5mm, and depth 40-80mm, using a honing head with four ABKh 5 x 4 x 50 stones, and M1 bond, and a diamond concentration of 100%.

The grain size of the natural diamond in the stones was 20/14-200/160, and of the ASR synthetic diamonds 100/80-125/100. The honing process took

20-30 seconds, and when using AM40/28-20/14 stones 60 seconds. A 3% aqueous solution of calcined soda was used as coolant.

The bushes were machined on a Model OF-38A vertical honing machine. The pressure p of the stones on the hole surface[1] was varied from 1,5 to 27,0kgf/cm^2, the peripheral head speed v_h being 25,8-62,8m/min and the reciprocation rate v_r being 3-7m/min.

The wear resistance of the diamond honing stones was determined after removing more than 150g of material. The glass bushes were machined at a pressure of $p = 3{,}6$ kgf/cm^2, v_h = 25,8m/min, v_r = 7m/min, the corresponding values for the ceramic bushes being 15kgf/cm^2, 62,8m/min and 7m/min.

The machining output was determined from the specific amount of material removed per second (g/s), while the wear resistance of the diamond honing stones was assessed from the specific consumption of diamonds per unit weight of the material removed (mg/g). The surface finish was evaluated with a Model 201 profilograph-profilometer, the hole dimensions being measured with a telescopic internal gauge having 0,01mm divisions.

To determine the effect of the stone characteristics and the machining conditions on the productivity of the process and the surface finish, the internal surfaces of the ceramic bushes were always machined at the same stone pressure, i.e., p = 3,0kgf/cm^2.

The method of clamping is an important factor when honing thin-walled bushes made from brittle materials. Bush-type components are generally clamped with rigid clamping plates and plates with a spherical pivot. Experience has shown that this method of clamping can be applied to glass or ceramic bushes having a ratio of height to wall thickness up to 11-13 or 14-16 respectively, provided the ratio of the external diameter to the wall thickness does not exceed 12-16. When this method is used, hard rubber plates (85-90 hardness units on a TIR instrument) should be fixed on the load carrying surfaces of the clamping attachment. The use of these plates ensures distribution of the axial load on the ends of the bushes and eliminates the possibility of their rotation during machining. When machining bushes in which the ratio of height and external diameter to wall thickness is high, it has been shown experimentally that attachments must be used in which the bushes are clamped on the external surface with an elastic material. It should be noted that the machining of chamfers on the bore eliminates the possibility of forming chips and cracks at the points of tool presentation and withdrawal.

The presence of significant initial geometrical profile errors in the bores of up to 0,7-0,9mm is a special feature of ceramic blanks. It has been shown experimentally that blanks having a profile error of less than 0,1mm can be honed at a constant radial stone pressure. Honing blanks having larger profile errors always involves fracture even at low pressures.

It is known that machines with a stepped, pulsating transverse stone feed[2] can be recommended for honing metal components with large initial profile errors in the bores at an initial stone pressure of 5-8kgf/cm^2. When honing non-metallic materials, pressures as high as this cannot be used because of fracture of the blanks. Accordingly, experiments on rectifying the initial blank errors were carried out as well as experiments on the basic criteria of the honing process, to determine the characteristics of the honing cycle and to bring out the feasibility of using honing machines at constant pressure. In the initial honing cycle the stones were therefore subjected at intervals to short term clamping with due allowance for the scope of the machine.

It was shown experimentally that the original profile errors are best rectified in two stages, the stone pressure on the bores never exceeding 0,9-1,5kgf/cm^2 in either stage.

The clamping time during the first stage should be 0,5-1,5s, the honing time in one run being 12-16s, which decreases the profile error in the bore by

0,07-0,1mm. In the second stage the clamping time should be increased to 2-3s, and the honing time per run should be decreased to 8-12s, which decreases the profile error in the bore by 0,2-0,25mm.

Knowing the initial blank error it is thus possible to determine the required number of honing stone clampings for each stage of correcting the bore profile.

The effect of stone grain size and honing conditions on the rate of correcting the bore profile was also studied. The data obtained, provides a basis for recommending stones of grain size 125/100-200/160 and a peripheral honing head speed of 40-60m/min.

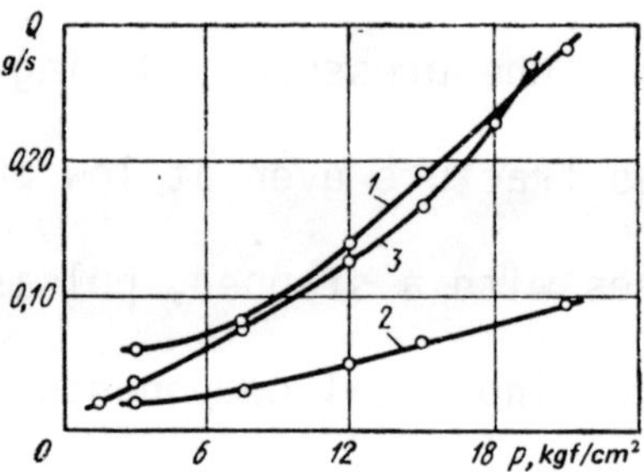

Fig. 1.
Output in diamond honing Q as a function of the pressure $p(v_r = 7m/min)$:
1 – ceramic A200/160, v_h = 62,8m/min;
2 – ceramic, A200/160, v_h = 25,8m/min;
3 – glass, ASR125/100, v_h = 62,8m/min.

The honing output Q depends on the pressure p at the zone of contact between the diamond honing stones and the workpiece. The data in Fig. 1 show that the output of the process is directly related to the stone pressure.

The experimental results showed that when honing non-metallic components the strength of the components is the only limiting factor. The effect of stone pressure on output also depends on the hardness of the workpiece material. Thus, increasing the pressure from 3,0 to 15kgf/cm^2 increases the output when honing glass by a factor of 4-6, and when honing ceramics by a factor of 3-4.

This can be explained by the decrease in depth of penetration of the cutting grains into the workpiece surface with the harder material.

The honing output increases with an increase in peripheral head speed, the effect being more marked at higher pressures. For example, when honing a ceramic at a pressure of 21kgf/cm^2 an increase in the peripheral speed from 25,8 to 62,8m/min increases the output by 0,18g/s and when honing at a pressure of 6kgf/cm^2 by 0,05g/s.

Experience shows that increasing the rate of reciprocation of the honing head increases output when honing non-metals. The relation between the honing output when honing a ceramic and the honing conditions for A200/160 stones and the experimental conditions used can be expressed by the equation[2]:

$$Q = 0{,}112 \cdot 10^{-3}\ p^{0{,}758}\ v_h^{1{,}148}\ v_r^{0{,}345}$$

The maximum deviations of the Q values from the experimental values obtained with this equation never exceeds 6%.

The cutting properties of honing stones are also a function of the grade of diamond used. Thus, when machining a ceramic with ASR 100/80 stones the honing output is less by a factor of 1,6-2,0 than when honing with natural diamond honing stones of the same grain size. When honing glass bushes, natural diamond honing stones of grain size A125/100-100/80 increase the output compared with ASR 125/100-100/80 diamonds only a factor of 1,3-1,5.

It was shown experimentally that the grain size of diamond honing stones has a significant effect on the material removal rate. Thus, with an increase in the dimensions a of the diamond grains from 35μm (AM48/28) to 280μm (A200/160) the rate of removal of ceramics at a pressure of p = 9,0kgf/m^2 increased from 0,0025g/s to 0,09g/s (Fig. 2). In Fig. 2 the horizontal axis gives the grain dimensions of the main fractions of the diamond powders[4]. The effect of grain size is even more obvious at high stone pressures. On increasing the dimensions of the diamond grains within the given limits at a pressure of

$p = 21{,}5 kgf/cm^2$ the honing output increased from 0,0025g/s to 0,09g/s (Fig. 2). In Fig. 2 the horizontal axis gives the grain dimensions of the main fractions of the diamond powders[4]. The effect of grain size is even more obvious at high stone pressures. On increasing the dimensions of the diamond grains within the given limits at a pressure of $p = 21{,}5 kgf/cm^2$ the honing output increased from 0,0045g/s to 0,28g/s, i.e., by a factor of more than 60.

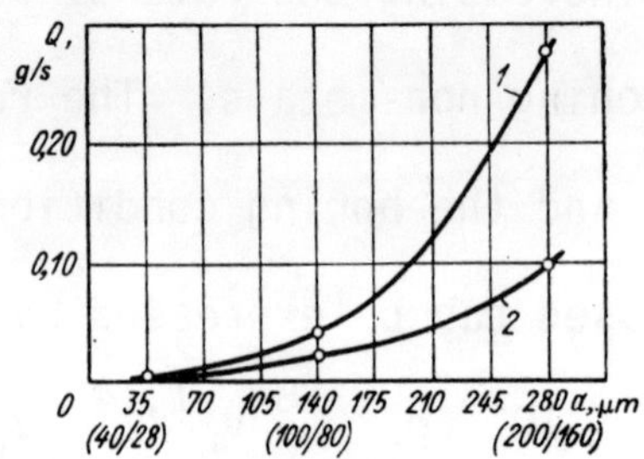

Fig. 2.
Honing output Q as a function of the grain size of diamond honing stones ($v_h = 62{,}8 m/min$, $v_r = 7 m/min$):
1 — ceramic, $p = 21{,}5 kgf/cm^2$;
2 — ceramic $p = 9{,}0 kgf/cm^2$.

Analysis of the results obtained and comparison with the outputs obtained when honing metals[2] shows that when honing non-metallic materials the honing output is significantly higher. Diamond honing of non-metallic materials provides a means of increasing honing output by a factor of 15-25 compared with internal grinding with diamond grinding wheels.

The experimental results on determining the wear of honing stones are given in the Table.

As can be seen from the data in the Table when machining a ceramic, large grained honing stones suffer significantly less wear than small grained stones, whilst when machining glass the wear is approximately the same with both. The data also show the effect of the properties of the bond on the wear resistance of honing stones. Thus, when machining glass using stones with an M5-9 bond the specific consumption of diamonds is more than double that with an M1 bond.

It is interesting to note that similar results for the consumption of synthetic diamonds were observed when grinding the spherical surfaces of K8 glass components with a circular tool[4]. The consumption of diamonds was directly related to the hardness of the workpiece material. For example, when grinding glass the specific consumption was less by a factor of 2-4 than when grinding a ceramic.

Stone characteristics	Workpiece material	Specific consumption of diamonds q,mg/g
AM40/28-M1	Ceramic 22KhS	0,65
A100/80-M1	Ceramic 22KhS	0,50 0,12
ASR100/80-M1	Glass K8	0,15
A200/160-M1	Ceramic 22KhS	0,17 0,11
A200/160-M5-9	Glass K8	0,26

The data in the Table also show that when honing glass with ASR 100/80 diamond honing stones the wear rate is 25% higher than with natural diamond stones.

Analysis of the data obtained shows that the specific consumption of diamonds when honing non-metallic materials is significantly lower than when honing quenched steels, and is the same as when internal diamond grinding non-metallic materials.

Measurement of the diameters of the honed bores was part of the experimental program. These measurements indicated high geometrical profile accuracy of the bores in the case of glass and ceramic bushes, the deviation from the cylindrical never exceeding 0,01mm.

The experimental results showed that changing the stone pressure and the honing head reciprocation rate over the range of parameters investigated did not significantly affect the height of the microroughnesses. It was shown that the stone grain size (Fig. 3) had a significant effect on the surface roughness: the height of the surface roughnesses was directly related to the diamond grain dimensions a.

The effect of the bond properties on the surface roughness was well brought out by comparative rests with A200/160 stones with M1 and M5-9 bonds. When honing glass with stones having an M1 bond the mean arithmetic profile deviation R_a = 3,4-3,6μm, and with an M5-9 bond R_a = 5,6-6,2μm.

It is known that the structure and hardness of the workpiece material affect the surface roughness. The studies described provided a means of showing that the surface roughness of ceramic components was somewhat less than with glass components. For example, when honing a ceramic with A100/80 stones R_a = 0,6-0,8μm, as against R_a = 1,6-2,0μm with glass.

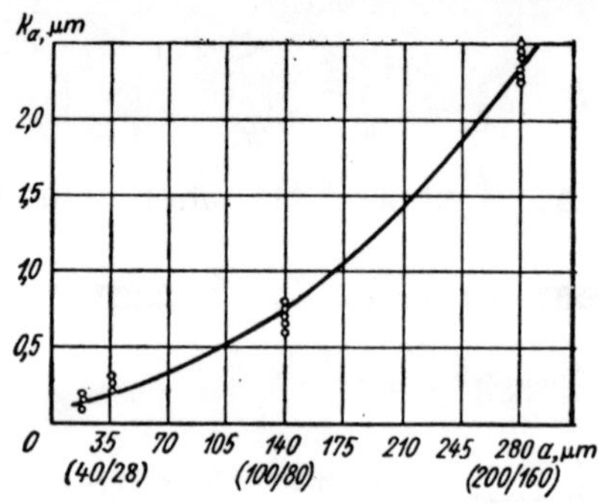

Fig. 3
Effect of diamond honing stone grain size on surface roughness of ceramic bushes: p = 9,0kgf/cm²; v_h = 62,8m/min; v_r = 7m/min.

The surface quality of components manufactured from hard, non-metallic materials depends on the depth of disturbance of the layer, which has a ramified network of microcracks, as well as on the surface roughness. It was found that the depth of the disturbed layer was less with ceramic bushes than with glass ones. For example, when honing with AM40/28 stones at a pressure of $9kgf/cm^2$ and a peripheral stone speed of 62,8m/min the depth of disturbance of the layer on ceramic bushes was 9-10μm, and on glass bushes 12-14μm.

The data obtained on diamond honing of non-metallic materials enables the cost effectiveness of this process to be estimated. Comparison of the specific cost[5] of honing a ceramic bush of internal diameter and depth 40mm by diamond honing as agains internal grinding showed that the specific cost per unit of mass removed (g) when grinding was 4,92copecks/g, as against 0,28copecks/g when honing, i.e., less approximately by a factor of 15. The effectiveness of honing is directly related to the depth to diameter ration of the component.

The significant increase in output when honing non-metallic materials compared with internal grinding and lapping gives justification for applying this method in batch production.

CONCLUSIONS

1. Diamond honing can be used for honing thin-walled components made from hard, non-metallic materials. In clamping the components (with a ratio of depth and external diameter to wall thickness up to 13-16) on the end faces the load carrying elements of the clamping attachment should be made from very hard rubber. Bushes having larger ratios of depth to wall thickness should be clamped on the external face with an elastic material.

2. To eliminate geometrical profile errors on the bore, stones of grain size 125/100-200/160 should be used. The honing cycle should include periodic, short term clamping of the stones at a pressure of $p = 0{,}9\text{-}1{,}5 kgf/cm^2$, honing for 12-16s in the first stage and 8-12s in the second stage. This separation of the cycle into two stages is essential when the profile error exceeds 0,3mm.

3. Rational honing conditions in which the soundness of the bushes is retained and chips on the edges avoided are: $p = 6\text{-}18 kgf/cm^2$, $v_h = 20\text{-}50 m/min$, $v_r = 5\text{-}10 m/min$ when machining glass, and $p = 9\text{-}30 kgf/cm^2$, $v_h = 40\text{-}80 m/min$, and $v_r = 5\text{-}12 m/min$ when machining ceramic.

4. The output in diamond honing as applied to preliminary finishing of glass is 0,1-0,5g/s and with ceramic 0,05-0,3g/s. In finish honing of glass the output is 0,005-0,02g/s, and in finish machining ceramic 0,002-0,006g/s.

REFERENCES

1. FRAGIN, I.E., BLIZHEVSKAYA, I.L. Determination of stone-pressure during honing. Russian Engineering Journal, No. 11, 1973.
2. NAERMAN, M.S., POPOV, S.A. Precision machining of components with diamond and abrasive stones. Mashinostroenie, M., 1971.
3. NALIMOV, V.V., CHERNOVA, N.A. Statistical methods of planning and extremal experiments. Nauka, M., 1965.
4. ROZNO, N.A. (editor). Synthetic diamonds when machining metals and glass. Mashinostroenie, M., 1968.
5. SEMKO, M.F., UZUNYAN, M.D. Economic basis for the selection of a diamond wheel. Prapor, Khar'kov, 1971.

INDEX

A

Abrasive flow machining, 3, 14, 42
Abrasive jet machining, 3, 14, 42
Abrasive processes guidelines, 49
Active nontraditional material removal processes, 3
Advanced composite materials, **54-55**, 59, 61
Aerospace, 124
Altered material zones, 33-34
Aluminum oxide wheels, 199
Aluminum, 61, 62
AMZ, See: Altered material zones
Ancient cutting stones, 159
Apparel, **53-54**, 58
Authors
- Alworth, R., 104-109
- Aspinwall, D., 95-103
- Ault, W. N., 198-211
- Belikov, A. M., 212-216
- Bellows, Guy, 3-12, 13-52
- Bhattacharyya, S. K., 95-103
- DeAngelis, William M., 125-128
- Deutschendorf, Hank, 187-197
- Frehn, F., 219-226
- Grieb, Paul, 182-186
- Hanson, Willard E., 53-63
- Hegland, Donald E., 79-82
- Herzog, Daniel E., 114-116
- Hujer, H., 217-218
- Kangun, V. R., 231-240
- Koepke, B. G., 145-157
- Kohls, John B., 13-52
- Loecy, Roger J., 174-178
- Loh, E., 227-230
- McLaughlin, Howard B., 158-167
- Mercincavage, C. Michael, 64-78
- Moir, R. A., 198-211
- Nicolls, M. O., 168-173
- Notter, A. T., 168-173, 219-224
- Pokhorovskii, A. D., 212-216
- Ramalingam, S., 104-109
- Stokes, R. J., 145-157
- Takeyama, Hidehiko, 110-113
- Tsypkin, R. Z., 231-240
- Tyrrell, William R., 83-92
- Van Groenou, A. Broese, 131-144
- Veldkamp, J. D. B., 131-144
- Wilson, G. F., 104-109
- Zsolnay, Laszlo, M., 179-181

Automotive, 124

B

Bearings, 124
Bond retention, 175
Boring, 118
Brittle ceramics, **147-148**
Brittle failure, 101-102
Brittle materials, **131-144, 147-148**
Broach, 120

C

Carbide hole drilling, 126
Carbides, **168-173, 174-178, 179-181, 219-224**
Cast iron, 121
Castings, 123
CBN wheels, **200-201**
Centering, **226-227**
Ceramics, 131, 146, 231
Chamfers, 233
Chemical machining, 3, 14, **18-23**
Chemical milling, 18
Chemical polisher, 167
Chemical polishing, 18
Chip formation, **148-150**
Chipping, **139-140**
Clamping, **232-233**
Coated diamond wheels, 199-200
Cold rolled steel, 127
Comparative abrasive mesh sizes, 185
Complex radial shape profiling, 73
Computer controlled lasercutting system, 63
Contouring wire saw, 164
Controlled fragmentation, 175
Coolants, 125, 126, 186, 232
Costs
- comparison, 68
- cutting, **64-78**
- dressing devices, 214
- drilling holes, 128
- effective surface grinding, **182-186**
- grinding, 179
- hardened steels, **182-186**
- honing, **239**
- superalloys, 182-186

Crack formation, **141-143**
Cracking, **139-143**
Crater wear, 98
Cutter sharpening, 211
Cutting speed, 4

D

Damage, **145-157**
Devitrified glass, 231
Diamond abrasive, **168-173**
Diamond honing, **231-240**
Diamond powders, 229-230
Diamond wire table, 161
Drill damage, 128
Drilling, 85, 126, 127, 128
Dry grinding, **172-173, 174-178**

E

EBM, See: Electron beam machining
EDM, See: Electrical discharge machining
Elastic springback, 214
Electrical discharge grinding, 3, 14, 24
Electrical discharge machining, 3, 5, **8-9**, 14, 24, **64-78**
Electrical discharge sawing, 3, 14, 24
Electrical discharge wire cutting, 3, 14, 24
Electrical material removal, 33
Electro gel machining, 18
Electro-steam, 3, 14, 33
Electrochemical deburring, 3, 14, 33
Electrochemical discharge grinding, 3, 14, 33
Electrochemical grinding, 3, 14, 33
Electrochemical honing, 3, 14, 33
Electrochemical machining, 3, 4, **6-7**, 14, 33
Electrochemical polishing, 3, 14, 33
Electrochemical processes guidelines, 41
Electrochemical turning, 3, 14, 33
Electrocorundum wheels, 213
Electroforming, 127
Electron beam machining, 3, 5, 14, 24
Electronic bandsaw, **79-82**
Electropolish, 3, 14, 18

F

Face turning, 119
Facetting, **226-227**
Facing, 118, 119
Flame-sprayed alloys, 117
Flank wear, 101
Forming, 125

G

Gas turbine, 119
Gear and pinion, 122
Glass industry, 125, 131
Glass, **231-240**

H

Hack saws, 125
Hard ceramics, **110-113**
Hard flame-sprayed alloys, 117
Hard material profiling, 73
Hardenable steel bonded carbide alloys, **219-224**
Hardened steels, 95, 182-186, **219-224**
Hardness of tool material, 111
Heat dissipation, **175-176**
High speed tool steels, **198-211**
High-performance carbide grinding, **179-181**
High-speed turning, 114
Hob grinding, 211
Hydrodynamic machining, 3, 14, 42

I

I.D. boring, 120
Impact shock, 176
Insert selection, 116

J

Jig grinding, 191

K

Kerosene, 128

L

Lapping, **229-230, 225-226**
Laser beam machining, 3, 5, **9-10**, 14, 24
Laser beam torch, 3, 14, 24
Laser cutting, **53,-63**
Laser drilling, 27
Liners, 123
Low stress grinding, 3, 14, 42
Lubrication, 128

M

Machine selection, 182-183
Machine tool control system, 119
Machining turbine parts, **217-218**
Mandrels, 214
Material removal rates, 4
Mechanical material removal, 42
Metal removal rates, 4
Milling, **87-88**, 225
Monocorundum wheels, 213
Multi-ribbed grinding, 214

N

Narrow slot machining, 73
NC, See: Numerical Control
Nickel alloy, 61
Non-metallic materials, **231-240**
Numerical Control, **64-78**, 82

O

O.D. turning, 118, 119, 120
Optics industry, **225-228**

P

Photochemical machining, 3, **11-12**, 14, 18
Piston body, 120
Plasma beam machining, 3, 14, 24
Ploughing, 134-139
Polishing, 158-167, 226, 228
Polycrystalline diamond tools, **104-109**
Polycrystalline tooling, 95-103, **104-109**
Precision grinding, **187-197**
Pressworking tools, 73
Programming, 81
Pump impellers, 123
Pumps, 124
Punch and die set, 81
Punch axes, 119
Punch rings, 120

Q

Quartz, 126
Quenched steels, 237

R

Relief grinding, 211
Residual stress, 21
Ring tool, 227
Roller bearing, 118
Rotary ultrasonic maching, **83-92**, 125
Rough turning parameters, 115
Routing operations, 73
Rust, 128

S

Selective etching, **107-108**
Semibrittle ceramics, **146-147**
Shaped tube electrolytic machining, 3, 14, 33
Silicon aluminum alloy castings, 96
Silicon aluminum alloys, 95
Single crystal eroding, 73
Sintered diamond tools, **110-113**
Six-wheel polishing machine, 167
Slot grinding, 186
Small part for profile contouring, 72
Stainless steel tubing, 9
Static force, **102-103**
Steel mill, 124
Steel rollers, 214
Steels, 114-116
Stone cutting saw, 159
Superabrasives, **187-197**
Superalloys, **182-186**
Superhard cutting edges, 114
Surface finish ranges, 14
Surface finish, 102-103
Surface integrity, 13-52
Surface roughness, 238
Surgical needle, 10

T

Tape controlled jig grinder, 193
Thermal material removal, 24-32
Thermochemical machining, 3, 14, 18
Thread cutting, 92
Thread grinding, **212-216**
Threading, 88-90
Tight tolerances, **127**
Tolerances, 126, 127
Tool evaluation, 96
Tool life, 126
Tool profiles, 73
Tool temperature, **100-101**
Tool wear measurement, 97
Tool wear pattern, 111
Transverse stone feed, 233
Truing, 201
Turbine parts, **217-218**
Turning, 105, 117, 118

U

Ultrasonic drilling, 126
Ultrasonic machining, 3, **5-6**, 14, 42, **83-92**, 125

V

Valve seat turning, 116

W

Wear resistance, 232
Wet surface grinding, **170-173, 223-224**
Wheels
  aluminum-epoxy, 190
  aluminum-oxide, 199
  CBN, **200-201**, 208
  characterizations, 180
  coated diamond, **199-200**
  conditioning, **184-185**
  diamond, **219-224**
  dressing, **201-202**
  elastic springback, 214
  electrocorundum, 213
  high speeds, 133
  impregnated, **188-189**
  moncorundum, 213
  mounting, **184-185**
  normal speed, 186
  plated, **189-190**
  resin bonded, 189
  selection, 180, **183-184**
  silicon carbide, 199
  truing, **184-185**
  types, **188-191**
Wire EDM, 79-82
Wire electrodes, **64-78**
Wood, 61